Ashford Books

Six Sense Strategy LTD.

IEE ELECTROMAGNETIC WAVES SERIES 36

Series Editors: Professor P. J. B. Clarricoats
Professor Y. Rahmat-Samii
Professor J. R. Wait

Propagation, scattering and dissipation of electromagnetic waves

Other volumes in this series:

Volume 1 **Geometrical theory of diffraction for electromagnetic waves** G. L. James
Volume 2 **Electromagnetic waves and curved structures** L. Lewin, D. C. Chang and E. F. Kuester
Volume 3 **Microwave homodyne systems** R. J. King
Volume 4 **Radio direction-finding** P. J. D. Gething
Volume 5 **ELF communications antennas** M. L. Burrows
Volume 6 **Waveguide tapers, transitions and couplers** F. Sporleder and H. G. Unger
Volume 7 **Reflector antenna analysis and design** P. J. Wood
Volume 8 **Effects of the troposphere on radio communications** M. P. M. Hall
Volume 9 **Schumann resonances in the earth-ionosphere cavity** P. V. Bliokh, A. P. Nikolaenko and Y. F. Flippov
Volume 10 **Aperture antennas and diffraction theory** E. V. Jull
Volume 11 **Adaptive array principles** J. E. Hudson
Volume 12 **Microstrip antenna theory and design** J. R. James, P. S. Hall and C. Wood
Volume 13 **Energy in electromagnetism** H. G. Booker
Volume 14 **Leaky feeders and subsurface radio communications** P. Delogne
Volume 15 **The handbook of antenna design, Volume 1** A. W. Rudge, K. Milne, A. D. Olver, P. Knight (Editors)
Volume 16 **The handbook of antenna design, Volume 2** A. W. Rudge, K. Milne, A. D. Olver, P. Knight (Editors)
Volume 17 **Surveillance radar performance prediction** P. Rohan
Volume 18 **Corrugated horns for microwave antennas** P. J. B. Clarricoats and A. D. Olver
Volume 19 **Microwave antenna theory and design** S. Silver (Editor)
Volume 20 **Advances in radar techniques** J. Clarke (Editor)
Volume 21 **Waveguide handbook** N. Marcuvitz
Volume 22 **Target adaptive matched illumination radar** D. T. Gjessing
Volume 23 **Ferrites at microwave frequencies** A. J. Baden Fuller
Volume 24 **Propagation of short radio waves** D. E. Kerr (Editor)
Volume 25 **Principles of microwave circuits** C. G. Montgomery, R. H. Dicke, E. M. Purcell (Editors)
Volume 26 **Spherical near-field antenna measurements** J. E. Hansen (Editor)
Volume 27 **Electromagnetic radiation from cylindrical structures** J. R. Wait
Volume 28 **Handbook of microstrip antennas** J. R. James and P. S. Hall (Editors)
Volume 29 **Satellite-to-ground radiowave propagation** J. E. Allnutt
Volume 30 **Radiowave propagation** M. P. M. Hall and L. W. Barclay (Editors)
Volume 31 **Ionospheric radio** K. Davies
Volume 32 **Electromagnetic waveguides: theory and application** S. F. Mahmoud
Volume 33 **Radio direction finding and superresolution** P. J. D. Gething
Volume 34 **Electrodynamic theory of superconductors** S.-A. Zhou
Volume 35 **VHF and UHF antennas** R. A. Burberry

Propagation, scattering and dissipation of electromagnetic waves

A. S. Ilyinsky,
G. Ya. Slepyan
and A. Ya. Slepyan

Peter Peregrinus Ltd. on behalf of the Institution of Electrical Engineers

Published by: Peter Peregrinus Ltd., on behalf of the Institution of Electrical Engineers, London, United Kingdom

Peter Peregrinus Ltd.,
The Institution of Electrical Engineers,
Michael Faraday House,
Six Hills Way, Stevenage,
Herts. SG1 2AY, United Kingdom

British Library Cataloguing in Publication Data

A CIP catalogue record for this book
is available from the British Library

ISBN 0 86341 283 1

Printed in England by Antony Rowe Ltd., Wiltshire

Contents

Preface

Textbooks and monographs on microwave theory contain very little material on methods of calculating microwave absorption. However, problems do exist in this field which is confirmed by the vast amount of journal publications on the issue, especially during the last decade. There are some mutually contradictory views concerning a number of aspects and arguments about resolved questions regularly arise.

This is largely due to the absence of a monograph dealing with the problems of electromagnetic wave absorption in waveguides, resonators and periodic structures. This book attempts to fill the gap.

The introduction reviews possible ways of reducing losses in waveguides and resonators, it formulates boundary-value problems of electromagnetics for cases when losses are considered, and gives analytical and numerical methods of analysis. Chapter 2 deals with the formulation of impedance boundary conditions for a high skin effect in various conducting media.

Chapter 3 presents the theory of regular waveguides with finitely conducting walls, and Chapter 4 the theory of cavities when the surfaces and the medium inside are absorbing. These chapters present the derivation of general formulae for the attenuation coefficient and Q-factor and consider concrete practical applications. The principal errors associated with the solution of these problems are also discussed.

Chapters 5 and 6 deal with wave diffraction by imperfectly conducting gratings in both free space and inhomogeneous anisotropic media. The main emphasis is on the description of dissipative resonance effects, in particular the effect of the abnormally low dissipation.

Chapter 7 considers the attenuation of eigenmodes in periodic waveguides. It presents results obtained for low-loss flexible corrugated waveguides and high-quality corrugated resonators of the millimetre waveband etc.

A number of theoretical results confirmed by experiment are included.

The book is intended for physicists and engineers performing theoretical research and designing microwave and millimetre-wave devices, and for students and postgraduates involved in these issues. It can be used by a design engineer to select a suitable type of a waveguide (resonator) and methods of computing its parameters.

As similar problems arise in acoustics, much of the book's material can be directly used for sound waves; this makes it useful for specialists in acoustics.

The authors are grateful to Prof. E.A. Alkhovsky, Prof. A.A. Kuraev, Prof. V.A. Cherepenin, Dr. T.N. Galishnikova, Dr. V.V. Zarubanov and Mrs. V.N. Rodionova who made valuable contributions to the research.

The authors greatly appreciate the support and encouragement given by Prof. P.J.B. Clarricoats which has enabled the book to be written.

List of notations

SI units are used

$\boldsymbol{E}, \boldsymbol{H}$ electric and magnetic intensities

$\exp(-j\omega t)$ time dependence

ω cyclic frequency

$j = \sqrt{(-1)}$

$k = 2\pi/\lambda$ wave number

λ wavelength

ε, μ permittivity and permeability of substances

$\varepsilon = \varepsilon' + j\varepsilon''$

$\tan\delta = \varepsilon''/\varepsilon'$

ε_0, μ_0 permittivity and permeability of vacuum

σ_0 static metal conductivity

$\tilde{\varepsilon} = \varepsilon/\varepsilon_0$ relative permittivity

$W_0 = (\mu_0/\varepsilon_0)^{1/2}$ impedance of free space

Z_S surface impedance

$\boldsymbol{n}_0$ interior unit normal

x, y, z Cartesian co-ordinates

$\hat{i}_x, \hat{i}_y, \hat{i}_z$ unit vectors in the x, y and z directions

$\psi(x, y)$ longitudinal component of the electric (or magnetic) field in 2-dimensional problems

$J_m(x), N_m(x)$ Bessel's and Neumann's functions

ν_{mn} nth root of $J_m(x)$

μ_{mn} nth root of $J'_m(x)$

a^* complex conjugate of a

I identity operator

A^{-1} operator inverse to A $(A^{-1}A = I)$

δ_{mn} Kronecker delta

∇^2 Laplacian

Chapter 1

Introduction

1.1 Loss reduction in microwave waveguides and resonators

An important feature of microwave waveguides and resonators is heat losses. The latter determine, in particular, the transient time, knowledge of which is indispensable for designing digital microwave devices (Kiang, 1991). Waveguides with low attenuation are required to create highly effective radio-relay, space and tropospheric-scatter communication facilities and to design measuring equipment (including that operating at the millimetre waveband and infrared). High-quality resonators are widely used in radio engineering (tunable oscillators with high frequency stability), in applied physics (wavemeters, equipment for measuring dielectric coefficients and permeabilities of substances and for microwave heating etc.), in electronics (gyrotrons, orotrons, free electron lasers etc.) and in unique physical experiments (for instance, experiments on detecting gravitational waves).

Therefore, heat-loss control and loss reduction in waveguides and resonators, in particular, are of great importance. Certainly, the most radical way is to use the superconductivity phenomenon. New possibilities in that direction have arisen through the discovery of high-T_C superconductors (Winters and Rose, 1991). Superconducting waveguides and resonators have unique attenuation parameters (so that, for instance, the unloaded Q-factor of a superconducting cavity resonator may reach 10^8–10^9). However, the use of the cryogenic technique is sometimes undesirable or even unacceptable. Moreover, in unique physical experiments, when an extremely high sensitivity of the measuring apparatus is required, the values of Q actually obtained are often insufficient (Braginsky *et al.*, 1981). This points to the need for further research in this area.

The use of low-temperature techniques in microwave devices is not confined to the superconductivity phenomenon. For example, experiments with cooled high-Q-factor ring-dielectric sapphire resonators without superconducting films (Braginsky and Vyatchanin, 1980) are of considerable interest.

Measures of a technological character, such as polishing of inner surfaces and deposition of special coatings, are also important for loss reduction. We shall not touch upon these questions here; we merely mention that the possibilities available in this direction have been largely exhausted. Therefore, optimal selection of the waveguide (resonator) type, configuration and working mode is of paramount importance. It is necessary to emphasise that this formulation of the problem is by no means an alternative to the low-temperature technique. On the contrary, the results obtained in this direction may widen the opportunities provided by cryogenic-engineering facilities.

To solve this problem, adequate methods of loss calculation for waveguides and resonators are needed. These methods form the main subject of this book, along with physical models, the most important numerical results and comparison of the latter with experimental data. In these introductory remarks we shall not try to summarise the contents of the book but merely dwell on the historical aspects of the development of the subject [see also the review material in publications by Ilyinsky and Slepyan (1983, 1990)].

Reports about allegedly abnormal low losses in some electrodynamic systems of a complex configuration have occurred from time to time; these have not been confirmed (see below). If we take into account difficulties related both to the calculation and to accurate measurement of low losses, such a situation is not surprising. Two directions of research are possible when trying to reduce losses through changing the electrodynamic-system configuration. The first ('direct') is to use waveguides and resonators with fundamental modes. Mode selection causes no problems and shape optimisation is used directly to reduce the losses. Rectangular waveguides with rounded corners, double-ridge waveguides (Kotayama *et al.*, 1979, Alkhovsky *et al.*, 1986) can be mentioned as examples. It is difficult to expect a significant gain in this way, but if the shape is not too complex even a minor gain (by 25–50%) is valuable. Besides, the question is of principal interest.

The use of multimode systems, where effective rarefaction of the eigenmodes spectrum and suppression of lower (fundamental) modes are provided by shape optimisation, is another method (let us call it 'indirect'). In this case the working mode has a high Q-factor and is higher-order with respect to its field structure, being, in fact, fundamental (or one of the fundamental modes).

The use of attenuation abnormality for the TE_{0i} modes in circular waveguides is a classical example of this. By taking a large enough waveguide radius it is possible to obtain the TE_{01}-mode attenuation as low as desired (Okress, 1968). However, though attractive, this result can not really be used: it occurs in essentially multimode regimes. The situation is made worse by the fact that the TE_{0i} and TM_{1i} modes are degenerate at any value of working frequency. The operational TE_{01} mode will be transformed into parasitic modes at the waveguide discontinuities (bends, junctions, technological deformations of the cross-section etc.). The attenuation of these parasitic modes is much more significant than that of the TE_{01} mode. The transformation effects are essentially enhanced by modes degeneration. As a result, the total losses increase and attenuation reduction may turn out to be illusive.

A similar situation is observed in cylindrical cavity resonators. The TE_{0in} modes have the highest Q but they are degenerate with the TM_{1in} modes for any relationship between the cavity dimensions, and can only be used in an essentially multimode regime. It leads to difficulties in excitation and tuning of the resonator.

A problem of parasitic-mode suppression therefore arises. In other words, it is a problem of rarefaction of the eigenmodes spectrum of a circular waveguide. One of the ways to solve it is to use spaced-disc or helix waveguides. Spaced-disc waveguides (circular waveguides with narrow ring slits) are open systems where a spectrum rarefaction takes place because of a strong parasitic wave radiation. If the parameters are chosen properly, the losses for the TE_{01} mode increase somewhat, compared with a smooth circular waveguide, but considerable suppression of parasitic modes is thus achieved. However, such waveguides in low-loss regime are also multimode ones, and are therefore highly sensitive to the waveguide irregularities.

To remove the degeneration of the TE_{0i} and TM_{1i} modes, a corrugation of the internal surface of a circular waveguide can be used. Considering cases when the period and depth of corrugations are small compared with the wavelength, Gent (1959) predicted an essential reduction (by an order) of attenuation of the TE_{0n} mode. However, a more careful theoretical analysis (Katsenelenbaum, 1959, Nefedov and Sivov, 1977) has not confirmed this highly promising result. According to the rigorous solution obtained by Katsenelenbaum (1959), the TE_{0n}-mode attenuation in a corrugated waveguide grows but not significantly. Nefedov and Sivov (1977) have analysed the mistaken conclusions drawn by Gent (1959) in detail. The idea had been forgotten for some time.

Fundamental investigations by Clarricoats and his coworkers (Clarricoats and Saha, 1970, Parini *et al.*, 1977, Clarricoats *et al.*, 1975a, 1975b) have revived the idea of using corrugated (comb-shaped) surfaces for loss reduction. But there it was related to the HE_{11} mode and the period and depth of corrugations were great enough—about a quarter of the wavelength. It was shown theoretically and experimentally that in such waveguides attenuation of the HE_{11} mode is approximately equal to that of the TE_{01} mode in conventional circular waveguides (~ 0.01 dB/m at $f \simeq 3$–20 GHz). However, unlike the latter, in corrugated waveguides the weakly attenuated mode can be fundamental and nondegenerate.

In such a way the effect of abnormally small dissipation of electromagnetic waves in periodical structures was discovered. Further work in this direction was continued by the authors of this book. The results are presented in Chapters 5 and 7. The emphasis was laid on the physical nature of the phenomenon, namely on the specific mechanism of the decrease in the induced surface currents and also the loss-calculation methods for electrodynamic systems with corrugated (comb-shaped) surfaces.

This effect has already been implemented in practice. Flexible low-loss corrugated waveguides based on this effect with various (rectangular, circular, elliptical, double-ridge) cross-sections are manufactured in the UK, USA, France, Russia, Germany, Canada and Japan. These waveguides are convenient for use in both stationary and mobile communication facilities because it is possible to bend them and to reel them on a drum many times. The theory, design, manufacture technology and examples of application of flexible corrugated waveguides were described by Alkhovsky *et al.* (1986). A broadband multi-mode corrugated low-loss waveguide ($\gamma'' < 0.01$ dB/m at $f \simeq 75$–535 GHz) was used to measure the parameters of electron cyclotron radiation in the TFTR Tokamak (Carallo *et al.*, 1990).

High-quality tunable resonators for the millimetre waveband (Luk *et al.*, 1988, Rodionova and G. Slepyan, 1989, Rodionova *et al.*, 1990) are another example of application of an abnormally small dissipation effect. To reduce the losses in the walls, the internal surface is partially corrugated. The TM_{0mn} modes of the resonator are operational. To rarefy the eigenmodes spectrum, special filtering slits are provided. The unloaded Q-factor of 10^5 is achieved for $f \simeq 32$–53.57 GHz without any cryogenic-engineering means. The results of theoretical and experimental investigations of such resonators are presented in Section 7.9.

The corrugated surfaces have another important application in antenna engineering. Using them as a basis, highly effective feeds for reflector antennas have been developed (Clarricoats and Olver, 1984).

Another possible way of reducing the losses in metallic walls is to apply metal–dielectric structures (Kazantsev *et al.*, 1974, Kazantsev and Kraftmakher, 1979). The idea is to introduce a dielectric insert of a special shape which leads to a reduction in the currents induced on the metallic surfaces. Although dielectric inserts cause some attenuation, the total losses in a metal–dielectric system may be lower than those in a similar system having no dielectric filling. A characteristic example is a hollow circular waveguide with a coaxial dielectric bushing (Kazantsev *et al.*, 1974). Another example is a metal–dielectric resonator described in Section 4.7. The physical nature of the loss reduction in this case is similar to the abnormally small dissipation effect in corrugated surfaces. However, in periodic structures there is also a highly effective mechanism of parasitic-mode filtration due to transmission and stop bands.

Of other types of highly effective transmission lines, we mention single groove-guides, coupled groove-guides and dielectric ribbon waveguides. According to Meissner (1990), attenuation in a single groove-guide at the band $f \simeq 40$–120 GHz is about 0.1–0.4 dB/m, whereas in a conventional single-mode rectangular waveguide it is more than 4 dB/m. A coupled groove-guide has similar characteristics: 0.1 dB/m for the even mode and 0.2 dB/m for the odd mode. Optimisation of the configuration (Yeh *et al.*, 1990) has permitted an attenuation of 0.02 dB/m to be obtained in a dielectric ribbon waveguide at millimetre and submillimetre wavebands using a material with $\tan\delta \sim 10^{-4}$.

1.2 Maxwell's equations; constitutive equations; boundary conditions

Problems of calculation of electromagnetic fields in microwave systems are solved on the basis of macroscopic electrodynamics. The behaviour of macroscopic electromagnetic phenomena is governed by Maxwell's equations

$$\left.\begin{aligned} \operatorname{curl} \boldsymbol{H} &= \frac{\partial \boldsymbol{D}}{\partial t} + \boldsymbol{J} + \boldsymbol{J}^{ext} \\ \operatorname{curl} \boldsymbol{E} &= -\frac{\partial \boldsymbol{B}}{\partial t} \end{aligned}\right\} \qquad (1.1)$$

where $\boldsymbol{E}$ and $\boldsymbol{H}$ are vectors of electric and magnetic intensities, $\boldsymbol{D}$ and $\boldsymbol{B}$ are vectors of electric and magnetic flux densities, $\boldsymbol{J}$ is the macroscopic-conduction current density in the medium and $\boldsymbol{J}^{ext}$ is the current density induced by external sources.

In the electromagnetic theory, time-harmonic fields like

$$\begin{bmatrix} \boldsymbol{E}(\boldsymbol{r}, t) \\ \boldsymbol{H}(\boldsymbol{r}, t) \end{bmatrix} = \mathrm{Re}\left\{ \begin{bmatrix} \boldsymbol{E}(\boldsymbol{r}, \omega) \\ \boldsymbol{H}(\boldsymbol{r}, \omega) \end{bmatrix} \exp(-j\omega t) \right\} \tag{1.2}$$

are of fundamental importance. Similar relationships can be written for the rest of the quantities in eqn. 1.1 (ω is a real number called cyclic frequency). The quantities $\boldsymbol{E}(\boldsymbol{r}, \omega)$ and $\boldsymbol{H}(\boldsymbol{r}, \omega)$ are called complex amplitudes of the corresponding vectors.

From now on, we shall study electromagnetic fields described by eqn. 1.2 and shall operate with complex amplitudes only, with no special notations for the latter; the only exception will be the problem of free oscillations in a resonator with a magnetodielectric absorbing body (see Section 4.3). Harmonic fields expressed by eqn. 1.2 are good models for the description of steady-state narrowband processes and, in fact, they have a much more general meaning. The fact is that an electromagnetic field with an arbitrary time dependence in a linear system can be expanded in the Fourier series or integral in terms of time-harmonic fields.

Maxwell's equations for complex amplitudes can be written as

$$\left. \begin{aligned} &\mathrm{curl}\, \boldsymbol{H} = -j\omega \boldsymbol{D} + \boldsymbol{J} + \boldsymbol{J}^{ext} \\ &\mathrm{curl}\, \boldsymbol{E} = j\omega \boldsymbol{B} \end{aligned} \right\} \tag{1.3}$$

Within the scope of macroscopic electrodynamics, eqns. 1.3 are completed by constitutive equations which, for example, can be in the form

$$\left. \begin{aligned} \boldsymbol{D}(\boldsymbol{r}) &= \varepsilon(\boldsymbol{r}, \omega)\, \boldsymbol{E}(\boldsymbol{r}) \\ \boldsymbol{B}(\boldsymbol{r}) &= \mu(\boldsymbol{r}, \omega)\, \boldsymbol{H}(\boldsymbol{r}) \\ \boldsymbol{J}(\boldsymbol{r}) &= \sigma(\boldsymbol{r}, \omega)\, \boldsymbol{E}(\boldsymbol{r}) \end{aligned} \right\} \tag{1.4}$$

The quantities $\varepsilon(\boldsymbol{r}, \omega)$, $\mu(\boldsymbol{r}, \omega)$ and $\sigma(\boldsymbol{r}, \omega)$ are called the permittivity, permeability and conductivity of the medium. They are scalars for isotropic media and tensors for anisotropic media.

Constitutive equations given by eqns. 1.4 arise in electromagnetic theory as a result of macroscopic averaging, i.e. field averaging over a physically infinitesimal volume.[1] This averaging leads to smoothing of microscopic details of the field structure: sharp field oscillations are averaged at an atom-molecular scale. As experience shows, these details are completely insignificant for the determination of macroscopic parameters of electromagnetic processes such as power absorbed, diffraction patterns, eigenmode spectra etc. But, in this case the problem is essentially simplified. The parameters of substances ε, μ and σ are assumed to be specified within the scope of macroscopic electrodynamics. The problem of their determination for various media is solved irrespective of the field-calculation problem. These parameters can be calculated from some structural models of the media or measured in experiments. The physical funda-

---

[1] By physically infinitesimal volume we mean a volume with dimensions much larger than atom-molecular dimensions but small compared with the wavelength, body size and other macroscopic quantities (Tamm, 1976).

mentals of macroscopic electrodynamics and the macroscopic averaging technique are described in Tamm's book (1976).

Still more general forms of constitutive equations are possible; for example, nonlocal connections may occur between $\boldsymbol{D}$ and $\boldsymbol{E}$, $\boldsymbol{B}$ and $\boldsymbol{H}$, $\boldsymbol{J}$ and $\boldsymbol{E}$:

$$\boldsymbol{D}(\boldsymbol{r}) = \int \varepsilon(\boldsymbol{r}, \boldsymbol{r}', \omega)\, \boldsymbol{E}(\boldsymbol{r}')\, \mathrm{d}^3\boldsymbol{r}'$$

$$\boldsymbol{B}(\boldsymbol{r}) = \int \mu(\boldsymbol{r}, \boldsymbol{r}', \omega)\, \boldsymbol{H}(\boldsymbol{r}')\, \mathrm{d}^3\boldsymbol{r}'$$

$$\boldsymbol{J}(\boldsymbol{r}) = \int \sigma(\boldsymbol{r}, \boldsymbol{r}', \omega)\, \boldsymbol{E}(\boldsymbol{r}')\, \mathrm{d}^3\boldsymbol{r}'$$

The nonlocality mentioned reflects a phenomenon of spatial wave dispersion (Agranovich and Ginzburg, 1979). Further, we shall not take into account spatial dispersion in magnetodielectrics. In Section 2.3 we shall consider in detail one of the spatial dispersion effects in metals, i.e. the anomalous skin effect.

The quantities ε and μ are real numbers for lossless media. For lossy magnetodielectrics they are complex: $\varepsilon = \varepsilon' + j\varepsilon''$, $\mu = \mu' + j\mu''$. The form of the constitutive equations in this case is the same as for lossless media. Because of this, analysis of electrodynamic systems with lossy magnetodielectrics can be reduced to replacement of real ε and μ by complex ones in the final expressions obtained for nonabsorbing structures.

A perfect conductor is characterised by the boundary condition

$$(\boldsymbol{n}_0 \times \boldsymbol{E})|_S = 0 \tag{1.5}$$

where S is the surface of the conductor and $\boldsymbol{n}_0$ is a unit vector normal to S. We shall take into account finite conductivity of the metal on the basis of the impedance boundary conditions. It means that instead of eqn. 1.5 the boundary condition

$$\boldsymbol{n}_0 \times \boldsymbol{E} = -Z_S\{\boldsymbol{n}_0 \times (\boldsymbol{n}_0 \times \boldsymbol{H})\} \tag{1.6}$$

should be used. Here $\boldsymbol{n}_0$ is the unit inward normal to the metal surface and Z_S is the given surface impedance. Chapter 2 deals with the derivation and substantiation of the impedance boundary conditions for various cases.

Certainly, a conducting medium could be formally considered as a 'conventional' dielectric characterised by the permittivity $\varepsilon = \sigma/j\omega$. Unfortunately, the use of such an approach is complicated by the need to describe the field inside the metal volume, although it has a significant magnitude in the thin near-boundary layer only (skin effect). Use of the impedance boundary conditions in the form of eqn. 1.6 exempts us from this necessity. The field within the conducting medium is not considered, but the full information about this field is contained in the magnitude of the surface impedance and its frequency dependence.

However, variations in the boundary-condition form compared with the case of perfect conductivity fundamentally change the mathematical statement of the problem: some methods giving excellent results for perfectly conducting structures become unacceptable for impedance structures; others require a considerable modification. Therefore, a major part of this book will be dedicated to solution techniques for electrodynamic problems with impedance boundary conditions.

It should be noted that ζ_S is a small parameter which is in many cases interesting from a practical point of view. This leads to the idea of applying the perturbation method to calculate the dissipation characteristics. However, the small size of ζ_S means that this technique cannot yet be applied. In many cases its direct application leads to nonsense, although sometimes the fact is masked. We analyse the causes of failures in detail and suggest correct modifications of this approach. Finally, within the framework of the perturbation technique various schemes are possible. They are equivalent in principle but differ in computational efficiency. We therefore pay special attention to the comparative analysis of various modifications of the perturbation method and the use of the latter for solving specific applied problems.

1.3 Solution techniques for mathematical problems of electromagnetics

The purpose of the following introductory remarks is to characterise the problems of electromagnetics studied in this book and the solution techniques used. From the mathematical point of view, these problems are boundary-value ones for partial differential equations. We shall consider both diffraction problems (excitation of electrodynamic systems by fields or currents given) and spectral problems (eigenmodes of smooth-wall and periodic waveguides, free oscillations of resonators). Figures 1.1 and 1.2 show typical examples of such structures.

The diffraction problem can be presented in an operator form as

$$A\,\boldsymbol{u}(\boldsymbol{r}) = \boldsymbol{f}(\boldsymbol{r}) \tag{1.7}$$

where $\boldsymbol{r} \in V$, A is the differential operator, $\boldsymbol{u}$ is the function sought and $\boldsymbol{f}$ is a given free term. Let us assume that the boundary condition

$$B\,\boldsymbol{u}(\boldsymbol{r}) = 0 \tag{1.8}$$

is imposed on the surface S bounding the volume V ($\boldsymbol{r} \in S$, B is some differential operator).

In the general case A is Maxwell's operator

$$A = \begin{bmatrix} \text{curl} & -j\omega\mu \\ j\omega\varepsilon & \text{curl} \end{bmatrix}$$

and

$$\boldsymbol{u}(\boldsymbol{r}) = \begin{bmatrix} \boldsymbol{E} \\ \boldsymbol{H} \end{bmatrix}$$

is an unknown vector of the electromagnetic field depending on three space variables.

For finitely conducting structures, the impedance boundary condition given by eqn. 1.6 plays the role of eqn. 1.8.

For a wide class of problems the electromagnetic field does not depend on one of the Cartesian co-ordinates and can be expressed in terms of a scalar

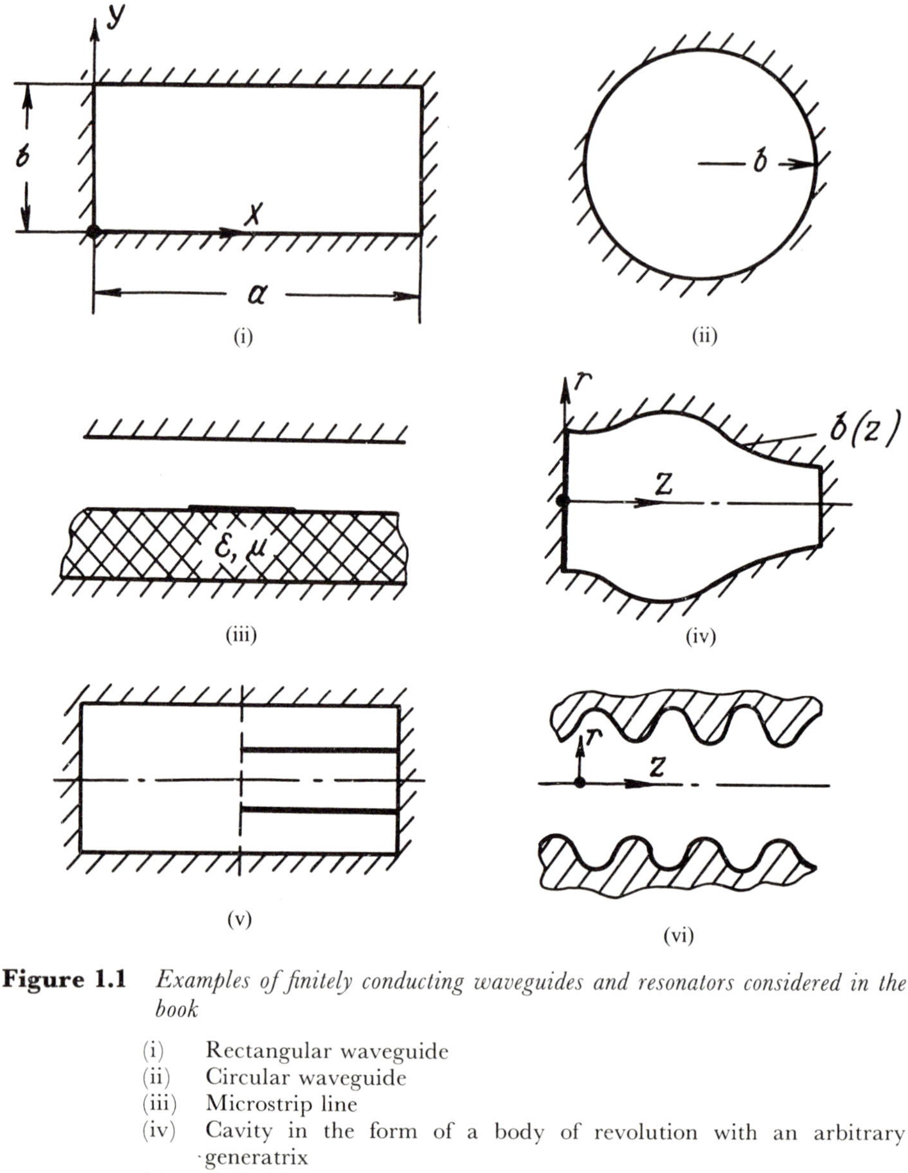

Figure 1.1 *Examples of finitely conducting waveguides and resonators considered in the book*

(i) Rectangular waveguide
(ii) Circular waveguide
(iii) Microstrip line
(iv) Cavity in the form of a body of revolution with an arbitrary generatrix
(v) Cylindrical resonator with coaxial insert
(vi) Corrugated waveguide

function.[2] In this case V is a plane area, S is the contour bounding this area, $A = \nabla^2 + k^2$ is the Helmholtz 2-dimensional operator and $\boldsymbol{u} = \psi$ is the unknown

[2] This also takes place when the field depends on this co-ordinate but this dependence is known from the symmetry properties.

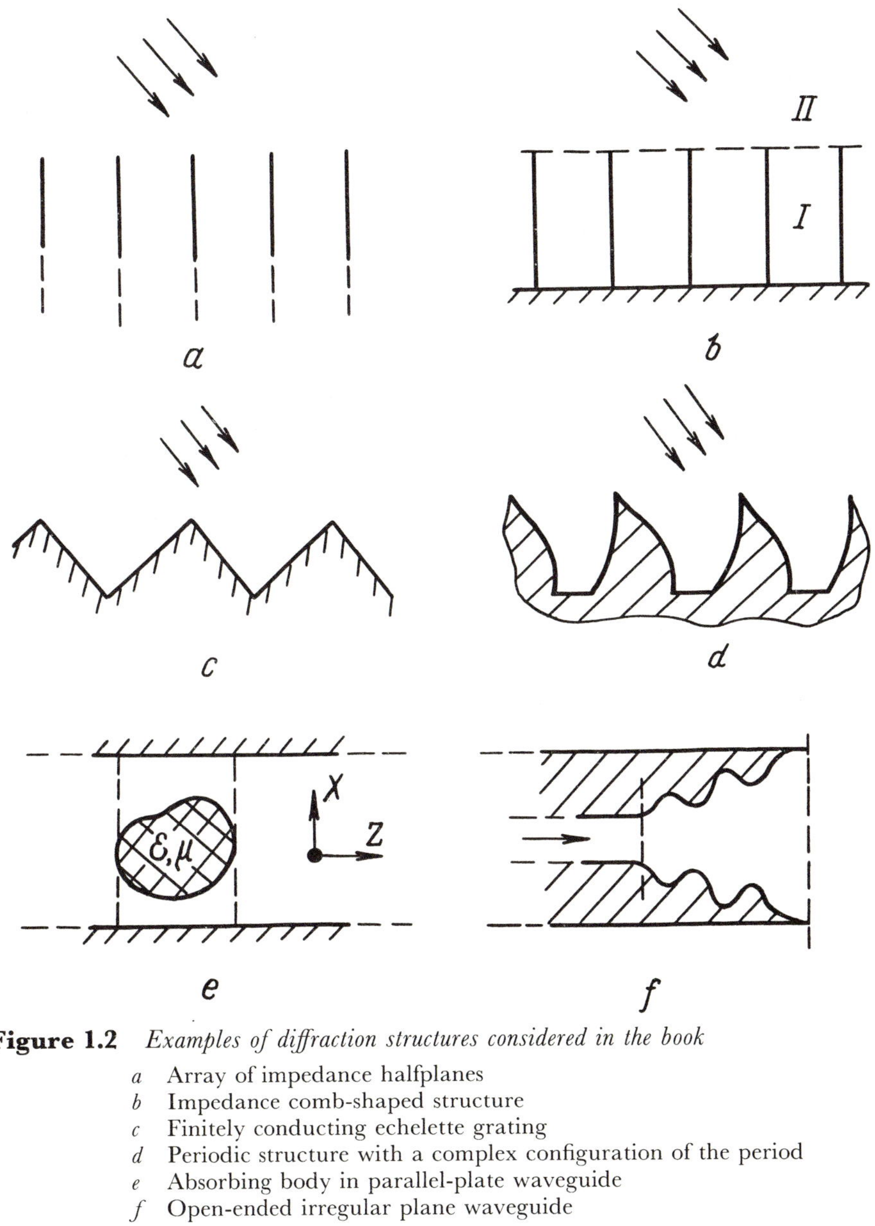

Figure 1.2 *Examples of diffraction structures considered in the book*

a Array of impedance halfplanes
b Impedance comb-shaped structure
c Finitely conducting echelette grating
d Periodic structure with a complex configuration of the period
e Absorbing body in parallel-plate waveguide
f Open-ended irregular plane waveguide

longitudinal-field component. Eqn. 1.6 is then transformed into a boundary condition of the third kind:

$$\left[\frac{\partial\psi}{\partial\boldsymbol{n}} + j\eta\psi\right]\Bigg|_S = 0$$

i.e. $B = (\partial/\partial\boldsymbol{n}) + j\eta$, where η is a given complex number.

The area V can be either bounded or unbounded (external in relation to S). In the latter case the solution of the boundary-value problem under discussion should satisfy the radiation condition at infinity.

The formulation of a spectral problem is different. Instead of eqn. 1.7, the following homogeneous operator equation is considered:

$$A\,\boldsymbol{u}(\boldsymbol{r}) = \lambda q\,\boldsymbol{u}(\boldsymbol{r}) \tag{1.9}$$

Here q is a given weight operator and λ is a constant. It is necessary to find nontrivial solutions of eqn. 1.9 (they are called eigenfunctions). The values of λ, when such solutions do exist, are called eigenvalues. These nontrivial solutions have to satisfy the required boundary condition on S.

A specific feature of systems with dissipation (i.e. when ε, μ and Z_S are complex numbers) is that the problems in question are non-self-adjoint due to the structure of the operators A and B. The theory of non-self-adjoint operators (Keldysh, 1951, Neimark, 1954) is much more complicated than that of self-adjoint ones. Very unusual properties may be typical for non-self-adjoint operators. It is enough to say that they may have no eigenfunctions at all, and that when eigenfunctions do exist their systems may not form the basis in $L_2(V)$. However, as it will be further shown in this book, many electrodynamic problems for systems with dissipation which are interesting from a practical point of view lead to non-self-adjoint operators which are small perturbations of self-adjoint ones. Their spectral properties are very close to those of self-adjoint operators: the specific character of non-self-adjoint problems is shown either partially or not at all. This makes it possible to use the conventional solution techniques.

In this book we apply mathematically rigorous solution methods, i.e. methods with no assumptions based on physical intuition. The approximate analytical expressions obtained in the book are a result of some parameters being small and the use of the perturbation method. The applicability limits for these expressions are indicated and their accuracy is evaluated. In this book we apply exact analytical, direct numerical and analytical–numerical methods. Of a variety of exact analytical methods, the method of separation of variables (Jones, 1986) and the Wiener-Hopf technique (Noble, 1958, Mittra and Lee, 1971) are used. Examples of structures to which the two methods are applied are shown in Figures 1.1(i) and (ii) and Figure 1.2*a*, respectively. The capabilities of these approaches are limited by very specific configurations of the objects.

The use of direct numerical methods provides far more opportunities. In the framework of these methods the original boundary-value problem can be reduced to a standard mathematical procedure performed by a computer. We can list four such procedures:

(i) finite-order matrix inversion;
(ii) determination of eigenvalues and eigenvectors of a finite-order matrix;
(iii) solution of a transcendental equation;
(iv) solution of a 2-point boundary-value problem for a linear system of ordinary differential equations by the shooting technique.

Let us dwell upon the statement of problem (iv). It can be written in the form

$$\left.\begin{aligned} \frac{\mathrm{d}\boldsymbol{y}}{\mathrm{d}t} - A(t)\,\boldsymbol{y} &= \boldsymbol{f}(t) \\ B\,\boldsymbol{y}(0) &= \boldsymbol{b} \\ D\,\boldsymbol{y}(T) &= \boldsymbol{d} \end{aligned}\right\} \tag{1.10}$$

where $0 < t < T$, $\boldsymbol{y}(t)$ is an unknown vector of m components, $\boldsymbol{f}$, $\boldsymbol{b}$ and $\boldsymbol{d}$ are given column vectors of m, $m - r$ and r components, respectively, and A, B and D are matrices of the orders $m \times m$, $(m - r) \times m$ and $r \times m$. The ranks of the matrices B and D are $m - r$ and r. There are a few modifications of the shooting method which differ in their computational characteristics. Their description and comparative analysis are given in Appendix 1 (see also Roberts and Shipman, 1972).

Of direct methods, we use the moment method (Harrington, 1968), Galerkin's incomplete method (Sveshnikov, 1969) and the integral-equation technique (Poggio and Miller, 1973).

Let us consider an elementary scheme of the moment method for the spectral problem given by eqns. 1.9 and 1.8. Let the set of basis functions $\{\boldsymbol{u}_s\}$ ($s = 1, 2, \ldots, \infty$) complete in $L_2(V)$ be specified. In addition we assume that each of the functions $\boldsymbol{u}_s$ satisfies the required boundary condition. The solution of the spectral problem is sought in the form of a series

$$\boldsymbol{u}(\boldsymbol{r}) = \sum_{p=1}^{\infty} c_p\,\boldsymbol{u}_p(\boldsymbol{r}) \qquad (\boldsymbol{r} \in V) \tag{1.11}$$

where c_p are unknown coefficients. Introducing another set of functions $\{\boldsymbol{v}_s\}$ complete in $L_2(V)$, we replace eqn. 1.9 by a system of orthogonality relations

$$(A\boldsymbol{u} - \lambda q\boldsymbol{u}, \boldsymbol{v}_s) = 0 \tag{1.12}$$

($s = 1, 2, \ldots, \infty$). Eqn. 1.12 results from the fact that a function orthogonal to all functions of a complete set is identical to zero (Kantorovich and Akilov, 1977). The parentheses in eqn. 1.12 mean the scalar product in $L_2(V)$ defined as $(\mathrm{f}, \phi) = \int_V \boldsymbol{f}(\boldsymbol{r})\phi^*(\boldsymbol{r})\,\mathrm{d}V$.

Substituting eqn. 1.11 into eqn. 1.12, we come to an eigenvalue problem for infinite-dimensional matrices like

$$\sum_{p=1}^{\infty} c_p(A\boldsymbol{u}_p, \boldsymbol{v}_s) = \lambda \sum_{p=1}^{\infty} c_p(q\boldsymbol{u}_p, \boldsymbol{v}_s) \tag{1.13}$$

The particular case $\boldsymbol{v}_s = \boldsymbol{u}_s$ corresponds to Galerkin's method. At the final stage, eqn. 1.13 is reduced to a set of a finite order: N first terms in the series and N equations with $s = 1, 2, \ldots, N$ are taken into account. Then, eqn. 1.11 is transformed into

$$\boldsymbol{u}(\boldsymbol{r}) = \sum_{p=1}^{N} c_p^{(N)}\,\boldsymbol{u}_p(\boldsymbol{r}) \tag{1.14}$$

Eqn. 1.14 is approximate due to the truncation of the highest terms with $p > N$ in the series, and since the first N coefficients are calculated approximately from a finite set of equations (this fact is emphasised by the superscript N in the coefficients in eqn. 1.14). An important question arising here is that of convergence of the approximate eigenvalues $\lambda^{(N)}$ to the exact ones and the convergence of the representations, given by eqn. 1.14, to eigenfunctions, when $\mathcal{N}\to\infty$. A number of works has been devoted to mathematical proofs of this problem (e.g. Mikhlin, 1970).

This elementary scheme is used in Section 4.6, in particular, for the structure shown in Figure 1.1(iv). The case where the basis functions do not satisfy the required boundary conditions is more typical for systems with dissipation. In this case the orthogonality relations given by eqn. 1.12 have to be modified so that the expansion in eqn. 1.11 would provide both the fulfilment of eqn. 1.9 and the boundary condition given by eqn. 1.8. Examples of such modifications are presented in Sections 3.2 and 7.6.

The idea of Galerkin's incomplete method is to expand the function sought $\boldsymbol{u}(q_1, q_2, q_3)$ in terms of basis functions $\{\boldsymbol{u}_s(q_1, q_2, q_3)\}$ forming a complete set in each section of V by the surface $q_3 = \text{constant}$:[3]

$$\boldsymbol{u}(q_1, q_2, q_3) = \sum_{p=1}^{\infty} c_p(q_3)\, \boldsymbol{u}_p(q_1, q_2, q_3)$$

Here, in contrast with eqn. 1.11, the coefficients c_p are functions of one of the co-ordinates. The orthogonality relations now are written on a set of sections of the domain V by the surface $q_3 = \text{constant}$. A system of differential equations for $c_p(q_3)$, which can be reduced to eqns. 1.10, results from these relationships. In Sections 5.8, 7.6 and 7.10 Galerkin's incomplete method is applied to structures shown in Figures 1.1(vi), 1.2*d* and 1.2*f*.

The integral-equation method is applied to a 2-dimensional problem of wave diffraction by a finitely conducting complex-shaped grating (Section 5.7). Using the methods of potential theory, the following Fredholm integral equation of the second kind was obtained:

$$\mu(P) + \frac{1}{2\pi}\int_C K(P, P')\mu(P')\,\mathrm{d}l_{P'} = F(P) \qquad (1.15)$$

where $\mu(P)$ is an unknown function specified at the contour C of one period of the structure. The essential property of this integral equation is that the kernel $K(P, P')$ has a logarithmic singularity when the arguments P and P' coincide. To solve it numerically the Krylov–Bogolyubov method (Kantorovich and Krylov, 1958) is used. The unknown function is expanded in a series in terms of pulse functions, and eqn. 1.15 is reduced to a set of linear algebraic equations. An important advantage of this method is a high degree of universality as regards the shape of the contour C.

The methods mentioned above place practically no limitations on the configuration of the areas. However, there are classes of problems related to areas of special forms, which are of great importance for applications. To solve these problems it is possible to employ the techniques leading to specialised computational algorithms which, however, have a higher computational efficiency. These are so-called 'co-ordinate' structures where the boundaries of areas considered consist of parts of co-ordinate lines (surfaces) of any classical system of orthogonal co-ordinates (Cartesian, cylindrical, spherical etc.). In this case the area under consideration may be divided into subareas for which general solu-

[3] In particular cases $\boldsymbol{u}_s$ may be completely independent of the variable q_3.

tions of the equations being solved are written in a convenient form.[4] The fields matching at the interfaces between the subareas results in sets of linear algebraic equations or dual integral (series) equations. A microstrip line [Figure 1.1(iii)] is an example of such a structure. Its cross-section is decomposed into two subareas in the form of layers over and under the strip. Solutions for each layer are presented as Fourier integrals. The imposition of the boundary conditions results in dual integral equations, which, unlike eqn. 1.9, are nonlinear in the spectral parameter λ. A numerical solution of these equations is carried out by the moment method (see Sections 3.8 and 3.9).

For regions of special form the numerical–analytical methods (Mittra and Lee, 1971, Shestopalov *et al.*, 1984) are of great importance. Their essence is that the analytical techniques are not applied to find the solution in a closed form but to transform the equation solved to a form allowing a more effective application of numerical algorithms.

Thus, for wave diffraction by a comb-shaped structure (Figure 1.2*b*) it is convenient to separate the area under consideration into a halfspace over the structure and internal cell areas. In the first subarea the field is expanded in a series in terms of Floquet's harmonics; in others it is expanded in terms of forward and backward eigenmodes of the corresponding plane waveguide. The field matching leads to the infinite set of linear algebraic equations

$$A\boldsymbol{x} = \boldsymbol{f} \tag{1.16}$$

where $\boldsymbol{x}$ is an unknown infinite-dimensional vector, $\boldsymbol{f}$ is a known infinite-dimensional vector and A is the matrix operator (see Section 5.3).

It is often ineffective to solve such sets numerically by the truncation method since eqn. 1.16 belongs to a class of ill-posed problems: operator equations of the first kind (Tikhonov and Arsenin, 1977). In this case the limit, to which the truncation method converges, depends on the method of truncation and may not coincide with the solution required [a phenomenon of 'relative convergence' studied by Mittra and Lee (1971), Shestopalov *et al.* (1984)]. The correct method of truncation is unknown *a priori* but even if it is found the convergence will be slow. In addition, the sets obtained are ill-conditioned.

Therefore, an analytical regularisation of eqn. 1.16 is carried out. The operator A is presented in the form $A = A_0 + A_1$, where A_0 is an analytically invertible 'main part' of the operator A causing difficulties in numerical solution of eqn. 1.16. For wave diffraction by a comb-shaped structure, A_0 is a convolution matrix operator[5] of the form

$$A_{0,mn} = \frac{1}{\Gamma_m - \gamma_n}$$

where $m = 0, \pm 1, \pm 2, \ldots, \pm\infty$, $n = 0, 1, 2, \ldots, \infty$, $\Gamma_m = O(|m|)$ and $\gamma_n = O(n)$. If A_0^{-1} operates on the left-hand side and the right-hand side of eqn. 1.16, the result will be

$$(I + A_0^{-1}A_1)\boldsymbol{x} = A_0^{-1}\boldsymbol{f} \tag{1.17}$$

---

[4] These solutions are in the form of Fourier series, Fourier–Bessel series with unknown coefficients, or Fourier, Hankel, Mellin, Mehler–Fock and other integrals with unknown spectral amplitudes.

[5] The procedure of its analytical inversion is described by Mittra and Lee (1971) and Shestopalov *et al.* (1984).

Eqn. 1.17 reflects the essence of regularisation from the left or semi-inversion being developed by Shestopalov *et al.* (1984).

Another possible approach is a method of regularising substitution or regularisation from the right. The idea of this method is to introduce the linear substitution of the form $\boldsymbol{x} = A_0^{-1}\boldsymbol{g}$, where $\boldsymbol{g}$ is a new vector of the unknowns. Then, the following equation is obtained for $\boldsymbol{g}$:

$$(I + A_1 A_0^{-1})\boldsymbol{g} = \boldsymbol{f} \tag{1.18}$$

In some cases a linear–fractional substitution of the unknown quantities is expedient to simplify the operator structure in eqn. 1.18 (see, for example, Sections 5.1, 5.3).

The efficiency of a regularisation scheme depends on the particular features of the problem considered. However, if the 'main part' is reasonably defined, both of these provide the convergence of the approximate solution, obtained by the truncation method, to the accurate solution. The rate of convergence is very high.

These methods of analytical regularisation were applied to various classes of operator equations: matrix, integral, dual integral and dual series equations. Certainly, the possibility of extracting an analytically invertible 'main part' is created by their special form imposing restrictions on the configuration of structures under consideration. Normally the 'main part' has a clear physical meaning corresponding to the extreme value of frequency or one of the geometrical parameters of the problem. To invert it, various methods, such as the Wiener–Hopf technique (Noble, 1958), the method of the Riemann–Hilbert problem (Shestopalov, 1970) or the residue-calculus technique (Mittra and Lee, 1971) are applied.

In this book the numerical–analytical methods are developed in two directions, one for structures with impedance-boundary conditions, and the other for structures with a complex non-co-ordinate shape of the surface (Figures 1.2*d* and 1.2*e*). The latter case is notable because the diffraction problem is reduced to a 2-point boundary-value problem for an infinite set of ordinary differential equations like eqns. 1.10, and the semi-inversion procedure is carried out at one of its boundary conditions (see Sections 5.8 and 7.10).

1.4 Accuracy control and computational instabilities

In the previous Section we characterised briefly the mathematical methods used in the book. These methods are based on different ideas and differ in generality, computational efficiency and labour content of analytical transformations. The survey gives rise to the following questions. What criteria should be used when choosing a method to solve a particular problem? How should we monitor the computational accuracy obtained?

One of the most fundamental aspects in all methods under consideration is the truncation, i.e. an approximation of an infinite set of algebraic or differential equations by a finite set. The elements of matrices and free-term vectors involve integrals, series and infinite products which often cannot be calculated in a closed form. Errors due to rounding and other factors invariably occur when calculating such elements and solving these sets numerically. Despite the errors being small in each particular step of the algorithm, they may accumulate and, as a consequence, considerably distort the final result.

In the light of this we formulate the following key criteria for method efficiency:

(i) high rate of convergence;
(ii) simplicity of calculation of matrix and free-term-vector elements; and
(iii) stability against computational errors.

Let us suppose that as a result of application of any numerical method a boundary-value problem has been reduced to a matrix equation of the Nth order:

$$A\boldsymbol{x} = \boldsymbol{f} \tag{1.19}$$

Assuming that we have the following equation due to computational errors

$$(A + \delta A)\tilde{\boldsymbol{x}} = \boldsymbol{f} + \delta\boldsymbol{f} \tag{1.20}$$

instead of eqn. 1.19. The solution of eqn. 1.20 can be presented as $\tilde{\boldsymbol{x}} = \boldsymbol{x} + \delta\boldsymbol{x}$, where $\boldsymbol{x}$ is the solution of the unperturbed linear system (eqn. 1.19). We can evaluate the error $\delta\boldsymbol{x}$ (Fadeev and Fadeeva, 1963) as

$$\|\delta\boldsymbol{x}\| \leq C_1\|\delta A\| + C_2\|\delta\boldsymbol{f}\|$$

where $C_{1,2}$ are some constants.[6] The question is: how do C_1 and C_2 behave when N grows? If they are independent of N, then the algorithm will be stable. If $C_{1,2}$ still grow when N increases, then there will be no stability against computational errors. This question should be cleared up before the computer program is worked out. If an algorithm is unstable, it is impossible to provide a desirable accuracy, even when the approximate solution converges to an exact one at $N \to \infty$. In fact, any improvement in accuracy caused by an increase in N may be neutralised by a greater influence of computational errors. Starting from some N_0 the latter factor may prevail, in which case the accuracy of the solution may deteriorate when $N > N_0$.

Two kinds of computational instability should be distinguished. The first is caused by the fact that the original mathematical problem is ill-posed (e.g. a Fredholm integral equation of the first kind). In this case the inversion operator, as known, is unbounded (Tikhonov and Arsenin, 1977), and an attempt to approximate the latter by means of a sequence of finite-dimensional matrix operators A^{-1} causes $C_{1,2}$ to increase. To overcome instability here, it is necessary to use a special technique, e.g. Tikhonov's regularisation method.

Another case is when the problem is well-posed, and the instability arises because of poor algebraisation, for example, inappropriate choice of basis functions in Galerkin's method.[7] To provide computational stability, the basis-function sets should have special properties, e.g. 'strong minimality' (Mikhlin, 1966).

---

[6] Here the following norms are used: $\|\boldsymbol{x}\| = \left(\sum_{j=1}^{N} |x_j|^2\right)^{1/2}$ for vectors, and $\|A\| = \left(\sum_{i=1}^{N}\sum_{j=1}^{N} |A_{ij}|^2\right)^{1/2}$ for matrices.

[7] On the other hand, the contrary can be illustrated, i.e. when an ill-posed problem is solved by Galerkin's method. A special choice of basis functions may play the role of analytical regularisation.

To control the accuracy of numerical results obtained the following means are applied:

(i) rigorous mathematical substantiation of the method;
(ii) 'intrinsic-convergence' check (stabilisation of the results when the truncation order $\mathcal{N}$ is increased);
(iii) check that fundamental physical laws are satisfied (energy conservation, reciprocity theorem etc.); and
(iv) comparison of particular calculational results with those obtained by other methods.

Mathematical substantiation of a method (proof of convergence to a real solution of the problem) is very important. In any case, it can not be substituted by reference to 'common sense'. A good illustration is the phenomenon of 'relative convergence' when a certain class of infinite algebraic systems is solved by the truncation method (Mittra and Lee, 1971): when the method of truncation is arbitrary there is a convergence, but the limit is not a real problem solution.

Mathematical proofs are rather voluminous and we therefore do not present them in full in this book. We only demonstrate a general idea of the proof and refer the interested reader to works where this proof is described in detail.

Normally, a mathematical proof states that there is a truncation order $\mathcal{N}$ allowing one to achieve a computational accuracy as high as desired. This, however, does not produce a specific technique to select $\mathcal{N}$ providing any required accuracy. The dependence of $\mathcal{N}$ on frequency and geometrical parameters of the electrodynamic system makes this selection still more difficult. The common practice here is to perform a computational experiment solving the problem at various $\mathcal{N}$ and to compare the results obtained. Verification of the intrinsic convergence is an indispensable element of accuracy control but by itself it does not guarantee anything.[8]

The approximate solutions often identically satisfy the energy balance and reciprocity theorem, though it is not always obvious. In particular, the power is totally conserved for fictitious solutions obtained as a result of relative convergence (Shestopalov *et al.*, 1984). The energy conservation and reciprocity theorem can therefore be used only for detecting casual errors.

Thus, each of the enumerated means of accuracy control alone does not provide any reliability of the data obtained. They can only be relied upon when taken in combination. When presenting results of mathematical simulation we shall pay special attention to the problem of computational-accuracy control and reliability of calculation results.

In the following discussion infinite sets of linear algebraic equations of special form will also be of importance:

$$x_n + \lambda \sum_{m=1}^{\infty} A_{mn} x_m = f_n \tag{1.21}$$

where $n = 1, 2, \ldots, \infty$ and λ is a preset number. Let us assume that the matrix

[8] We emphasise that it is impossible to replace the mathematical proof of convergence by an 'intrinsic convergence' check. This was also confirmed in the paper by Heitkamper and Hienrich (1991), where the diverging algorithm was described though its divergence was very slow [as $O(\ln \mathcal{N})$]. In that case $\mathcal{N}$ varying within definite limits does not lead to any considerable change in computation results which, as Heitkamper and Hienrich pointed out, may be mistaken for convergence.

elements and free terms satisfy the conditions

$$\sum_{m=1}^{\infty} \sum_{n=1}^{\infty} |A_{mn}|^2 < \infty \qquad \sum_{n=1}^{\infty} |f_n|^2 < \infty \tag{1.22}$$

Then the Fredholm alternative is valid for the set given by eqn. 1.21: either λ is the eigenvalue for a corresponding homogeneous set (when $f_n = 0$), or eqn. 1.21 has a unique solution. This solution satisfies the condition

$$\sum_{n=1}^{\infty} |x_n|^2 < \infty$$

and may be obtained by the truncation method. Stability against errors in computation of A_{mn} and f_n is provided (Kantorovich and Akilov, 1977). Such sets are called 'sets of the second kind' or Fredholm sets. This reflects a similarity of their theory to that of Fredholm integral equations of the second kind. The conditions given by eqns. 1.22 do not cover a class of Fredholm sets as a whole. For example, infinite sets with a spectrum within the unit circle are also related to the latter.

Fredholm sets often arise as a result of application of numerical–analytical methods to boundary-value problems of electromagnetics. Typically, inequalities given by eqns. 1.22 are comparatively easy to check. It is important to note that they contain a full mathematical substantiation of the method. When the sets are obtained by numerical–analytical methods, there are very reliable ways of choosing the truncation order N according to the initial data of the problem (Shestopalov *et al.*, 1984). We have already pointed out that these methods provide very fast convergence and that therefore the machine time required is very short. This is an important factor but the development of computers gradually reduces its role. A much more important factor is that numerical–analytical methods have a considerably higher reliability than do numerical procedures because of the above-mentioned characteristic features. Therefore, these results, in particular, give nontrivial tests for software debugging and accuracy control when working with direct numerical methods. All these points justify the complex analytical transformations involved when numerical–analytical methods are used.

Chapter 2

Surface-impedance technique for the study of dissipation processes in bodies with finite conductivity

Dissipation of electromagnetic waves in metal structures, due to their finite conductivity, is usually investigated employing impedance boundary conditions of the form of eqn. 1.6. The use of these boundary conditions allows the problem to be simplified, as the field inside the metal is not considered, which is the essence of the surface-impedance method. Penetration of microwave fields into the metals is accompanied by a marked skin effect and this stipulates the following special features of eqn. 1.6, which are important for the conditions to be used in the future:

(i) surface impedances of metals in ordinary situations are small: $|\mathcal{Z}_S|/W_0 \ll 1$;
(ii) the impedance $\mathcal{Z}_S$ calculated for a plane wave does not in practice depend on the angle of incidence [this means that the conditions given by eqn. 1.6 can be used for the field of arbitrary (unspecified in advance) spatial structure].

It is important to note that the surface-impedance method is a fairly universal technique which covers practically all interesting cases when dissipation characteristics of microwave and millimetre-wave components are to be determined. In this Chapter, expressions of the surface impedance for such cases will be considered.

2.1 The Leontovich impedance boundary condition

When deriving the Leontovich boundary condition the simplest constitutive equation of metal is used:

$$\boldsymbol{J}(\boldsymbol{r}, \omega) = \sigma_0 \boldsymbol{E}(\boldsymbol{r}, \omega) \tag{2.1}$$

where σ_0 is the conductivity of metal for static electric fields. The key task to be examined is the incidence of a plane wave on the infinite flat interface 'vacuum–metal'. In fact, it is sufficient to characterise the metal by a conditional complex permittivity $\varepsilon = j\sigma_0/\omega$ and use Fresnel's formulae (Weinstein, 1988). At $\sigma_0 \gg \omega\varepsilon_0$ we find from the latter that eqn. 1.6 is fulfilled at

$$\zeta_S = (\omega\mu/2\sigma_0)^{1/2}(1-j) = W_0 k \Delta_0 (1-j)/2 \tag{2.2}$$

where $\Delta_0 = (2/\omega\mu\sigma_0)^{1/2}$ is the field penetration depth into the metal (we would like to underline once again that the angle of incidence is not involved in eqn. 2.2).

Though impedance boundary conditions are usually derived for a plane infinite surface, they are also applicable to curvilinear surfaces, if only the radii of curvature are large compared with Δ_0 (in this case the boundary can be considered as locally flat).

Physically the validity of this assertion is rather obvious though it can be proved quite rigorously if we solve the problem for a body with a high conductivity and curvilinear boundary by the integral-equation technique, as has been done by Mitzner (1967). In this case, more general boundary conditions can be obtained which contain corrections corresponding to the curvature of the surface. In the particular case of $\Delta_0 \ll R$ these conditions are reduced to the impedance condition expressed by eqns. 1.6 and 2.2.

Tisher (1974, 1976, 1978, 1979) in his works conducted experimental research about the Leontovich boundary conditions' correctness for the calculation of attenuation in waveguides. An attenuation coefficient in rectangular waveguides was measured over a rather broad frequency range ($f \simeq 25$–200 GHz) and the results of these measurements were compared with those of the calculations (Tisher, 1979). The value σ_0 was estimated from measurements at direct current (for copper at $T = 20°$C the value $\sigma_0 = 5.73 \times 10^7$ S/m has been received). Secondary factors, such as admixtures, temperature influence on the attenuation during hardening of the waveguide, surface roughness etc., were also taken into consideration. A satisfactory coincidence of theoretical and experimental data has been obtained. Thus, for example, the highly precise measurement carried out by Tisher (1978) has showed that for copper at the frequency $f = 35$ GHz the measured surface resistance is 1.129 ± 0.02 times that calculated.

The measurement of the influence of metal surface roughness carried out by Tisher (1978) is of some interest. The roughness was controlled by a profilometer gauged using the photomicrographic method. It was discovered that the influence of roughness is equivalent to a certain increase in the surface impedance: $\mathrm{Re}\,\zeta_S = \chi(\omega\mu/2\sigma_0)^{1/2}$, where χ is the surface impedance increase factor. Table 2.1 represents the results of measurements of χ from Tisher's work for $f = 30$ GHz.

Theoretical research in roughness influence is somewhat difficult. The surface microrelief is determined, in the first place, by technological factors and should

Table 2.1 *Results of measurements of χ (Tisher, 1978)*

Mean-square dimensions of inhomogeneity (μm)	Surface-impedance increase factor χ
0.2	1.1
0.5	1.2
1.0	1.3

be described by a random function. The problem of wave diffraction by a random rough surface was considered, for example, by Bass and Fuks (1979) and De Santo and Brown (1986). A simpler model to take the roughness into account is suggested in works by Mende and Spitsyn (1985) and Baryshnikov *et al.* (1988). It presents the surface microrelief as a periodic structure with a period and depth equal to a mean-square length and height of inhomogeneity, respectively. This is the very model which leads to impedance ζ_S being independent of the angle of incidence with the numerical factor χ (Mende and Spitsyn called it a roughness factor).

Methods of calculation of χ differ significantly for two cases: when geometrical parameters of microrelief greatly exceed Δ_0, and when they are comparable with Δ_0. In the first case we have a problem of wave reflection from the finitely conducting periodic structure, on the surface of which the Leontovich condition is imposed. As microrelief parameters are significantly less than λ, this problem can be solved in a quasistatic approximation which considerably simplifies it. For a rectangular microrelief profile, Mende and Spitsyn (1985) obtained approximate formulae for χ which give results of the same order as those obtained by Tisher (1978). In the second case the problem is significantly more complicated: it is necessary also to solve Maxwell's equations for the electromagnetic field inside the metal. For this case Mende and Spitsyn (1985) presented numerical results obtained by the finite-difference method. Note that the microrelief can be anisotropic, for example consisting of long thin rulings. In this case, instead of the numerical factor χ, a tensor of roughness should be introduced.

The local constitutive equation given by eqn. 2.1 and, consequently, the Leontovich impedance obtained on its basis, are correct at the condition $l \ll \Delta_0$ (l being the mean free path of electrons in the metal), which corresponds to the frequencies $f \ll f^0_{cr}$, where

$$f^0_{cr} \simeq (\sigma_0 \mu l^2 \pi)^{-1} \tag{2.3}$$

(Lifshitz and Pitaevskii, 1979, Abrikosov, 1987). Note that f^0_{cr} depends indirectly (through σ_0 and l) on the temperature (f^0_{cr} corresponds to the condition $l = \Delta_0$).

If l were comparable to, or exceeded Δ_0, then the constitutive equation given by eqn. 2.1 would be replaced by a more general nonlocal relationship

$$\boldsymbol{J}(\boldsymbol{r}, \omega) = \int \hat{\sigma}(\boldsymbol{r}, \boldsymbol{r}' \mid \omega)\, \boldsymbol{E}(\boldsymbol{r}', \omega)\, \mathrm{d}^3\boldsymbol{r}' \tag{2.4}$$

in which the metal conductivity has the form of an integral operator. In this case the skin effect has a number of specific features and is called anomalous skin effect. The surface impedance in eqn. 1.6 is also expressed differently. The next Section deals with this problem.

2.2 The surface impedance of normal metals for the anomalous skin effect

The nonlocal connection of the current density and the electric-field intensity in eqn. 2.4 indicates the presence of a spatial dispersion of the medium. For a homogeneous infinite metal $\hat{\sigma}(\boldsymbol{r}, \boldsymbol{r}' \mid \omega) = \sigma_\infty(\boldsymbol{r} - \mathbf{r}' \mid \omega)$, i.e. the kernel in eqn. 2.4

depends only on the arguments difference. Then

$$\boldsymbol{J}(\boldsymbol{r}, \omega) = \int \sigma_\infty(\boldsymbol{r} - \boldsymbol{r}' \,|\, \omega)\, \boldsymbol{E}(\boldsymbol{r}', \omega)\, \mathrm{d}^3\boldsymbol{r}' \tag{2.5}$$

Presenting $\boldsymbol{J}$ and $\boldsymbol{E}$ as plane-wave spectra

$$\begin{Bmatrix} \boldsymbol{J}(\boldsymbol{r}, \omega) \\ \boldsymbol{E}(\boldsymbol{r}, \omega) \end{Bmatrix} = \int \begin{Bmatrix} \boldsymbol{J}(\boldsymbol{q}, \omega) \\ \boldsymbol{E}(\boldsymbol{q}, \omega) \end{Bmatrix} \exp(j\boldsymbol{qr})\, \mathrm{d}^3\boldsymbol{q}$$

and using the convolution theorem for eqn. 2.5, we obtain for the spectral amplitudes the relationship

$$\boldsymbol{J}(\boldsymbol{q}, \omega) = \sigma_\infty(\boldsymbol{q}, \omega)\, \boldsymbol{E}(\boldsymbol{q}, \omega) \tag{2.6}$$

where $\sigma_\infty(\boldsymbol{q}, \omega)$ is the Fourier transform of the kernel σ_∞ in eqn. 2.1.[9]

The specific feature of eqn. 2.6 (unlike, for example, eqn. 2.1) is dependence of σ_∞ on the wave vector $\boldsymbol{q}$ (spatial dispersion). The nonlocal connection between $\boldsymbol{J}$ and $\boldsymbol{E}$ can be physically explained, for example, by the fact that electrons of conductivity, acquiring a considerable velocity component in the direction normal to the metal surface as a result of electron collisions, rapidly leave the skin and thereby do not contribute to the microwave current (the concept of 'ineffectness' by Pippard).

The partial conductivity $\sigma_\infty(\boldsymbol{q}, \omega)$ in eqn. 2.6 can be found from the self-consistent solution of Maxwell's equations and the kinetic equation for free electrons. Next, well developed methods of electrodynamics for media with spatial dispersion can be used to calculate the impedance Z_S (Agranovich and Ginzburg, 1979).

Suppose that the E-polarised plane wave $\boldsymbol{E} = \hat{i}_y \exp\{jk(z \sin\vartheta - x\cos\vartheta)\}$ is incident from $x > 0$ on an infinite plane boundary vacuum–metal (the axes are chosen so that the interface is the $x = 0$ plane). Considering the spatial dispersion while solving boundary-value problems, we need to impose the so-called additional boundary conditions (Agranovich and Ginzburg, 1979). This results from the boundary being responsible for the difference between the kernels $\hat{\sigma}$ in eqn. 2.4 and σ_∞ (corresponding to an infinite medium). Moreover, the kernel $\hat{\sigma}$ remains indefinite even after σ_∞ has been calculated. Additional boundary conditions are physical assumptions *a priori* which allow us to specify the relationship between $\hat{\sigma}$ and σ_∞. To obtain these conditions, the assumption of a mirror character of the electrons reflection from the boundary can be made. Then, for E_y inside the metal we receive the integral–differential equation

$$\nabla^2_{xz} E_y(\boldsymbol{R}) + j\omega\mu \int \sigma_\infty(\boldsymbol{R} - \boldsymbol{R}') E_y(\boldsymbol{R}')\, \mathrm{d}^2\boldsymbol{R}' = -2\delta(x - 0)\, \frac{\partial E_y}{\partial x} \tag{2.7}$$

where $\sigma_\infty(\boldsymbol{R} - \boldsymbol{R}')$ corresponds to a homogeneous infinite metal and thus depends only on the difference between the arguments, and $\boldsymbol{R}$ and $\boldsymbol{R}'$ are radii-vectors in the plane Oxz.

Solution of eqn. 2.7, using general methods for problems of electrodynamics

---

[9] We consider only isotropic metals, where the tensor $\sigma_\infty(\boldsymbol{q}, \omega)$ becomes a scalar; the surface impedance Z_S in eqn. 1.6 is therefore also a scalar.

of media with spatial dispersion, allows us to obtain the following expression for the reflection coefficient:

$$R = -\frac{1 + jk\cos\vartheta \mathcal{Z}(k\sin\vartheta)}{1 - jk\cos\vartheta \mathcal{Z}(k\sin\vartheta)} \tag{2.8}$$

where $\mathcal{Z}(k\sin\vartheta)$ is the partial impedance corresponding to the angle of incidence ϑ and is expressed by

$$\mathcal{Z}(\nu) = \frac{1}{\pi}\int_{-\infty}^{\infty} \frac{d\beta}{\nu^2 + \beta^2 - j\omega\mu\sigma_\infty(\nu, \beta)} \tag{2.9}$$

where $\sigma_\infty(\nu, \beta) = \sigma_\infty((\nu^2 + \beta^2)^{1/2}) = \sigma_\infty(|\boldsymbol{q}|)$.

From eqn. 2.9 it follows that for good metals the partial impedance $\mathcal{Z}(\nu)$, with a high degree of accuracy, does not depend on ν (and hence the angle of incidence ϑ). This means that the impedance $\mathcal{Z}$ in eqn. 2.8 can be identified with the surface impedance $\mathcal{Z}_S$ to be found for the impedance boundary condition given by eqn. 1.6. Assuming $\nu = 0$ in eqn. 2.9 and substituting the specific function $\sigma_\infty(|\boldsymbol{q}|)$, we can obtain the required expression for $\mathcal{Z}_S$.

In case of an extreme anomalous skin effect when $l \gg \Delta_0$, according to Abrikosov (1987), we can write

$$\sigma_\infty(|\boldsymbol{q}|) \simeq (l|\boldsymbol{q}|)^{-1}\sigma_0$$

Then from eqn. 2.9 we obtain

$$\mathcal{Z}_S \cong W_0 k\zeta^{-1/3}(1 - j\sqrt{3})/2 = k\Delta_{eff} W_0(1 - j\sqrt{3})/2 \tag{2.10}$$

where $\zeta = \omega\mu\sigma_0 l^{-1}$ and $\Delta_{eff} = (l/\omega\mu\sigma_0)^{1/3}$ is the effective depth of field penetration into the metal.[10]

The qualitative character of frequency dependence of Re $\mathcal{Z}_S$ on f is shown in Figure 2.1. The region $f \ll f_{cr}^0$ (f_{cr}^0 is evaluated by eqn. 2.3) is the region of normal skin effect described by the Leontovich impedance expressed by eqn. 2.2. The region $f \gg f_{cr}^0$ corresponds to an extreme anomalous skin effect (the impedance is defined by the eqn. 2.10). Most of the difficulties arise in the intermediate region, where $l \sim \Delta_0$ and where it is rather difficult to obtain simple analytical relationships.[11] However, it is necessary to take into consideration the fact that the transition from normal to anomalous skin effect takes place fairly smoothly [this is also confirmed by experiments (see Abrikosov, (1987)]. This already allows us to consider the surface impedance in the intermediate region as also independent of the angle of incidence. The value of $\mathcal{Z}_S$ can be estimated by means of a simple approximation using two extreme cases given by eqns. 2.2 and 2.10 (broken line in Figure 2.1).

It is important to note that the surface impedance for the anomalous skin effect turns out to be independent of the angle of incidence in spite of the nonlocality of the constitutive equation. Because of this, the frequency dependence of the surface impedance differs from that in eqn. 2.2.

---

[10] For the anomalous skin effect the microwave field when penetrating further into the metal diminishes nonexponentially. This is why the value Δ_{eff} ceases to have a direct physical meaning, unlike the value Δ_0 in the normal skin effect.

[11] The analysis of this case has been carried out by Kittel (1963), but the expression obtained for $\mathcal{Z}_S$ is rather complicated.

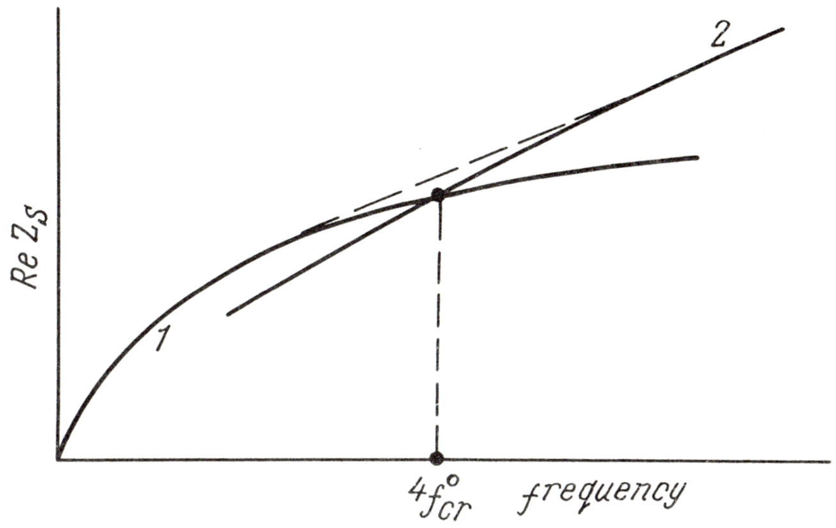

Figure 2.1 *Sketch of surface-impedance active component against frequency*

1 Normal-skin-effect region
2 Extremely anomalous-skin-effect region
– – – – approximation in intermediate region

The theory of anomalous skin effect is much more complicated if the character of the electron reflection from the boundary is a diffuse one.[12] However, the final expression for the impedance does not depend greatly on the character of the electron reflection. Thus instead of eqn. 2.10, at $l \gg \Delta_0$ we obtain an expression that differs from eqn. 2.10 by the constant numerical factor $\frac{9}{8}$ (Abrikosov, 1987).[13]

Numerical evaluations show that the above-mentioned Pippard mechanism of anomalous skin effect takes place in the microwave and millimetre wavebands only at very low temperatures; for room temperatures it is essential only at very high frequencies (infra-red and optical band).

In the work by Wang (1978) an attempt has been made to explain why the experimental data (Tisher, 1978) differ from the results of calculations according to eqn. 2.2. Wang interpreted this as a result of anomalous character of the skin effect. He also suggested that the mechanism of anomalous skin effect involves a Coulomb's screening of the donor potentials by the electrons of conductivity. The attempt to calculate the surface impedance for this case was also made by Slepyan (1984). In this respect it is necessary, however, to note the following: in the presence of the spatial dispersion even an isotropic metal becomes anisotropic due to the wave propagation; the anisotropic properties are defined by

[12] Under diffuse reflection we understand such a reflection when electrons do not preserve a 'memory' about their original states.

[13] Physically the assumption about a mirror character of the electron reflection appears to be more realistic. This is because for the electrons moving inside the skin for a long time the reflection law is close to a mirror one and these are the very electrons which contribute most to the microwave current (Lifshitz and Pitaevskii, 1979).

the orientation of the wave vector $\boldsymbol{q}$. That is why one should distinguish between longitudinal and transverse conductivities of the metal (Landau *et al.*, 1984). In eqn. 2.6 and, hence, in eqn. 2.9 just the transverse conductivity appears, whereas Coulomb's screening affects the longitudinal conductivity of the metal.[14] The expressions for Z_S offered by Wang (1978) and Slepyan (1984) are therefore erroneous, as is the suggestion of the influence of Coulomb's screening on the character of the skin effect.

2.3 Surface impedance of superconductors

As is well known, microwave engineering and low-temperature physics are closely related. On the one hand, the physics of superconductors finds new materials with unique physical characteristics for microwave applications; on the other hand, microwave methods are highly effective for measurements of superconductor parameters. Undoubtedly, the discovery of high-temperature superconductivity in barium ceramics (see, for example, Ekholm and McKnight, 1990) will contribute even further to their mutual influence.

Superconducting waveguides, resonators and microstrip lines are widely used in microwave and millimetre-wave devices (powerful highly stable tunable oscillators, wideband delay lines, charged-particle accelerators etc.). Thus the problem of electrodynamic modelling of these components is important. Certainly, our task here is not to give a systematic description of the high-frequency properties of superconductors: that has been done, for example, by Lifshitz and Pitaevskii (1979), Abrikosov *et al.* (1958, 1959), Ginsberg (1966), Mattis and Bardeen (1958) and Mende and Spitsyn (1985). We wish only to demonstrate that these properties are described by means of the surface-impedance concept. Hence, all the information in this book applies equally to electrodynamic systems with normal metals and to superconducting systems.

The constitutive equation for a homogeneous infinite superconductor is usually written as a linear relationship between Fourier transforms (with respect to time and all space variables) of the current density and the vector potential $\boldsymbol{A}$

$$\boldsymbol{J}(\boldsymbol{q}, \omega) = -Q(\boldsymbol{q}, \omega)\,\boldsymbol{A}(\boldsymbol{q}, \omega) \tag{2.11}$$

where $Q(\boldsymbol{q}, \omega)$ is the so-called generalised susceptibility, calculated on the basis of a superconductor physical model. Eqn. 2.11, strictly speaking, is written for the transverse components $\boldsymbol{J}$ and $\boldsymbol{A}$, and $Q(\boldsymbol{q}, \omega)$ is transverse susceptibility.

In the theory of superconductivity the displacement current inside the superconductor can be neglected, i.e. $\operatorname{div}\boldsymbol{J} = 0$. This means that $\boldsymbol{q}\cdot\boldsymbol{J}(\boldsymbol{q}, \omega) = 0$ and, hence $\boldsymbol{q}\cdot\boldsymbol{A}(\boldsymbol{q}, \omega) = 0$. Then the vector potential $\boldsymbol{A}$ is related to the electric-field intensity by the formula $\boldsymbol{A} = \boldsymbol{E}/j\omega$. Then eqn. 2.11 is reduced to eqn. 2.6, where the partial conductivity and the generalised susceptibility relationship is $\sigma_\infty(\boldsymbol{q}, \omega) = jQ(\boldsymbol{q}, \omega)/\omega$. Therefore if $Q(\boldsymbol{q}, \omega)$ is determined from the physical model, the scheme described in Section 2.2 can be applied to find the surface impedance. For this, as for normal metals, the impedance does not depend on the angle of incidence of the plane wave.

To determine $\sigma_\infty(\boldsymbol{q}, \omega)$ one can use simple phenomenological theories of

[14] The authors' attention was drawn to this fact by Prof. F.G. Bass.

superconductivity (London's model, Pippard's model) and the microscopic BCS theory as well. For example, in London's model a superconductor is presented as a 2-component electron liquid (one component corresponds to 'normal' electrons, another to 'superconducting' ones). The 'superconducting' current density $\boldsymbol{J}_s$ (current caused by the motion of 'superconducting' electrons) is described by the formula

$$\boldsymbol{J}_s = -N_s e^2 \boldsymbol{A}/\mathrm{m}$$

where N_s is the density of 'superconducting' electrons and e and m are the charge and mass of the electron. Within the framework of this model

$$\sigma_\infty = \sigma_0\left(\frac{N_n}{N_s + N_n}\frac{1}{\{1+(\omega\tau)^2\}} + \frac{j}{\omega\tau}\frac{N_s}{N_s+N_n}\left[1 + \frac{N_n}{N_s}\frac{(\omega\tau)^2}{\{1+(\omega\tau)^2\}}\right]\right) \tag{2.12}$$

where σ_0 is the normal static conductivity, N_n is the density of 'normal' electrons, and $\tau = l/v_F$ where v_F is the velocity of electrons on the Fermi surface. The specific feature of eqn. 2.12 is independence of σ_∞ from the wave vector $\boldsymbol{q}$ (this means that within the framework of London's model there is no spatial dispersion). From eqn. 2.9 there immediately follows an expression for the surface impedance:

$$Z_S = (1-j)(\omega\mu_0/2\sigma_0)^{1/2}(g_1 + jg_2)^{-1/2} \tag{2.13}$$

where

$$g_1 = \frac{1}{\{1+(\omega\tau)^2\}}\frac{N_n}{N_s+N_n}$$

$$g_2 = \frac{1}{\omega\tau}\frac{N_s}{N_s+N_n}\left[1 + \frac{N_n}{N_s}\frac{(\omega\tau)^2}{\{1+(\omega\tau)^2\}}\right]$$

The impedance expressed by eqn. 2.13 depends on the temperature through σ_0 and the ratio N_s/N_n. There is an approximate formula

$$\frac{N_s}{N_s+N_n} \sim 1 - \left(\frac{T}{T_C}\right)^4$$

where T_C is the critical temperature of the superconductor measured in kelvins.

It is far more complicated to determine the surface impedance on the basis of a rigorous microscopic theory. We shall present here only some final results (Ginsberg, 1966, Mattis and Bardeen, 1958). For the surface impedance Z_S in the extreme local limit, eqn. 2.13 is valid, but the coefficients $g_{1,2}$ are expressed, according to BCS weak-coupling theory, as

$$g_1 = \frac{2}{\hbar\omega}\int_\Delta^\infty \frac{\{f(\xi) - f(\xi+\hbar\omega)\}(\xi^2+\Delta^2+\hbar\omega\xi)\,d\xi}{(\xi^2-\Delta^2)^{1/2}\{(\xi+\hbar\omega)^2-\Delta^2\}^{1/2}}$$

$$+\frac{1}{\hbar\omega}\int_\Delta^{s(\hbar\omega,\Delta)} \frac{\{1-2f(\hbar\omega-\xi)\}(\hbar\omega\xi-\xi^2-\Delta^2)\,d\xi}{(\xi^2-\Delta^2)^{1/2}\{(\hbar\omega-\xi)^2-\Delta^2\}^{1/2}}$$

$$g_2 = \frac{1}{\hbar\omega}\int_{-s(\hbar\omega,\Delta)}^{\Delta} \frac{\{1-2f(\xi+\hbar\omega)\}(\xi^2+\Delta^2+\hbar\omega\xi)\,d\xi}{(\xi^2-\Delta^2)^{1/2}\{(\xi+\hbar\omega)^2-\Delta^2\}^{1/2}}$$

where

$$S(\hbar\omega, \Delta) = \begin{cases} \hbar\omega - \Delta & (\hbar\omega \leq 2\Delta) \\ \Delta & (\hbar\omega > 2\Delta) \end{cases}$$

Δ is the energy gap, $f(\xi)$ is the Fermi distribution and $\hbar$ is Planck's constant (its presence in the formulae points to a quantum nature of superconductivity). Through the energy gap Δ and the function $f(\xi)$, coefficients $g_{1,2}$ depend on the temperature.

Qualitative dependencies $g_{1,2}(\omega)$ at various temperatures are shown in Figure 2.2 for YBCO ceramics (Ekholm and McKnight, 1990).

As in Section 2.2, we can introduce the effective depth of penetration of the microwave field Δ_{eff}, determining it from the condition Re $Z_S = k\Delta_{eff} W_0/2$. Then

$$\Delta_{eff} = \Delta_0 (g_1^2 + g_2^2)^{-1/4} \{\cos(\phi/2) - \sin(\phi/2)\}$$

where $\phi = \tan^{-1}(g_2/g_1)$ and $\Delta_0 = (2/\omega\mu\sigma_0)^{1/2}$ is the normal depth of penetration.

A more detailed description of the high-frequency properties of superconductors is presented in the monograph by Mende and Spitsyn (1985). Apart from the physical models and formulae for Z_S, it also contains extensive numerical and experimental data for various superconductors used (lead, niobium etc.).

2.4 Surface impedance modification for structures with edges

Many theoretical models of electrodynamic systems contain infinitely thin unclosed metal surfaces. Typical examples of these systems are microstrip and

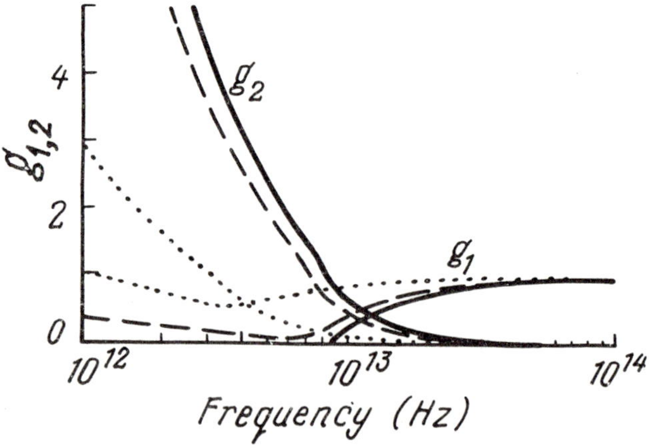

Figure 2.2 $g_{1,2}$ *factors for YBCO (Ekholm and McKnight, 1990)*

$\Delta(T)/\pi\hbar = 8.5$ THz

———	$T = 4.2$ K
- - - -	$T = 60$ K
·······	$T = 85$ K

Reproduced by permission of the IEEE © IEEE 1990

slot lines and their modifications, some types of open resonator, diffraction gratings etc. In general, it is impossible to take the metal losses into consideration by imposing the conventional impedance boundary conditions (when $\mathcal{Z}_S$ is determined by any of eqns. 2.2, 2.10 and 2.13) on both sides of the unclosed surface. The reason for this is that the local field and current-density distributions near the edge are described inaccurately by these impedance boundary conditions (Zhurav, 1987, Weinstein *et al.*, 1986, Ilyinsky and Slepyan, 1986). This can be seen from the fact that, in reality, field and current-density distributions near the screen edge are determined by its thickness and geometry and such parameters are not involved at all in the model of an infinitely thin screen with the impedance conditions examined above.

If the field is electrically polarised along the edge (***H*** is parallel to the edge), the use of impedance boundary conditions considered in Sections 2.1–2.3, to calculate such an integral characteristic as dissipated power, is relevant. This is because the current density near the edge in this case is rather small and its contribution to the full dissipated power is insignificant. Therefore, the error in calculation of the current density is not important.

The situation becomes more complicated when there is a substantial component of magnetic polarisation (***E*** is parallel to the edge). In this case the usual impedance-boundary conditions near the edge are not applicable, so it is necessary to take into account directly the screen thickness and the actual edge geometry; this significantly complicates the construction of effective numerical algorithms. For small thicknesses some methods (for example, the integral-equations technique) result in algebraic systems with poorly conditioned matrices and, as a consequence, large computational errors (Galishnikova and Ilyinsky, 1987).

This Section presents an approach, based on the replacement of the actual surface of finite thickness by an 'effective' infinitely thin surface, on both sides of which the boundary condition expressed by eqn. 1.6 with a modified impedance $\mathcal{Z}_S$ is imposed. The modified impedance is introduced in such a manner that it should describe satisfactorily both the dissipative properties of the metal and the induced currents in the edge vicinity. The modified impedance then involves parameters describing the edge geometry (screen thickness, radius of curvature near the edge).

We consider two cases: the rounded-edge geometry (Figure 2.3*a*) and the rectangular-edge geometry (Figure 2.3*b*). Halfplanes with such edge configurations are the key structures to determine the modified impedance. The material below is based on works by Ilyinsky and Slepyan (1986, 1990) and Slepyan (1988). To describe correctly the dissipative properties of the metal we have to consider $\mathrm{Re}\,\mathcal{Z}_S$ having the value which follows from the skin-effect theory ($\mathrm{Re}\,\mathcal{Z}_S = k\Delta_{eff} W_0/2$). This theory gives the value of $\mathrm{Im}\,\mathcal{Z}_S$ but, as $\mathrm{Im}\,\mathcal{Z}_S/W_0 \ll 1$, the influence of $\mathcal{Z}_S$ on waveguide mode dispersion, resonance frequencies and diffraction characteristics of electrodynamic systems is negligibly small and is not usually considered. It will not therefore introduce a serious error in the calculation of these characteristics if we consider $\mathrm{Im}\,\mathcal{Z}_S$ as arbitrary while the condition $\mathrm{Im}\,\mathcal{Z}_S/W_0 \ll 1$ is satisfied. The 'degree of freedom' thus obtained can be used to calculate correctly the losses within the framework of the infinitely thin impedance model. For this purpose, $\mathrm{Im}\,\mathcal{Z}_S$ should be chosen so that the current-density distribution near the edge would be the same as in

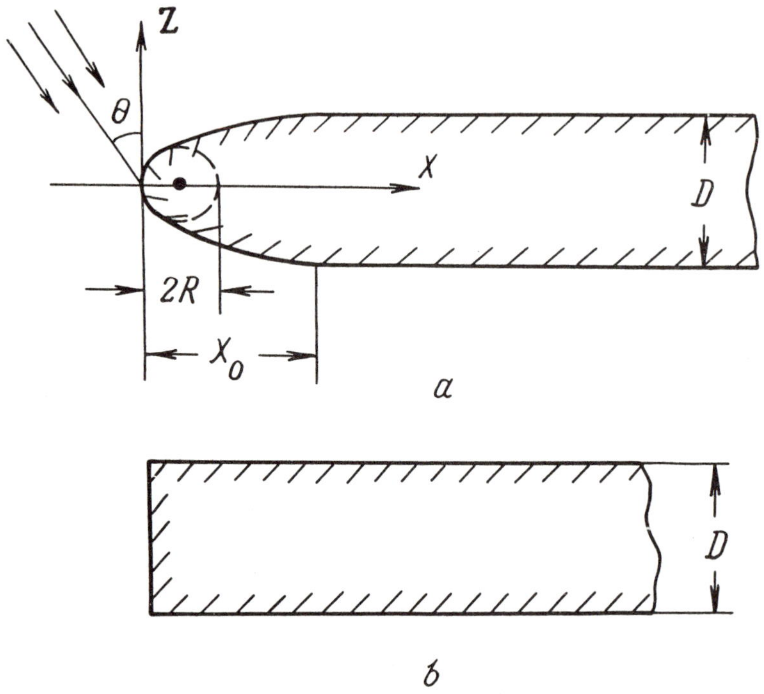

Figure 2.3 *Geometry of thin unclosed screen*

a Rounded edge
b Rectangular edge

an actual structure of finite thickness (we shall describe the methods of choosing this later).

Most important for this procedure is the fact that the current density on the edge of an infinitely thin impedance halfplane has no Meixner's singularity.

Let us consider the solution of the problem of E-polarised plane-wave diffraction by an infinitely thin impedance halfplane (Noble, 1958). The current-density distribution is given by the expression

$$\boldsymbol{J}(x) = \frac{\hat{i}_y}{\omega\mu_0} \frac{\{(1+\cos\vartheta)k\}^{1/2}}{2\pi K_-(k\cos\vartheta)} \int_{j\tau-\infty}^{j\tau+\infty} \frac{(w-k)^{1/2}\exp(-jwx)\,\mathrm{d}w}{(w-k\cos\vartheta)K_+(w)} \tag{2.14}$$

where $K_\pm$ are split functions regular in $\mathrm{Im}\, w > -k''$ and $\mathrm{Im}\, w < k''$, respectively, and $K(w) = K_+(w)K_-(w) = 1 + j\mathcal{Z}_S(w^2-k^2)^{1/2}/W_0 k$.

Let us evaluate the leading asymptotic term ($\mathcal{Z}_S \to 0$) of the integral

$$\mathcal{J} = \int_{j\tau-\infty}^{j\tau+\infty} \frac{(w-k)^{1/2}\exp(-jwx)\,\mathrm{d}w}{(w-k\cos\vartheta)K_+(w)} \tag{2.15}$$

at $x \to +0$. We assume that $k = k' - jk''$ ($k'' > 0$) and choose the brunch cuts in the plane of complex variable w, as shown in Figure 2.4.

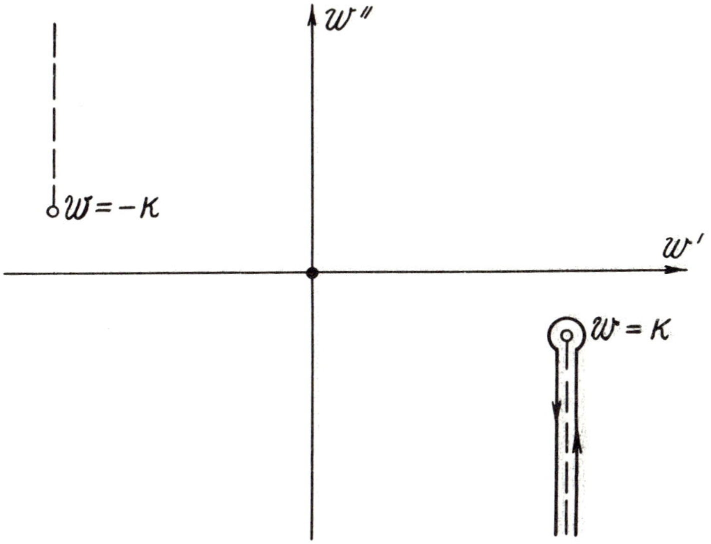

Figure 2.4 *Brunch cuts and contour of integration in the complex w-plane*

Deforming the integration path in the lower halfplane we can show that the evaluation sought is the same as that for the integral of the same function along the contour embracing the brunch cut. In this case eqn. 2.15 can be rewritten at $k'' \to 0$ as

$$\tilde{J} = \exp(j\pi/4) \int_0^\infty \frac{(\sqrt{u}) \exp(-ux) S(u) \, du}{(k - k \cos \vartheta - ju)} \tag{2.16}$$

where

$$S(u) = \frac{2K_-(k - ju)}{1 - (Z_S/W_0 k)^2 (u^2 + 2jku)}$$

Assuming that $Z_S = Z'_S - jZ''_S$, $Z'_S, Z''_S > 0$, $Z'^2_S \ll Z''^2_S$ and $x = 0$ in eqn. 2.16 and introducing the variable v according to the formula $v = jZ''_S u/W_0 k$, we obtain the relationship

$$|\tilde{J}| = \pi |G_1| (2W_0 k/|Z_S|)^{1/2} \tag{2.17}$$

where

$$G_1 = \frac{\sqrt{2}}{\pi} \int_0^\infty \frac{(\sqrt{v}) K_+(jW_0 kv/Z_S - k) \, dv}{(1 + v^2 + 2jZ''_S v/W_0)(v + Z''_S(1 - \cos \vartheta)/W_0)}$$

To obtain the leading asymptotic term we have to carry out the limiting transition $Z''_S \to 0$ in the expression for G_1. Then

$$|G_1| \to G = \frac{\sqrt{2}}{\pi} \int_0^\infty \frac{f(v) \, dv}{(\sqrt{v})(1 + v^2)} \tag{2.18}$$

where

$$f(v) = \lim_{\bar{\varepsilon}\to 0} \bar{K}_+(jv)\ (\bar{\varepsilon} = \bar{\varepsilon}' + j\bar{\varepsilon}'',\ \bar{\varepsilon}',\ \bar{\varepsilon}'' > 0)$$

$\bar{K}_+(\alpha)$ is a split function regular in

$$\operatorname{Im}\alpha > -\bar{\varepsilon}'' \qquad \text{and} \qquad \bar{K}(\alpha) = \bar{K}_+(\alpha)\bar{K}_-(\alpha) = 1 + (\alpha^2 - \bar{\varepsilon}^2)^{1/2}$$

Finally, we have

$$f(v) = (1 + v^2)^{1/4} \exp\left(-\frac{1}{\pi} \int_0^v \frac{\ln \zeta \, d\zeta}{\zeta^2 + 1} \right) \tag{2.19}$$

The integral in eqn. 2.18 converges: G is a finite number independent of ϑ. The integrals in eqns. 2.18 and 2.19 cannot be expressed analytically. As a result of numerical integration, we obtain $G = 2.3392$.

Thus, it follows from eqns. 2.14 and 2.17 that

$$|\boldsymbol{J}|_{|x=0} = \frac{kG}{\omega\mu_0} \left\{ \frac{(1 + \cos\vartheta)}{2} \frac{W_0}{|Z_S|} \right\}^{1/2} \tag{2.20}$$

to an accuracy of $O(|Z_S|^{3/2})$.

We now deal with the current-density distribution in the quasistatic vicinity of the edge determined by the condition $kx \ll 1$. Using eqn. 2.14, we obtain

$$|\boldsymbol{J}(x)| = \frac{\hat{i}_y}{\omega\mu_0} \frac{\{k(1 + \cos\vartheta)\}^{1/2}}{2\pi} F(x) \tag{2.21}$$

Unfortunately the integral representation for $F(x)$ resulting from eqn. 2.14 is rather complicated for further use. We shall therefore construct a simple approximation of $F(x)$ on the basis of the following *a priori* information: first $F(0)$, as follows from eqn. 2.20, is finite; secondly, from the physical point of view it is evident that $F(x)$ diminishes monotonously, and that $F(x) \to 2(\pi/x)^{1/2}$ at $x \to \infty$. The simplest approximation of the function satisfying all conditions mentioned above has the form

$$F(x) \approx 2\pi^{1/2}(x + 2|Z_S|/\pi G^2 k W_0)^{-1/2} \tag{2.22}$$

This approximation is quite sufficient for calculation of the dissipated power.

Let us now analyse the current-density distribution for a structure with a rounded edge. We assume that

(i) the effective penetration depth Δ_{eff}[15] is small compared with all geometrical dimensions of the structure; and
(ii) $kR \ll 1$, $kD \ll 1$.

From assumption (i) it follows that the structure material can be considered as a perfect conductor. The scheme of analysis is as follows: beyond the bounds of a certain small vicinity of the edge ($x \lesssim \delta$, $k\delta \ll 1$) the induced current-density distribution does not differ from the corresponding distribution for an infinitely thin halfplane. Inside this vicinity the current-density distribution has a quasi-

[15] For the Leontovich impedance, the value Δ_0 is used as Δ_{eff}.

static character and, after simple curve approximation of the edge shape, can be determined by the conformal-mapping method up to a constant factor. This factor is determined by means of matching of the solutions at $x \simeq \delta$.

According to the conformal-mapping technique, the current density in the edge vicinity can be expressed as

$$\boldsymbol{J} = C\hat{i}_y \left| \frac{\mathrm{d}f}{\mathrm{d}\zeta} \right|_{|f|=1} \tag{2.23}$$

where $f(\zeta)$ is a function which describes the conformal mapping of the preset region on the exterior of the unit circle and C is a constant.

The current density near the edge of an infinitely thin halfplane excited by the E-polarised plane wave is determined by the formula

$$\boldsymbol{J} = -\frac{\hat{i}_y}{j\omega\mu_0} \left\{ -\frac{jk(1+\cos\vartheta)}{\pi x} \right\}^{1/2} \tag{2.24}$$

The constant C can be found through matching of the representations given by eqns. 2.23 and 2.24 at large values of x.

We assume that the edge of a 'thick' halfplane has an elliptical shape (Figure 2.3*a*). The field structure near the edge will then be the same as that near the edge of the elliptical cylinder. The function $f(\zeta)$ is determined, according to Lavrentyev and Shabat (1987), by the expression

$$f(\zeta) = \frac{\zeta + (\zeta^2 - c^2)^{1/2}}{a+b}$$

where $\zeta = x - a + jz$, $c = (a^2 - b^2)^{1/2}$ and a and b are the large and small semi-axes of the ellipse. From eqn. 2.23 it follows that near the edge

$$\boldsymbol{J}|_{x\to 0} \simeq C\hat{i}_y \left(x + \frac{R}{2} \right)^{-1/2} \tag{2.25}$$

where $R = p$ and $p = b^2/a$ is the ellipse focal parameter. Comparing eqn. 2.25 with eqn. 2.24, it is easy to find C and obtain the expression for the current density near the edge

$$\boldsymbol{J}|_{x=0} = \hat{i}_y \frac{\exp(j\pi/4)}{\omega\mu_0} \left\{ \frac{k(1+\cos\vartheta)}{\pi(x+R/2)} \right\}^{1/2} \tag{2.26}$$

If the edge has a parabolic shape described by the equation $z^2 = 2px$, where p is a positive number, then $\zeta = x - p/2 + jz$ and $f(\zeta) = (\sqrt{\zeta}) - j\{\sqrt{(p/2)}\}$ which, taking into account eqn. 2.23, leads to eqns. 2.25 and 2.26.

Now we can obtain the expression for the impedance to be found. A relevant criterion of equivalence of an infinitely thin impedance screen and the actual screen of finite thickness is the equality of mean-square-current densities (and hence energy losses) in the edge vicinity of the order of the curvature radius R. For a parabolic cylinder, the dissipated power for both sides in this vicinity is given by the relationship

$$P = Z'_S \int_0^R |\boldsymbol{J}(x)|^2 \left(1 + \frac{p}{2x} \right)^{1/2} \mathrm{d}x$$

Determining the dissipated power in the edge vicinity $x < R$ for an infinitely thin impedance structure in a similar manner on the basis of eqns. 2.21 and 2.22 and comparing the results, we obtain

$$\mathcal{Z}_S \simeq k(\Delta_{eff}/2 - j\mathfrak{C}R)W_0 \qquad (2.27)$$

where $\mathfrak{C} = 1.12$. For the elliptic edge we obtain the same expression, but $\mathfrak{C} = 1.04$. Calculations for the edge of a rectangular shape are similar; in this case

$$\mathcal{Z}_S \simeq k(\Delta_{eff}/2 - j0.025D)W_0 \qquad (2.28)$$

It is noteworthy that the impedances expressed by eqns. 2.27 and 2.28 do not depend on the incident angle of the primary plane wave.

It is necessary to emphasise that the numerical coefficient in the imaginary part of $\mathcal{Z}_S$ is involved in the expression for the dissipated power in the argument of the 'large' logarithm (see Section 5.2), which makes its numerical specification (through considering fine details of the edge shape etc.) superfluous. The dissipated power is therefore only slightly sensitive to the dimension of the edge vicinity in which the comparison is drawn (the vicinity being determined in just the order of its dimension). For example, Ilyinsky and Slepyan (1986) treated structures with a rounded edge on the basis of a much rougher criterion of equivalence than that mentioned above: the equality of induced current densities on the edge of the actual structure with finite thickness and its infinitely thin impedance analogue. In this case $\mathfrak{C} = 4.0$, but even such a difference in $\mathfrak{C}$ does not change an order of dissipated power and the qualitative character of its dependencies on the parameters of the structures under study.

The proposed variant of impedance boundary conditions allows us to avoid solving mathematically complicated diffraction problems for screens of finite thickness when calculating the dissipated power. This method can be used in combination with the Wiener–Hopf technique and its modifications and also with direct numerical methods, e.g. integral-equation or dual-integral-equation techniques. To conclude, we emphasise that it cannot be used to take into account the influence of the screen thickness on the diffraction or dispersion characteristics of electrodynamic systems. The thickness is only considered when calculating the metal losses in the cases where diffraction or dispersion characteristics are described satisfactorily through the model of an infinitely thin surface.

Special attention has been paid to the choice of the sign of the imaginary part of the impedances expressed by eqns. 2.27 and 2.28. It may appear that this choice is of no importance as long as the lost power P is determined only by $\mathcal{Z}'_S$ and $|\mathcal{Z}_S|$. However, with the opposite sign for the imaginary part of the impedance, the plane may support the fictitious surface wave with dispersion properties defined by the edge geometry. In this case, when solving spectral problems it is necessary to detect and remove such fictitious modes. In diffraction problems the excitation of the latter may distort the distributions of the electromagnetic fields calculated.

2.5 The edge condition for an impedance halfplane located at media interface

In Section 2.4 we obtained expressions for an impedance halfplane, allowing us to consider the influence of edge effects on the absorbed power. However, the analysis was made for a halfplane in free space. A question arises: is it possible to generalise eqns. 2.27 and 2.28 for the case of a halfplane located at an interface of media with different permittivities? To answer this question we shall analyse the edge condition for this case, using an approach different from that in Section 2.4—the modified Meixner method (Braver *et al.*, 1988). This will allow us to illustrate and compare various approaches to this class of problems.

Figure 2.5 shows the geometry of the problem. We introduce the cylindrical-co-ordinate system r, ϕ, the origin of which coincides with the edge of the halfplane. Mathematical formulation of the problem is as follows: let $\psi_1 = E_z$ in region $\mathscr{A}$ $(0 < \phi < \pi)$, $\psi_2 = E_z$ in region $\mathscr{B}$ $(\pi < \phi < 2\pi)$. We wish to find the solution of the Helmholtz equations

$$\nabla^2_{r\phi}\psi_1 + k^2\psi_1 = 0 \tag{2.29}$$

$$\nabla^2_{r\phi}\psi_2 + k^2\tilde{\varepsilon}\psi_2 = 0 \tag{2.30}$$

satisfying the boundary conditions

$$\left.\left(\psi_1 - \frac{j\delta}{r}\frac{\partial\psi_1}{\partial\phi}\right)\right|_{\phi=0} = 0 \tag{2.31}$$

$$\left.\left(\psi_2 + \frac{j\delta}{r}\frac{\partial\psi_2}{\partial\phi}\right)\right|_{\phi=2\pi} = 0 \tag{2.32}$$

$$(\psi_1 - \psi_2)|_{\phi=\pi} = 0 \tag{2.33}$$

$$\left.\left(\frac{\partial\psi_1}{\partial\phi} - \frac{\partial\psi_2}{\partial\phi}\right)\right|_{\phi=\pi} = 0 \tag{2.34}$$

where $\delta = \mathcal{Z}_S/W_0k$.

The Meixner method (Meixner, 1972) is applicable for the case of perfectly conducting surfaces. The solutions sought are represented in the form of power series

$$\psi_{1,2} = \sum_{m=0}^{\infty}\begin{Bmatrix} a_m(\phi) \\ b_m(\phi)\end{Bmatrix} r^{m+\tau} \tag{2.35}$$

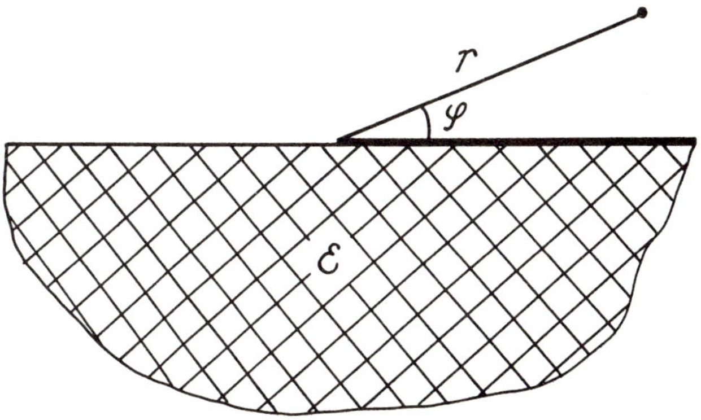

Figure 2.5 *Impedance halfplane located at media interface*

where $a_m(\phi)$ and $b_m(\phi)$ are unknown functions and τ an unknown positive number. Substitution of eqn. 2.35 into eqns. 2.29 and 2.30 leads to a recurrent chain of equations for $a_m(\phi)$ and $b_m(\phi)$. Their solution and imposition of boundary conditions results in a characteristic equation for τ, the smallest positive root of which defines the field structure near the edge.

But, as Braver *et al.* (1988) have shown, for impedance structures such an approach enables us to obtain only a trivial solution: $a_m(\phi) = b_m(\phi) = 0$. It means that at $\delta \neq 0$ the expansions for $\psi_{1,2}$ given by eqn. 2.35 are not applicable. Braver *et al.* (1988) proposed a modification of Meixner's method which uses the expansions

$$\psi_{1,2} = \sum_{m=0}^{\infty} \sum_{n=0}^{m} \begin{Bmatrix} a_{mn}(\phi) \\ b_{mn}(\phi) \end{Bmatrix} r^{m+\tau} \ln^n r \tag{2.36}$$

instead of eqn. 2.35. Recurrent chains of equations and boundary conditions

$$\begin{aligned} a''_{mn}(\phi) + (m+\tau)^2 a_{mn}(\phi) + 2(m+\tau)(n+1) a_{m,n-1}(\phi) \\ + (n+2)(n+1) a_{m,n-2}(\phi) + k^2 a_{m-2,n}(\phi) = 0 \end{aligned} \tag{2.37}$$

$$\begin{aligned} b''_{mn}(\phi) + (m+\tau)^2 b_{mn}(\phi) + 2(m+\tau)(n+1) b_{m,n-1}(\phi) \\ + (n+2)(n+1) b_{m,n-2}(\phi) + k^2 \tilde{\varepsilon} b_{m-2,n}(\phi) = 0 \end{aligned} \tag{2.38}$$

$$a_{mn}(0) - j\delta a'_{m-1,n}(0) = 0 \tag{2.39}$$

$$b_{mn}(2\pi) + j\delta b'_{m-1,n}(2\pi) = 0 \tag{2.40}$$

follow from eqns. 2.29–2.34, where the prime denotes differentiation with respect to ϕ. The coefficients with at least one negative index in eqns. 2.37–2.40 are assumed to be zero.

Let us write out the first few equations of the chain given by eqns. 2.37–2.40. These equations define the behaviour of the current density near the edge and have the form

$$\left.\begin{aligned} a''_{00}(\phi) + \tau^2 a_{00}(\phi) = 0 \\ b''_{00}(\phi) + \tau^2 b_{00}(\phi) = 0 \end{aligned}\right\} \tag{2.41}$$

$$\left.\begin{aligned} a''_{10}(\phi) + (1+\tau)^2 a_{10}(\phi) = 0 \\ b''_{10}(\phi) + (1+\tau)^2 b_{10}(\phi) = 0 \end{aligned}\right\} \tag{2.42}$$

$$a_{00}(0) = b_{00}(2\pi) = 0 \tag{2.43}$$

The current density $\boldsymbol{J}$ is expressed in terms of $\psi_{1,2}$ by

$$\left.\begin{aligned} \boldsymbol{J}_1 = -\frac{\hat{i}_z}{j\omega\mu_0 r} \frac{\partial \psi_1}{\partial \phi}\bigg|_{\phi=0} \\ \boldsymbol{J}_2 = \frac{\hat{i}_z}{j\omega\mu_0 r} \frac{\partial \psi_2}{\partial \phi}\bigg|_{\phi=2\pi} \end{aligned}\right\} \tag{2.44}$$

The equalities $a_{00}(\phi) = b_{00}(\phi)$ and $a_{10}(\phi) = b_{10}(\phi)$ result from eqns. 2.33 and 2.34. Solving eqns. 2.41 and 2.42 and taking into account eqn. 2.43 at $\phi = 0$ we have

$$a_{00}(\phi) = Q_{00} \sin(\tau\phi)$$

$$a_{10}(\phi) = P_{10}\cos\{(1+\tau)\phi\} + Q_{10}\sin\{(1+\tau)\phi\}$$

where Q_{00}, Q_{10} and P_{10} are the constants of integration. It follows from eqn. 2.43 that τ may be only integer or half-integer (for our purposes it is enough to consider $\tau = 0, \frac{1}{2}, 1$). Using eqns. 2.36, 2.39 and 2.40, we obtain the first terms of asymptotic expansions for the field in the vicinity of the edge:

$$\left.\begin{aligned} \tau &= 0, \quad \psi = Q_{10} r \sin\phi + \cdots \\ \tau &= \tfrac{1}{2}, \quad \psi = Q_{10} r^{3/2} \sin(3\phi/2) + \cdots \\ \tau &= 1, \quad \psi = Q_{00} r \sin\phi + \cdots \end{aligned}\right\} \tag{2.45}$$

The constants Q_{00}, Q_{10} remain indefinite in this analysis. This situation occurs because we did not specify field sources and consequently we did not solve any definite diffraction problem (the analysis carried out reveals the general structure of asymptotic expansions for the field near the edge of impedance halfplane).

Let us analyse eqn. 2.45. When $\tau = 0$ and $\tau = 1$ it follows from eqn. 2.44 that the longitudinal component of the current density J_z is finite at the edge of impedance halfplane, which is in agreement with results of Section 2.4. When $\tau = \frac{1}{2}$ the current density is zero (this case is of no interest to us). We can apply Meixner's approximation $J_z(r) \simeq Cr^{-1/2}$ (C is a constant defined by the incident field) at some distance from the edge within the quasistatic domain $kr \ll 1$. Thus we come to the approximation of the current density in the edge vicinity

$$J_z(r) = C\left\{r + \left(\frac{C}{J_z(0)}\right)^2\right\}^{-1/2}$$

which is similar to eqn. 2.22. A new element following from the analysis performed is that the approximation is quite applicable when there is an interface between media with different permittivities. The ratio $C/J_z(0)$ does not depend on the relative permittivity $\tilde{\varepsilon}$. Consequently, eqns. 2.27 and 2.28 may be also used for an unclosed screen placed at the media interface. That allows us, in particular, to apply them to microstrip lines and other planar structures on dielectric substrata.

Completing our consideration of impedance boundary conditions, we would like to mention the papers by Grinberg (1981), Schimert *et al.* (1991) and Das and Pozar (1991), where generalisation of the Leontovich boundary condition has been implemented. In these papers an infinite plane-metal layer of thickness d, characterised by the permittivity $\varepsilon = \varepsilon_0 - j\sigma_0/\omega$, is taken as the key structure. Then the boundary conditions which are imposed on the field at an infinitesimally thin surface are obtained and both the thickness d and the conductivity σ_0 enter into the expressions for the coefficients in the boundary conditions. In general, these conditions may not, however, be compared with those obtained in Section 2.4. They are intended for taking into account the field penetration through the metal layer and do not consider the edge effects (the key structure is unbounded). The use of such conditions is justified for rather thin metal layers when the thickness d becomes comparable with penetration depth Δ_0 (in this case the fields on both sides of the layer are interrelated). Attempts to apply these conditions for calculation of the dissipation characteristics in structures with unclosed screens, e.g. in microstrip lines (Das and Pozar, 1991), proved to be problematic, as did the use of the ordinary Leontovich condition for this purpose. Sufficient accuracy may be obtained, when the contribution of edge effects to the lost power is negligible (e.g. in a microstrip line with a very wide strip).

Chapter 3

Normal modes in waveguides with losses

This Chapter deals with the theory of regular waveguides with impedance walls. Spectral problems for normal modes have been formulated. General characteristics of different types of normal modes (eigenmodes, associated modes) have been considered.

Calculation of attenuation coefficients of the eigenmodes for the case of small losses is discussed. General expressions for attenuation coefficients are obtained with the aid of the perturbation technique. Applicability cases for different variants of the perturbation method are analysed in detail (energy-perturbation method, perturbation technique for solving the dispersion equation). Their modifications are described for cases when classic schemes of the method are not usable (degenerated modes of D-multiplicity, models of waveguides with unclosed infinitely thin metal surfaces). Typical principal errors in attenuation-coefficient calculation are analysed.

Computational peculiarities of different approaches are pointed out, as well as the expediency of their application for different cases. General principles are illustrated by specific examples (rectangular and circular waveguides, coaxial and microstrip lines).

3.1 Excitation of waveguides without losses

Here we consider the theory of excitation of regular lossless waveguides by given arbitrarily distributed currents. This theory appears to be important from the methodological point of view: similar ideas and methods will be used later for the analysis of waveguides with losses.

Suppose that we have a longitudinally regular waveguide in the form of an infinite metal tube filled with a homogeneous medium. The axis of the tube is parallel to the z axis. The side surface of the waveguide will be denoted by Σ, the cross-section in $z = \text{constant}$ by $S_{\perp}$, and its perimeter by C. Let the electric-current density

$$\boldsymbol{J}(x, y, z, t) = \mathrm{Re}\{\boldsymbol{J}^e(x, y, z)\exp(-j\omega t)\}$$

be described by a finite function ($\boldsymbol{J}^e \neq 0$ only in the interval $z_1 \leq z \leq z_2$). We have to find the solution of Maxwell's equations

$$\left.\begin{aligned} \mathrm{curl}\,\boldsymbol{H} &= -j\omega\varepsilon\boldsymbol{E} + \boldsymbol{J}^e \\ \mathrm{curl}\,\boldsymbol{E} &= j\omega\mu\boldsymbol{H} \end{aligned}\right\} \tag{3.1}$$

satisfying the boundary condition

$$(\boldsymbol{n}_0 \times \mathbf{E})|_\Sigma = 0 \tag{3.2}$$

where $\boldsymbol{n}_0$ is the unit normal to Σ. In addition, $\boldsymbol{E}$ and $\boldsymbol{H}$ should meet the radiation condition at $z \to \pm\infty$: that means that the fields at $z > z_2$ and $z < z_1$ are presented as superpositions of waves moving away from the source. As there are no losses, $\varepsilon'' = \mu'' = 0$.

The theory of excitation of waveguides is one of the applications of the spectral method, which is a rather general approach to solving a wide class of problems of mathematical physics. Its essence is as follows: solutions sought are expanded into series in terms of basis functions determined from specially chosen spectral problems (eigenvalue problems). In our case, eigenmodes of the waveguide under consideration are taken as basis functions. The expansion coefficients are determined from Maxwell's equations using such properties of basis functions as completeness and orthogonality.

There are different enunciations of the theory of waveguide excitation based on the spectral method. Weinstein (1988) solved eqn. 3.1 directly; Kisun'ko (1948) proceeded from the fields intensities $\boldsymbol{E}$ and $\boldsymbol{H}$ to scalar and vector potentials Φ and $\boldsymbol{A}$ and solved equations for them; Tikhonov and Samarsky (1977) constructed a solution introducing the electric and magnetic Hertz vectors. All these versions of the excitation theory are equivalent and can be transformed from one to another. We shall dwell here upon the most convenient variant offered by Weinstein (1988) and use it in our further analysis.

Let us introduce the Dirichlet's and Neumann's spectral problems in the domain $S_\perp$:

$$\nabla_{xy}^2 \Phi_s^{e,m} + (\chi_s^{e,m})^2 \Phi_s^{e,m} = 0$$

$$\Phi_s^e|_C = 0 \qquad \left.\frac{\partial \Phi_s^m}{\partial \boldsymbol{n}}\right|_C = 0$$

where $(\chi_s^{e,m})^2$ are the eigenvalues.

These problems are classical ones; their properties are studied and described in detail in many works on mathematical physics.

Let us form the vector functions

$$\left.\begin{aligned} \boldsymbol{E}_s^e(x, y, z) &= \left(\hat{i}_z \Phi_s^e + \frac{jh_s^e}{(\chi_s^e)^2}\nabla_{xy}\Phi_s^e\right)\exp(jh_s^e z) \\ \boldsymbol{H}_s^e(x, y, z) &= -\frac{j\omega\varepsilon}{(\chi_s^e)^2}(\nabla_{xy}\Phi_s^e \times \hat{i}_z)\exp(jh_s^e z) \end{aligned}\right\} \tag{3.3}$$

$$\left.\begin{aligned} \boldsymbol{E}_s^m(x,y,z) &= \frac{j\omega\mu}{(\chi_s^m)^2}(\nabla_{xy}\Phi_s^m \times \hat{i}_z)\exp(jh_s^m z) \\ \boldsymbol{H}_s^m(x,y,z) &= \left(\hat{i}_z\Phi_s^m + \frac{jh_s^m}{(\chi_s^m)^2}\nabla_{xy}\Phi_s^m\right)\exp(jh_s^m z) \end{aligned}\right\} \tag{3.4}$$

where $h_s^{e,m} = \{k^2 - (\chi_s^{e,m})^2\}^{1/2}$, $s = 1, 2, \ldots$ and $k = \omega(\varepsilon\mu)^{1/2}$. Alongside eqns. 3.3 and 3.4 let us introduce the functions $\boldsymbol{E}_{-s}^{e,m}$, $\boldsymbol{H}_{-s}^{e,m}$ obtained from eqns. 3.3 and 3.4 through the replacement of $h_s^{e,m}$ by $-h_{-s}^{e,m}$. Let us combine the subsystems $\boldsymbol{E}_s^e$, $\boldsymbol{H}_s^e$ and $\boldsymbol{E}_s^m$, $\boldsymbol{H}_s^m$ in the united system $\boldsymbol{E}_s$, $\boldsymbol{H}_s$ $(s = \pm 1, \pm 2, \ldots)$:

$$\{\boldsymbol{E}_s\} = \{\boldsymbol{E}_s^e\} \oplus \{\boldsymbol{E}_s^m\}, \qquad \{\boldsymbol{H}_s\} = \{\boldsymbol{H}_s^e\} \oplus \{\boldsymbol{H}_s^m\}$$

It is easy to show that the functions $\boldsymbol{E}_s$ and $\boldsymbol{H}_s$ are the eigenmodes of a given waveguide: they satisfy homogeneous (without sources) Maxwell's equations and the boundary conditions required. The functions expressed by eqns. 3.3 are TM modes of a given waveguide $(H_{sz}^e = 0)$; the functions expressed by eqn. 3.4 are TE modes $(E_{sz}^e = 0)$. If the cross-section $S_\perp$ is not single connected, we should also add TEM modes in $\{\boldsymbol{E}_s\}$, $\{\boldsymbol{H}_s\}$ (for N-connected domain $S_\perp$ there are N − 1 TEM modes). As is shown by Weinstein (1988), the orthogonality relation

$$\int_{S_\perp} \{(\boldsymbol{E}_s \times \boldsymbol{H}_{s'}) - (\boldsymbol{E}_{s'} \times \boldsymbol{H}_s)\}\hat{i}_z \, \mathrm{d}S_\perp = \delta_{s,-s'}\mathcal{N}_{s'} \tag{3.5}$$

is valid where $\mathcal{N}_s$ is the sth eigenmode norm.

At $z > z_2$ the excited field can be presented as

$$\begin{Bmatrix} \boldsymbol{E} \\ \boldsymbol{H} \end{Bmatrix} = \sum_{s>0} T_s \begin{Bmatrix} \boldsymbol{E}_s \\ \boldsymbol{H}_s \end{Bmatrix} \tag{3.6}$$

and at $z < z_1$

$$\begin{Bmatrix} \boldsymbol{E} \\ \boldsymbol{H} \end{Bmatrix} = \sum_{s>0} R_s \begin{Bmatrix} \boldsymbol{E}_{-s} \\ \boldsymbol{H}_{-s} \end{Bmatrix} \tag{3.7}$$

where T_s and R_s are complex coefficients which at this stage are unknown. Eqns. 3.6 and 3.7 are an analytical form of the radiation condition.

Validity of representations given by eqns. 3.6 and 3.7 follows from the completeness property of the systems $\{\boldsymbol{E}_s\}$, $\{\boldsymbol{H}_s\}$ $(s = \pm 1, \pm 2, \ldots)$ in a class of functions meeting homogeneous Maxwell's equations for an arbitrary cylindrical area (including the z-unbounded one) with the cross-section $S_\perp$. $\boldsymbol{E}$ and $\boldsymbol{H}$ are determined by the same systems of coefficients T_s and R_s. This can be explained by the fact that there are no sources in the area under consideration. As has been shown by Tikhonov and Samarsky (1977), the series given by eqns. 3.6 and 3.7 are uniformly and absolutely convergent.

The transverse components of the vectors $\boldsymbol{E}_s$ and $\boldsymbol{H}_s$ can be used as basis functions to expand the transverse components of the vectors sought inside the region containing the sources $(z_1 \leq z \leq z_2)$. Thus let us present

$$\boldsymbol{E} = \boldsymbol{E}^t + \boldsymbol{E}^l \qquad \boldsymbol{H} = \boldsymbol{H}^t + \boldsymbol{H}^l \qquad \boldsymbol{J}^e = \boldsymbol{J}^{et} + \boldsymbol{J}^{el}$$

superscripts t and l denoting the transverse and longitudinal components of a corresponding vector. It can be shown that the following formulae are valid:

$$\left.\begin{aligned} \operatorname{curl} \boldsymbol{E}^l + \left(\hat{i}_z \times \frac{\partial \boldsymbol{E}^t}{\partial z}\right) &= j\omega\mu \boldsymbol{H}^t \\ \operatorname{curl} \boldsymbol{H}^l + \left(\hat{i}_z \times \frac{\partial \boldsymbol{H}^t}{\partial z}\right) &= -j\omega\varepsilon \boldsymbol{E}^t + \boldsymbol{J}^{et} \end{aligned}\right\} \tag{3.8}$$

$$\left.\begin{aligned} \operatorname{curl} \boldsymbol{E}^t - \left(\hat{i}_z \times \frac{\partial \boldsymbol{E}^t}{\partial z}\right) &= j\omega\mu \boldsymbol{H}^l \\ \operatorname{curl} \boldsymbol{H}^t - \left(\hat{i}_z \times \frac{\partial \boldsymbol{H}^t}{\partial z}\right) &= -j\omega\varepsilon \boldsymbol{E}^l + \boldsymbol{J}^{el} \end{aligned}\right\} \tag{3.9}$$

Eqns. 3.8 are the transverse part of Maxwell's equations, and eqns. 3.9 the longitudinal part.

Let us expand the transverse components of the electromagnetic field into series in terms of the transverse vector functions $\boldsymbol{E}_s^t$ and $\boldsymbol{H}_s^t$:

$$\begin{Bmatrix} \boldsymbol{E}^t \\ \boldsymbol{H}^t \end{Bmatrix} = \sum_{s>0} \left[C_s(z) \begin{Bmatrix} \boldsymbol{E}_s^t \\ \boldsymbol{H}_s^t \end{Bmatrix} + C_{-s}(z) \begin{Bmatrix} \boldsymbol{E}^t_{-s} \\ \boldsymbol{H}^t_{-s} \end{Bmatrix} \right] \tag{3.10}$$

Here $C_{\pm s}(z)$ are unknown functions of the longitudinal co-ordinate z to be determined. Substituting eqn. 3.10 into eqns. 3.9, we obtain the expansions for the longitudinal components of the electromagnetic field:

$$\boldsymbol{E}^l = \sum_{s>0} \{C_s(z)\boldsymbol{E}_s^l + C_{-s}(z)\boldsymbol{E}^l_{-s}\} + \frac{\boldsymbol{J}^{el}}{j\omega\varepsilon} \tag{3.11}$$

$$\boldsymbol{H}^l = \sum_{s>0} \{C_s(z)\boldsymbol{H}_s^l + C_{-s}(z)\boldsymbol{H}^l_{-s}\} \tag{3.12}$$

Combining eqn. 3.10 with eqns. 3.11 and 3.12 we finally obtain the expansions

$$\boldsymbol{E} = \sum_{s>0} \{C_s(z)\boldsymbol{E}_s + C_{-s}(z)\boldsymbol{E}_{-s}\} + \hat{i}_z \frac{(\boldsymbol{J}^e \hat{i}_z)}{j\omega\varepsilon} \tag{3.13}$$

$$\boldsymbol{H} = \sum_{s>0} \{C_s(z)\boldsymbol{H}_s + C_{-s}(z)\boldsymbol{H}_{-s}\} \tag{3.14}$$

Our task will now be to determine the coefficients $C_{\pm s}(z)$. We shall use the Lorentz lemma for this purpose. Let the electromagnetic fields $\boldsymbol{E}_1, \boldsymbol{H}_1$ and $\boldsymbol{E}_2, \boldsymbol{H}_2$ be excited by the currents $\boldsymbol{J}_1^e$ and $\boldsymbol{J}_2^e$, respectively. Then the following identity is correct (Lorentz lemma):

$$\oint_S \{(\boldsymbol{E}_1 \times \boldsymbol{H}_2) - (\boldsymbol{E}_2 \times \boldsymbol{H}_1)\}\boldsymbol{n}_0 \, \mathrm{d}S = \int_V (\boldsymbol{J}_1^e \boldsymbol{E}_2 - \boldsymbol{J}_2^e \boldsymbol{E}_1) \, \mathrm{d}V \tag{3.15}$$

where V is an arbitrary closed volume and S is the boundary surface of this volume.

Let us apply eqn. 3.15 to the infinitely small volume inside the waveguide bounded by two sections $z = z$, $z = z + \mathrm{d}z$ and part of the side surface Σ enclosed between them. Taking into account that $\boldsymbol{E}_1$ and $\boldsymbol{E}_2$ satisfy eqn. 3.2, we obtain from eqn. 3.15

$$\frac{d}{dz}\oint_{S_\perp}\{(\boldsymbol{E}_1\times\boldsymbol{H}_2)-(\boldsymbol{E}_2\times\boldsymbol{H}_1)\}\hat{i}_z\,dS=\int_{S_\perp}(\boldsymbol{J}_1^e\boldsymbol{E}_2-\boldsymbol{J}_2^e\boldsymbol{E}_1)\,dV \tag{3.16}$$

This equation can be used for determining $C_{\pm s}(z)$. Assuming that $\boldsymbol{E}_1, \boldsymbol{H}_1$ in eqn. 3.15 is the field to be found, presented by eqns. 3.13 and 3.14 ($\boldsymbol{J}_1^e=\boldsymbol{J}^e$) and $\boldsymbol{E}_2=\boldsymbol{E}_{-s}$, $\boldsymbol{H}_2=\boldsymbol{H}_{-s}$, ($\boldsymbol{J}_2^e=0$) and using the orthogonality relation expressed by eqn. 3.5, we have

$$\frac{dC_s}{dz}=\frac{1}{N_s}\int_{S_\perp}\boldsymbol{J}^e\boldsymbol{E}_{-s}\,dS_\perp \tag{3.17}$$

where $s=\pm 1, \pm 2, \ldots, \pm\infty$.

Boundary conditions for eqn. 3.17 result from the radiation condition; in accordance with eqns. 3.6 and 3.7 they have the form

$$C_s(z_1)=0 \qquad C_{-s}(z_2)=0$$

$s=1, 2, \ldots, \infty$. The fields matching in the $z=z_{1,2}$ planes results in

$$R_s=C_{-s}(z_1) \qquad T_s=C_s(z_2)$$

Thus the problem of excitation of a waveguide by given currents is solved completely.

3.2 Excitation of waveguides with losses in the walls

The classical formulation for the excitation equations of a waveguide by given sources, examined in Section 3.1, is based on the expansion of the field sought into a series in terms of the eigenmodes of the waveguide considered. To substantiate this approach we have to prove the completeness (basisness) of the system of eigenmodes. For lossless waveguides it can be done with rather general suppositions concerning the form of the waveguide cross-section (see Section 3.1). In the presence of losses the situation is more complicated. Eigenmodes systems of waveguides with losses may be incomplete and not forming a basis in the required functional space. However, where we have a basisness, for waveguides with losses we can repeat the scheme described in Section 3.1.

It would therefore be quite natural to construct a theory of excitation for systems with losses using the expansions of the fields sought in terms of some other complete systems of functions, for example the eigenmodes system of a lossless waveguide of the same shape. In this case, the field sought and the basis functions used for its expansion meet different boundary conditions, so generalisation of the method is needed. Later we shall describe this generalisation based mainly on the paper by Arkadaksky and Tsykin (1975*a*).

We need to find the solution of nonhomogeneous Maxwell's equations satisfying the boundary condition

$$(\boldsymbol{n}_0\times\boldsymbol{E})+Z_S\{\boldsymbol{n}_0\times(\boldsymbol{n}_0\times\boldsymbol{H})\}=0 \tag{3.18}$$

on the side surface Σ. Here, the impedance Z_S is independent of the z co-ordinate. The field sought should satisfy the condition 'at infinity', that for waveguides with losses ($\mathrm{Re}\, Z_S > 0$) has the form

$$\lim_{z \to \pm\infty} \begin{Bmatrix} \boldsymbol{E} \\ \boldsymbol{H} \end{Bmatrix} = 0 \tag{3.19}$$

This condition is responsible for the uniqueness of the solution of the problem.

Before solving the boundary-value problem given by eqns. 3.1, 3.18 and 3.19, let us derive an auxiliary identity. We introduce two systems of fields $\boldsymbol{E}_1, \boldsymbol{H}_1$ and $\boldsymbol{E}_2, \boldsymbol{H}_2$ (excited by sources $\boldsymbol{J}_1^e$ and $\boldsymbol{J}_2^e$) satisfying the boundary condition given by eqn. 3.18 with the impedances Z_{S1} and Z_{S2}, respectively. Applying eqn. 3.15 to the infinitely small volume $\mathrm{d}V = S_\perp\, \mathrm{d}z$ we obtain for the given class of fields the formula

$$\frac{\mathrm{d}}{\mathrm{d}z} \int_{S_\perp} \{(\boldsymbol{E}_1 \times \boldsymbol{H}_2) - (\boldsymbol{E}_2 \times \boldsymbol{H}_1)\}\hat{i}_z\, \mathrm{d}S$$
$$- (Z_{S1} - Z_{S2}) \oint_C \{(\boldsymbol{n}_0 \times \boldsymbol{H}_1) \times \boldsymbol{H}_2\}\boldsymbol{n}_0\, \mathrm{d}C$$
$$= \int_{S_\perp} (\boldsymbol{J}_1^e \boldsymbol{E}_2 - \boldsymbol{J}_2^e \boldsymbol{E}_1)\, \mathrm{d}V \tag{3.20}$$

Eqn. 3.20 is a generalised version of eqn. 3.16 for the case where impedance boundary conditions are imposed.

As a basis for our analysis we shall use Galerkin's incomplete method (see Section 1.3). Let $\boldsymbol{E}_1, \boldsymbol{H}_1$ be the fields sought ($\boldsymbol{J}_1^e = \boldsymbol{J}^e$, $Z_{S1} = Z_S$) and $\boldsymbol{E}_2, \boldsymbol{H}_2$ be the fields of the pth eigenmode $\boldsymbol{E}_{-p}, \boldsymbol{H}_{-p}$ ($\boldsymbol{J}_2^e = 0$, $Z_{S2} = 0$, $p = \pm 1, \pm 2, \ldots$). Then eqn. 3.20 at various p forms a chain of orthogonality relations equivalent to the original boundary-value problem given by eqns. 3.1, 3.18 and 3.19. Let us present the fields sought as expansions in terms of the eigenmodes $\{\boldsymbol{E}_s\}$, $\{\boldsymbol{H}_s\}$ of a waveguide of the same shape but without losses (the expressions for $\boldsymbol{E}_s, \boldsymbol{H}_s$ are presented in Section 3.1). Corresponding expansions have the form of eqns. 3.13 and 3.14. It should be mentioned that in the case considered the expansions given by eqns. 3.13 and 3.14 (unlike those in Section 3.1) cannot be directly substituted into Maxwell's equations to determine the coefficients $C_{\pm s}(z)$ because it is impossible to apply the term-by-term curl operation to these expansions. However, this is not necessary; we avoid it through using eqn. 3.20.

Using eqns. 3.13 and 3.14 and considering the energy orthogonality of the eigenmodes (see Section 3.1), we obtain the following system of ordinary differential equations for the coefficients $C_{\pm s}(z)$:

$$\frac{\mathrm{d}C_p}{\mathrm{d}z} = \frac{1}{N_p} \int_{S_\perp} \boldsymbol{J}^e \boldsymbol{E}_{-p}\, \mathrm{d}S_\perp$$
$$- \frac{1}{N_p} Z_S \sum_{\substack{s=-\infty \\ s \neq 0}}^{\infty} C_s(z) \oint_C \{\boldsymbol{H}_{-p} \times (\boldsymbol{n}_0 \times \boldsymbol{H}_s)\}\boldsymbol{n}_0\, \mathrm{d}C \tag{3.21}$$

The last term in eqn. 3.21 describes a mutual coupling of different eigenmodes due to the surface impedance of the walls. Eqn. 3.21 is a formulation of excitation theory in terms of coupling modes.

To make further analysis easier we introduce new variables according to the formula

$$C_p = X_p \exp(-jh_p z)$$

Then for X_p we obtain the system of ordinary differential equations

$$\frac{dX_p}{dz} = R_p + \sum_{\substack{s=-\infty \\ s \neq 0}}^{\infty} X_s A_{sp}$$

where

$$R_p = \frac{1}{N_p} \int_{S_\perp} \boldsymbol{J}^e \boldsymbol{E}^0_{-p} \, dS_\perp$$

$$A_{sp} = j \left[\delta_{sp} h_p + j \frac{Z_S}{N_p} \oint_C \{ \boldsymbol{H}^0_{-p} \times (\boldsymbol{n}_0 \times \boldsymbol{H}^0_s) \} . \boldsymbol{n}_0 \, dC \right]$$

$$\boldsymbol{E}^0_p = \boldsymbol{E}_p \exp(-jh_p z) \qquad \boldsymbol{H}^0_p = \boldsymbol{H}_p \exp(-jh_p z) \tag{3.22}$$

Proceeding to vector–matrix designations, we have

$$\frac{d\boldsymbol{X}}{dz} = A\boldsymbol{X} + \boldsymbol{R} \tag{3.23}$$

where $\boldsymbol{X}$ and $\boldsymbol{R}$ are vectors of the coefficients $\{X_p\}$ and $\{R_p\}$, respectively, A is a matrix of the elements A_{sp} and the matrix A is independent of z.

The system given by eqn. 3.23 does not represent the full formulation of the theory of excitation of waveguides, as it has no unique solution. This has to do with the fact that eqn. 3.23 does not take into account the condition at infinity given by eqn. 3.19. This system should be solved under some auxiliary conditions on the vector $\boldsymbol{X}$ sought. However, the analytical form of these conditions in terms of coupling modes is rather complicated. Therefore we postpone their formulation until the next Section where the general solution of this system is examined.

3.3 Eigenmodes in waveguides; dispersion characteristics

Suppose that we have a nonsingular matrix B which reduces the matrix A to the diagonal form, i.e. implementing the transformation

$$\Gamma = BAB^{-1} \tag{3.24}$$

where Γ is the diagonal matrix

$$\left\| \begin{matrix} \ddots & & & & & \\ & j\gamma_{-2} & & & 0 & \\ & & j\gamma_{-1} & & & \\ & & & j\gamma_1 & & \\ & 0 & & & j\gamma_2 & \\ & & & & & \ddots \end{matrix} \right\|$$

and γ_i are some complex numbers. From the reciprocity theorem it follows that they can be numbered in such a way to fulfil the equation $\gamma_{-i} = -\gamma_i$.

Then, introducing new variables

$$\boldsymbol{g} = B\boldsymbol{X}$$

we obtain from eqn. 3.23 the equation for $\boldsymbol{g}$

$$\frac{d\boldsymbol{g}}{dz} = \Gamma \boldsymbol{g} + \boldsymbol{K} \tag{3.25}$$

where $\boldsymbol{K} = B\boldsymbol{R}$.

The field sought can be presented in terms of the new variables in the form (Arkadaksky and Tsykin, 1975*b*):

$$\left.\begin{aligned} \boldsymbol{E} &= \sum_{s>0} \{g_s(z)\hat{\boldsymbol{E}}_s(x, y) + g_{-s}(z)\hat{\boldsymbol{E}}_{-s}(x, y)\} + \hat{i}_z \frac{(\boldsymbol{J}^e \hat{i}_z)}{j\omega\varepsilon} \\ \boldsymbol{H} &= \sum_{s>0} \{g_s(z)\hat{\boldsymbol{H}}_s(x, y) + g_{-s}(z)\hat{\boldsymbol{H}}_{-s}(x, y)\} \end{aligned}\right\} \tag{3.26}$$

where $g_s(z)$ is the sth component of the vector $\boldsymbol{g}(z)$ and

$$\begin{Bmatrix} \hat{\boldsymbol{E}}_s(x, y) \\ \hat{\boldsymbol{H}}_s(x, y) \end{Bmatrix} = \sum_{p>0} \left[\tilde{b}_{p,s} \begin{Bmatrix} \boldsymbol{E}^0_p(x, y) \\ \boldsymbol{H}^0_p(x, y) \end{Bmatrix} + \tilde{b}_{p,-s} \begin{Bmatrix} \boldsymbol{E}^0_{-p}(x, y) \\ \boldsymbol{H}^0_{-p}(x, y) \end{Bmatrix} \right] \tag{3.27}$$

$\tilde{b}_{p,s}$ are elements of the matrix B^{-1}, and $\boldsymbol{E}^0_p(x, y)$ and $\boldsymbol{H}^0_p(x, y)$ are expressed by eqn. 3.22.

It can be easily seen that eqns. 3.26 are an expansion of the excited field in terms of the eigenmodes of the impedance waveguide. Eqn. 3.25 can be written in the form

$$\frac{dg_s}{dz} = j\gamma_s g_s + \frac{1}{n_s} \int_{S_\perp} \boldsymbol{J}^e \hat{\boldsymbol{E}}_{-s} \, dS_\perp \tag{3.28}$$

Eqn. 3.28 coincides with that obtained according to a classical scheme (see Section 3.1) when we introduce new variables: $g_s(z) = C_s(z)\exp(j\gamma_s z)$. By n_s and γ_s we mean the norm and the wave number of the sth mode of the impedance waveguide.

Thus if there is a transformation given by eqn. 3.24 for matrix A, then the excited field can be presented as a series in terms of eigenmodes expressed by eqn. 3.26 with coefficients determined from eqn. 3.28. This equation is easily integrated in quadratures. The functions $\hat{\boldsymbol{E}}_s$, $\hat{\boldsymbol{H}}_s$ defined by eqn. 3.27 are transverse field distributions of the sth eigenmode in the waveguide with losses (to be more precise, their representation in the basis $\{\boldsymbol{E}^0_s\}$, $\{\boldsymbol{H}^0_s\}$).

As was pointed out in Section 3.2, to determine the unique solution of the problem some auxiliary physical conditions are required. Suppose that the external sources are localised in some finite area $z_1 \leq z \leq z_2$. Then, if the transformation expressed by eqn. 3.24 exists, to meet the condition at infinity in the areas $z > z_2$ and $z < z_1$ we should reject the increasing waves, i.e. add the following boundary conditions to eqn. 3.25

$$g_{-s}(z_2) = 0, \qquad g_s(z_1) = 0 \tag{3.29}$$

In lossless systems the solution sought can be selected by different methods: with the aid of Sommerfeld's radiation condition, Mandelstam radiation energy principle, principles of limiting absorption and limiting amplitude. Analysis and comparison of such principles applied to the problems of the dynamic theory of elasticity, are contained in the book by Vorovich and Babeshko (1979). We would like to stress here an *a priori* and heuristic character of these principles which limits the area of their application. Only for the simplest problems are these principles equivalent. The main difficulties in their use are met when there are associated modes or waves with anomalous dispersion (see eqn. 3.45).

Thus, using eqn. 3.29, the electromagnetic field at $z > z_2$ and $z < z_1$ can be presented as

$$\begin{Bmatrix} \boldsymbol{E} \\ \boldsymbol{H} \end{Bmatrix} = \sum_{s>0} g_s(z_2) \begin{Bmatrix} \hat{\boldsymbol{E}}_s(x, y) \\ \hat{\boldsymbol{H}}_s(x, y) \end{Bmatrix} \exp\{j\gamma_s(z - z_2)\} \qquad (z > z_2)$$

$$\begin{Bmatrix} \boldsymbol{E} \\ \boldsymbol{H} \end{Bmatrix} = \sum_{s<0} g_s(z_1) \begin{Bmatrix} \hat{\boldsymbol{E}}_s(x, y) \\ \hat{\boldsymbol{H}}_s(x, y) \end{Bmatrix} \exp\{j\gamma_s(z - z_1)\} \qquad (z < z_1)$$

The eigenmodes of a waveguide with losses in the walls satisfy the orthogonality relations given by eqn. 3.5, and hence each of them in the area which is free of sources propagates independently from the left. Thus eigenmodes have a clear physical sense: these are the fields that can be excited in a waveguide outside the area occupied by sources.

Let us consider now the conditions of existence of the diagonalised transformation expressed by eqn. 3.24 for the matrix A. The theorems formulated here are proved in the book by Fadeev and Fadeeva (1963) for a matrix of finite order. Transition to infinite matrices is valid under certain conditions which justify the operations carried out. Comprehensive substantiation can be given only for specific cases [see, for example, Arkadaksky and Tsykin (1975*a*)] where it was made for a parallel-plate impedance waveguide. Below we shall consider the corresponding conditions fulfilled without mentioning specifically the difference between the finite and infinite matrices. The following theorem is correct:

Theorem 3.1: If all eigenvalues of the matrix A are different in pairs, then the transformation given by eqn. 3.24 exists and diagonal elements of the matrix Γ coincide with eigenvalues of the matrix A.

The characteristic equation of the matrix A

$$\mathrm{Det}\{A(\omega) - j\gamma I\} = 0 \qquad (3.30)$$

defines the dependencies $\gamma_s(\omega)$ at given waveguide parameters.

Thus, eqn. 3.30 is the dispersion equation of the waveguide with losses.

Eigenmodes propagating in waveguides can be divided into four groups:

(i) fast modes characterised by the condition $\mathrm{Re}\,\gamma_s(\omega) < k$ $(v_{Ph} > c)$ (for a given shape of a waveguide their spectrum is defined by the area of the waveguide cross-section);

(ii) slow (surface) modes characterised by the condition $\mathrm{Re}\,\gamma_s(\omega) > k$ $(v_{Ph} < c)$;

(iii) TEM modes (modes which do not have longitudinal field components) characterised by the condition $\gamma_s(\omega) = k = \omega/c$; and

(iv) near-boundary modes; fast waves, whose spectrum at a given shape of a waveguide is determined not by the area but by the perimeter of the waveguide cross-section (Nefedov and Rossiisky, 1978).

The modes of the first type are specifically conventional TM and TE modes of hollow rectangular and circular waveguides with a smooth metal surface.

The modes of the second type are typical for their field being concentrated near the guiding surface and rapidly transversely decreasing (without losses; exponentially). Such modes can exist in waveguides with impedance walls or in the presence of dielectric insertions.

The TEM modes do not undergo any dispersion and their field structure is strictly transverse. Such modes can exist in multiply connected lossless guiding systems (e.g. coaxial lines). In the presence of heat losses in the walls there are longitudinal field components and dispersion effects, but with small losses they are notable only slightly.

Near-boundary modes can exist both in impedance waveguides and in ideal waveguides of a sufficiently complex shape of the cross-section (in the latter case they should necessarily be TE modes). The theory of near-boundary modes is described by Nefedov and Rossiisky (1978). Thus, for example, the TE_{m1} modes of a coaxial waveguide and biconical horn and some TE modes in a cross-shaped waveguide can be considered as near-boundary. The important features of such modes are a rather rare spectrum and the ability to reflect totally from widening segments of a waveguide. Based on the same principle, effective selection of parasitic types of eigenoscillations in open resonators is possible.

One of the most important characteristics of eigenmodes is the dispersion characteristic: the dependence $\gamma_s(\omega)$. As well as calculation of dispersion characteristics of a specific waveguide system, qualitative investigation of dispersion laws, in the most general sense, is very significant. It permits discovery of general physical properties typical of different classes of eigenmodes and not connected with the special features of the waveguide shape.

An effective method of carrying out such an investigation, based on the theory of analytical functions of many complex variables, was suggested by Krasnushkin and Fedorov (1972). Below we shall touch upon the problem's formulation, methods and the main results of these investigations.

Let us assume that the waveguide under study is characterised by the parameters $W_1, W_2, \ldots, W_n$ (these parameters describe the waveguide geometry, characteristics of its surface and filling media etc.). Then eqn. 3.30 can be symbolically written as

$$\Delta(\gamma, W_1, W_2, \ldots, W_n) = \Delta(\gamma, \{W_i\}_n) = \Delta(\gamma, \boldsymbol{W}) = 0$$

Let W_i be the co-ordinates of a point in the n-dimensional complex space C^n and $(\gamma, \{W_i\}_n)$ the co-ordinates of the point $\Lambda \in C^{n+1}$. If the coefficients A_{sp} are analytical functions of $\{W_i\}$, then Δ is an analytical function of Λ. In this case the dependencies $\gamma_s(\boldsymbol{W})$ would be holomorphic functions of n variables $\{W_i\}_n$ excluding the degeneration points in which

$$\frac{\partial \Delta}{\partial \gamma} = \frac{\partial^2 \Delta}{\partial \gamma^2} = \cdots = \frac{\partial^m \Delta}{\partial \gamma^m} = 0 \qquad (m < n) \tag{3.31}$$

These points will be branch points of the mth order. When among eigenmodes are degenerated ones (with the same propagation coefficients), i.e. complying with eqn. 3.31, then the question on existence of diagonalised transformation expressed by eqn. 3.24 is still open. The answer is given by theorem 3.2.

Theorem 3.2: If some of the eigenvalues of the matrix A are multiple, the transformation given by eqn. 3.24 exists only when the multiplicity of each eigenvalue equals the number of linearly independent eigenvectors corresponding to this eigenvalue (Fadeev and Fadeeva, 1963).

If the conditions of theorem 3.2 are fulfilled, the degenerated modes have different field structures and their orthogonal linear combinations can be formed. Such degenerated modes are called modes of diagonalised multiplicity (D-multiplicity modes). If all the degenerated modes are of D-multiplicity, the excitation theory in terms of eigenmodes is also correct.

When affirmation of theorem 3.2 does not take place then the transformation expressed by eqn. 3.24 does not exist and transition from coupled waves to eigenmodes is impossible.

In the general case, this transformation can be used to reduce the matrix A to the Jordan canonical matrix form (Fadeev and Fadeeva, 1963):

$$\Gamma = \left\| \begin{array}{cccccccccc}
\ddots & & & & & & & & & \\
& \lambda_{-1} & & & 0 & & & & 0 & \\
& 1 & \lambda_{-1} & & & & & & & \\
& & 1 & \ddots & & & & & & \\
& 0 & & & \lambda_{-1} & & & & & \\
& & & & & \lambda_1 & 1 & & 0 & \\
& & & & & & \lambda_1 & 1 & & \\
& & & & & & & \ddots & & \\
0 & & & & & 0 & & & \lambda_1 & \\
& & & & & & & & & \ddots
\end{array} \right\|$$

Thus, if M_j-multiple degenerated jth eigenvalue corresponds to $m_j < M_j$ linearly independent vectors, $j = 1, 2, \ldots, N$, then the matrix Γ contains N Jordan blocks of the order of $M_j - m_j + 1$ belonging to the eigenvalues λ_j. Degenerated eigenmodes responsible for the Jordan blocks of the matrix Γ are called modes of Jordan multiplicity or, in short, J-multiplicity. It is worth pointing out that J-multiple eigenvalues can be found only with non-self-adjoint operators.

The dependencies $\gamma_s(\boldsymbol{W})$ are presented in C^{n+1} by a countable set of analytic surfaces. These dependencies are studied by Krasnushkin and Fedorov (1972) on sets of routes $\mathscr{L}(s, \mu)$, where s is a real parameter determining the location of a point on a curve presenting the route and μ is the real index selecting the given route out of the set. To study the behaviour of branches $\gamma_s(\boldsymbol{W})$ in the neighbourhood of branch points determined by eqn. 3.31 it is necessary to transfer from C^n to some Riemann manifold Σ on which the functions under examination are single-valued. To construct such a Riemann manifold we should introduce m copies of C^n, make cuts along their branch surfaces and glue different edges of branch cuts to each other in pairs.

If at the wave 'movement' in the space C^n a branch cut is crossed, then a transfer to another 'sheet' of Riemann manifold Σ takes place [to another branch of $\gamma_s(\boldsymbol{W})$]. A secondary crossing in the opposite direction restores the initial situation. Thus the global structure of branches of $\gamma_s(\boldsymbol{W})$ is determined by the location of the branches' intersection points and their behaviour in the neighbourhood of these points. If we take a frequency ω as one of parameters $\{W_i\}_n$, then we can also study a structure of dispersion characteristics of waveguides.

As an example, let us analyse eqn. 3.30 assuming that only two modes are closely coupled (these will be the modes with numbers s and p). Then eqn. 3.30 transfers into a 2nd-order equation and we can obtain the following expressions for the roots:

$$2\gamma_{p,s} = A_{ss} + A_{pp} \pm (\delta^2 + l^2)^{1/2} \tag{3.32}$$

where

$$\delta = A_{ss} - A_{pp} \qquad l = 2(A_{sp}A_{ps})^{1/2}$$

Fields' distributions of these modes are presented as

$$\boldsymbol{u}_s = C_{ss}\boldsymbol{V}_s + C_{sp}\boldsymbol{V}_p$$

$$\boldsymbol{u}_p = C_{pp}\boldsymbol{V}_p + C_{ps}\boldsymbol{V}_s$$

where

$$\boldsymbol{u}_s = \begin{Bmatrix} \hat{\boldsymbol{E}}_s \\ \hat{\boldsymbol{H}}_s \end{Bmatrix} \quad \boldsymbol{V}_s = \begin{Bmatrix} \boldsymbol{E}_s^0 \\ \boldsymbol{H}_s^0 \end{Bmatrix}$$

and C_{mn} $(m, n = s, p)$ are coefficients satisfying the formulae

$$\left.\begin{aligned} \frac{C_{ss}}{C_{ps}} &= -\frac{1}{2A_{ps}}\{\delta + (\delta^2 + l^2)^{1/2}\} \\ \frac{C_{sp}}{C_{pp}} &= -\frac{1}{2A_{ps}}\{\delta - (\delta^2 + l^2)^{1/2}\} \end{aligned}\right\} \tag{3.33}$$

Eqns. 3.32 and 3.33 are relatively similar to the corresponding formulae for plane-layered media studied through this method by Krasnushkin and Fedorov (1972). So, without going much into detail, let us formulate the main results.

In this approximation, it is evident that not more than a 2-fold degeneration of modes is possible. The points $\boldsymbol{W}^2$ (points of 2-fold degeneration) form in C^2 a set determined by equation

$$\delta^2 + l^2 = 0$$

Assuming that $W_1 = \delta$ and $W_2 = l$, we find from eqn. 3.24 that the points of J-multiplicity W_J lie on straight lines $\delta = \pm l$ and the point of D-multiplicity W_D is located at their intersection and thus is not isolated. Taking into account the above considerations let us introduce two copies of the space C^2 and cut them along the surface S_C spanned by the straight lines $\delta = \pm l$. Then we glue the copies of C^2 along both sides of S_C (Figure 3.1). In such space the function

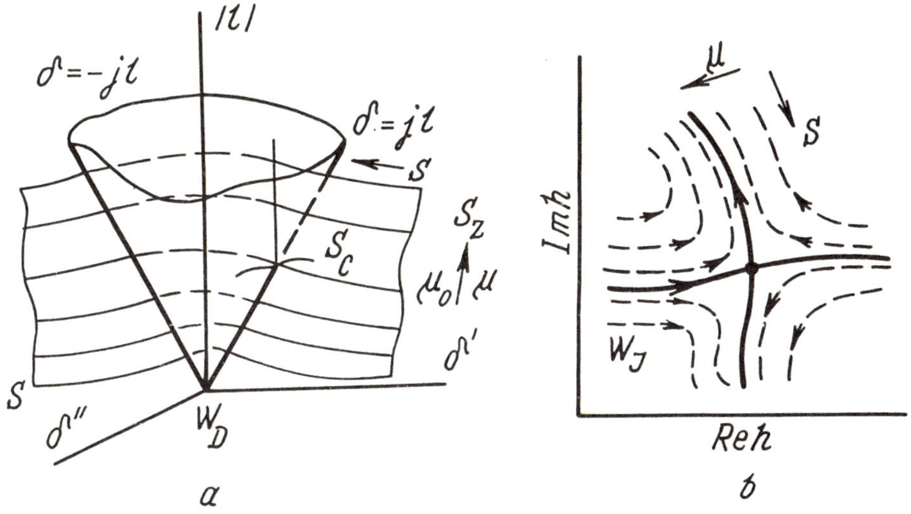

Figure 3.1 *Schematic representation of points of 2-fold degeneration in C^2 (Krasnushkin and Fedorov, 1972)*

a The set $\mathscr{L}(s, \mu)$
b The dependencies $\gamma(\boldsymbol{W})$

$\gamma(\boldsymbol{W})$ would be single-valued. Five types of the set $\mathscr{L}(s, \mu)$ have been examined by Krasnushkin and Fedorov (1972):

(i) $S_{\mathscr{L}}$ does not cross W_D and S_C (there are no wave transformations).

(ii) $S_{\mathscr{L}}$ does not pass through W_D but crosses S_C. In this case at $\mu = \mu_0$, when $S_{\mathscr{L}}$ crosses a point of S_C, the so-called exchange of partial systems by normal waves takes place. The essence of this phenomenon is as follows: at a transition through a point of J-multiplicity the character of dispersion dependencies (see Figure 3.1*b*) and the field structure of the eigenmodes change sharply because of the change of the sign before the square root in eqns. 3.32 and 3.33.

(iii) $S_{\mathscr{L}}$ passes through W_D but does not cross S_C [for example, $S_{\mathscr{L}}$ coincides with the plane (δ', δ'') and $s = \delta'$, $\mu = \delta''$]. In this case there is no field-structure alteration when passing through the degeneration point $\delta' = \delta'' = 0$, as at this point there is no coupling between the partial waves. However, there is an alternative method of construction of branches of dispersion characteristics, as in the previous case.

(iv) $S_{\mathscr{L}}$ passes through W_D and along one of the straight lines $\delta = \pm l$ (degenerated case). Then all routes passing through W_J (or W_D) will cause branches of dispersion characteristics that are separatrixes (solid lines in Figure 3.1*b*). In this case for each such route there is an alternative, as in case (iii) for $\mu = 0$.

(v) $S_{\mathscr{L}}$ passes through W_D, at once reaching the area S_C. In this case, as in case (ii), for routes at any small μ there is an exchange of partial systems by normal waves. It is caused by the fact that W_D is not isolated from the connected set of points W_J. This case, however, differs from case (ii) in that the wave shapes at $W = W_D$ remain different.

To conclude we wish once more to underline the fact that for the existence of wave-transformation effects not only an existence of degenerated points is necessary but also the presence of coupling between degenerated partial waves. When there is no coupling ($A_{sp} = 0$), the phenomenon of wave transformation is not observed and points of multiplicity do not differ from the ordinary ones. We shall later call such a case a trivial multiplicity. The term 'D-multiplicity' will be reserved for cases when there is a coupling between degenerated partial waves.

For example, let us consider modes of rectangular and circular impedance waveguides and analyse dependencies $\gamma_p(Z_S)$, where Z_S is the surface impedance of the walls. At $Z_S = 0$ for any operating frequency the TE_{mn} and TM_{mn} modes ($m, n \geq 1$) in a rectangular waveguide and the TE_{0n} and TM_{1n} modes in a circular waveguide are degenerate. In a rectangular waveguide, when $Z_S \neq 0$, the partial TE_{mn} and TM_{mn} modes are coupled (with the exception of a square waveguide $a = b$) because they are D-multiple modes. In a circular waveguide degenerated modes have different azimuthal dependency and are not coupled; therefore, the point $Z_S = 0$ in this case is that of trivial multiplicity.

3.4 Associated waves

Let us examine in more detail the specific case when there is only a 2-fold degenerated eigenvalue that is of J-multiplicity (let us assume γ_1 to be such an eigenvalue). Here we shall demonstrate the peculiar features of excitation equations caused by the presence of eigenmodes of J-multiplicity.

Jordan canonical form for this case can be written as

$$\Gamma = \left\| \begin{matrix} \ddots & & & & \\ & j\gamma_{-2} & & & \\ & & \Gamma_0 & & \\ & & & j\gamma_2 & \\ & & & & \ddots \end{matrix} \right\| \qquad (3.34)$$

where the matrix Γ_0 is presented as

$$\Gamma_0 = \left\| \begin{matrix} j\gamma_1 & 1 & & \\ & & 0 & \\ 0 & j\gamma_1 & & \\ & & -j\gamma_1 & 0 \\ & 0 & & \\ & & 1 & -j\gamma_1 \end{matrix} \right\|$$

The corresponding equation for the variables g_s has the form of eqn. 3.25, where the matrix Γ is given by eqn. 3.34. For all g_s except $s = 1$, equations have an ordinary form, so we shall in future limit our discussion to examination of the degenerated mode. It is characterised by a four-component vector

$$G = \left\| \begin{matrix} g_1^+ \\ G_1^+ \\ G_1^- \\ g_1^- \end{matrix} \right\|$$

satisfying the equation

$$\frac{dG}{dz} = \Gamma_0 G + K_0 \tag{3.35}$$

Eqn. 3.35, as is evident, cannot be reduced to excitation equations in the form of normal modes given by eqns. 3.28 and 3.29 but differs greatly from the latter. To understand the essence of these differences let us consider eqn. 3.35 in an area, free of sources ($K_0 = 0$). Writing eqn. 3.35 in a scalar form, we obtain

$$\left.\begin{aligned} \frac{dg_1^\pm}{dz} &= \pm j\gamma_1 g_1^\pm \\ \frac{dG_1^\pm}{dz} &= \pm j\gamma_1 G_1^\pm \pm g_1^\pm \end{aligned}\right\} \tag{3.36}$$

The solution to eqn. 3.36 has the form

$$g_1^\pm(z) = C \exp(\pm j\gamma_1 z) \quad G_1^\pm = \pm Cz \exp(\pm j\gamma_1 z) \tag{3.37}$$

where C is an arbitrary constant. The amplitude coefficients $g_1^\pm(z)$ describe the eigenmode of the given type, and $G_1^\pm(z)$ is the so-called associated wave. Thus, in the excitation theory, the mode of J-multiplicity is presented as the sum of the eigenmode and the associated mode. The latter, though it disappears at infinity, has a rather unusual dependency of the field on the longitudinal co-ordinate. When the associated wave occurs, the condition at infinity given by eqn. 3.19 cannot be defined in the form

$$\begin{Bmatrix} E \\ H \end{Bmatrix} = O\{\exp(-\tilde{C}|z|)\}$$

at $z \to \pm\infty$, where $\tilde{C}$ is a constant. This condition is too rigid and, as follows from the analysis presented, the problem stated in this manner may have no solution.

Determining the transformation matrix B in eqn. 3.24 (to be more precise, the $\tilde{b}_{1,p}$ coefficients), we can obtain a representation of the associated wave in the form of an expansion in terms of eigenmodes of a lossless waveguide of the same configuration.

Let us now examine equations for the electromagnetic field in the presence of modes of J-multiplicity. Assume that we have the eigenvalue problem

$$Lu - \lambda Qu = 0 \tag{3.38}$$

where $Q^{-1}L$ is a non-self-adjoint operator. As is known, if λ is an eigenvalue of multiplicity m, then the set of eigenfunctions and associated functions of the problem given by eqn. 3.38 satisfies the operator equation (Keldysh, 1951)

$$(L - \lambda Q)^m \boldsymbol{u} = 0$$

Taking $m = 2$ for simplicity, we obtain the chain of equations

$$\left.\begin{aligned} (L - \lambda Q)\boldsymbol{u}_{(0)} &= 0 \\ (L - \lambda Q)\boldsymbol{u}_{(1)} &= \boldsymbol{u}_{(0)} \end{aligned}\right\} \tag{3.39}$$

where $\boldsymbol{u}_{(0)}$ and $\boldsymbol{u}_{(1)}$ are eigenfunctions and associated functions, respectively.

Applying the general formulae given by eqns. 3.39 to waveguide problems, we can write for transverse field distributions of eigenmodes the equations

$$\left.\begin{aligned} \text{curl}_\perp \hat{\boldsymbol{E}}_{(0)} + j\gamma(\hat{i}_z \times \hat{\boldsymbol{E}}_{(0)}) &= j\omega\mu\hat{\boldsymbol{H}}_{(0)} \\ \text{curl}_\perp \hat{\boldsymbol{H}}_{(0)} + j\gamma(\hat{i}_z \times \hat{\boldsymbol{H}}_{(0)}) &= -j\omega\varepsilon\hat{\boldsymbol{E}}_{(0)} \end{aligned}\right\} \tag{3.40}$$

where the $\hat{\boldsymbol{E}}_{(0)}$ and $\hat{\boldsymbol{H}}_{(0)}$ satisfy the impedance-boundary condition of the form of eqn. 3.18, the symbol $\perp$ indicates that the derivatives $\partial/\partial z$ are replaced by zeros and $\lambda = -j\gamma$.

Transverse-field distributions of the associated modes satisfy the equations

$$\left.\begin{aligned} \text{curl}_\perp \hat{\boldsymbol{E}}_{(1)} + j\gamma(\hat{i}_z \times \hat{\boldsymbol{E}}_{(1)}) &= j\omega\mu\hat{\boldsymbol{H}}_{(1)} + \hat{\boldsymbol{E}}_{(0)} \\ \text{curl}_\perp \hat{\boldsymbol{H}}_{(1)} + j\gamma(\hat{i}_z \times \hat{\boldsymbol{H}}_{(1)}) &= -j\omega\varepsilon\hat{\boldsymbol{E}}_{(1)} + \hat{\boldsymbol{H}}_{(0)} \end{aligned}\right\} \tag{3.41}$$

and the same boundary condition. It is noteworthy that similar equations can also be written at $m > 2$.

In the area without sources, the forward and backward associated modes are expressed by

$$\begin{Bmatrix} \boldsymbol{E}^+ \\ \boldsymbol{H}^+ \end{Bmatrix} = g_1^+(z) \begin{Bmatrix} \hat{\boldsymbol{E}}_{(1)}(x, y) \\ \hat{\boldsymbol{H}}_{(1)}(x, y) \end{Bmatrix} - G_1^+(z) \begin{Bmatrix} \hat{\boldsymbol{E}}_{(0)}(x, y) \\ \hat{\boldsymbol{H}}_{(0)}(x, y) \end{Bmatrix} \tag{3.42}$$

$$\begin{Bmatrix} \boldsymbol{E}^- \\ \boldsymbol{H}^- \end{Bmatrix} = g_1^-(z) \begin{Bmatrix} \hat{\boldsymbol{E}}_{(1)}(x, y) \\ \hat{\boldsymbol{H}}_{(1)}(x, y) \end{Bmatrix} + G_1^-(z) \begin{Bmatrix} \hat{\boldsymbol{E}}_{(0)}(x, y) \\ \hat{\boldsymbol{H}}_{(0)}(x, y) \end{Bmatrix} \tag{3.43}$$

where $g_1^\pm(z)$ and $G_1^\pm(z)$ are expressed by eqns. 3.37, $\hat{\boldsymbol{E}}_{(0,1)}(x, y)$ and $\hat{\boldsymbol{H}}_{(0,1)}(x, y)$ are determined from eqns. 3.40 and 3.41.

Using eqns. 3.40 and 3.41, we can show that the associated modes expressed by eqns. 3.42 and 3.43 satisfy Maxwell's homogeneous equations.

Note that the associated wave is not such a rare and exotic phenomenon as it may seem at first. Associated waves occur in conventional hollow waveguides with perfectly conducting walls excited at a cutoff frequency. As $\gamma(\omega_C) = 0$, the waves travelling in opposite directions become indistinguishable (linearly dependent) and the points $\omega = \omega_C$ become of J-multiplicity. Therefore the classical excitation equations for waveguides at the cutoff frequencies contain the indeterminate forms. The modification of excitation theory has been made by Arkadaksky and Tsykin (1976) by means of the limiting transition $\omega \to \omega_C$; thus the associated wave is extracted in an explicit form.

3.5 Types of dispersion characteristics; a concept of anomalous dispersion; complex waves in lossless waveguides

Let us consider typical dispersion characteristics of waveguides. As mentioned above, their trend will be determined mainly by the specific points of the multiple-valued function $\gamma(\omega)$ that can be obtained from eqn. 3.30.

The most simple type of dispersion characteristics is observed in lossless waveguides. In this case forward and backward modes of the same type may be degenerated. It corresponds to the cutoff frequency ω_C $[\gamma(\omega_C) = h(\omega_C) = 0]$. This case is illustrated by Figure 3.2*a*. Branches BC and BC′ correspond to the propagation area $(\omega > \omega_C)$; in this area Im $h = 0$, Re $h > 0$ for the forward wave, and Re $h < 0$ for the backward one. Branches BA and BA′ correspond to the below-cutoff area $(\omega < \omega_C)$; in this area Re $h = 0$, Im $h > 0$ for the forward wave, and Im $h < 0$ for the backward one. In the presence of losses both branches CBA and C′BA′ become complex-valued at any ω and may have no contact points. However, if the losses are small, their trend remains approximately the same as in lossless systems (broken lines in Figure 3.2*a*).[16]

The trend of dispersion dependencies in the propagation area depends essentially on the type of waveguide. For waveguides with smooth perfectly con-

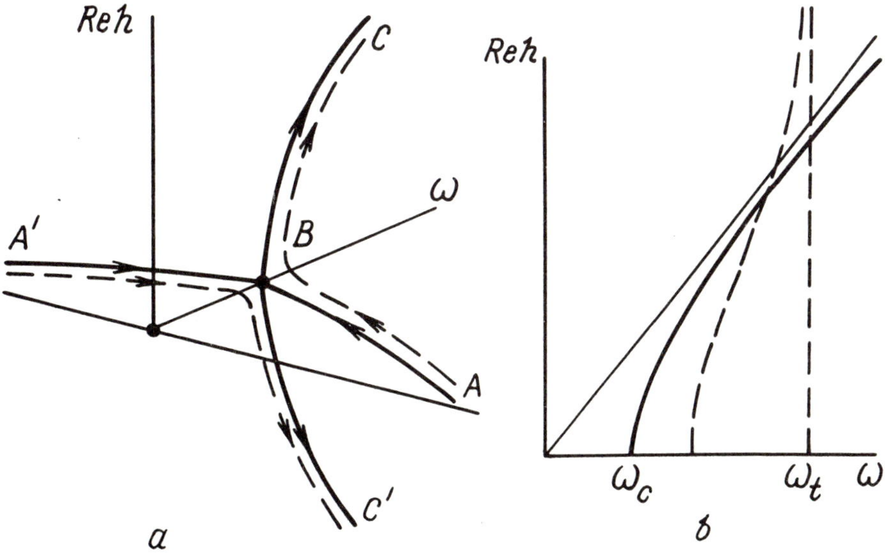

Figure 3.2 *Dispersion diagrams*

a Cutoff vicinity
b Propagation region

[16] This is why the concept of cutoff frequency for lossy waveguides is not uniquely determined and has no such direct meaning as for lossless systems. However, when the losses are small, it would be advisable to use this concept, as it permits differentiation between two essentially different operational regimes of a waveguide.

ducting walls the trend of the dependency $h(\omega)$ in the propagation area is shown by the solid line in Figure 3.2*b*. In this case the dispersion curve always lies under the straight line $\mathrm{Re}\, h = \omega/c$. In waveguides with impedance walls, another type of dispersion characteristic is possible (broken line in Figure 3.2*b*). At the intersection point of this curve with the straight line $\mathrm{Re}\, h = \omega/c$ transformation of the fast wave into a slow (surface) one occurs. At the point $\omega = \omega_t$ the surface wave undergoes a cutoff: in this case the function $h(\omega)$ has a pole owing to the fact that the function $\Delta(h, \omega)$ for the impedance waveguide may be nonanalytical. It distinguishes this system from, for example, a layered dielectric medium. Note that for the modes shown in Figure 3.2*b* the so-called normal dispersion is typical, i.e. the following condition is fulfilled:

$$\frac{\mathrm{d}}{\mathrm{d}\omega}\left(\frac{\mathrm{Re}\, h(\omega)}{\omega}\right) > 0 \qquad (3.44)$$

In complicated waveguide systems, for some types of modes an anomalous dispersion can be observed when the condition given by eqn. 3.44 is replaced by the opposite one

$$\frac{\mathrm{d}}{\mathrm{d}\omega}\left(\frac{\mathrm{Re}\, h(\omega)}{\omega}\right) < 0 \qquad (3.45)$$

In some types of waveguide even when there are no losses the branch points of dispersion characteristics can be located, unlike the case shown in Figure 3.2*a*, not at the origin of co-ordinates but on the co-ordinate axes of the plane of the complex variable h. In this case more complicated branch systems are formed which can cause couples of complex waves in a certain range of frequencies. For these waves the z-dependence is one of the forms

$$\left.\begin{array}{l} \exp(\pm jh'z)\exp(h''z) \\ \exp(\pm jh'z)\exp(-h''z) \end{array}\right\}$$

where $h' = |\mathrm{Re}\, h| > 0$ and $h'' = |\mathrm{Im}\, h| > 0$.

It may seem at first that such waves cannot be found in lossless waveguides. However, we have to take into account the fact that complex waves are excited in pairs with equal amplitudes[17] and form two reactive (i.e. not carrying the energy) waves, similarly to ordinary evanescent modes but having a sine spatial-amplitude modulation. For such waves, usual orthogonality conditions expressed by eqn. 3.5 are inapplicable (see Illarionov *et al.*, 1980).

Complex waves in lossless systems were found for the first time by Miller (1952) in a waveguide with anisotropic impedance walls and more recently in a number of electrodynamic systems: gyrotropic plasma (Gershman, 1955), circular waveguide with a dielectric rod (Clarricoats and Taylor, 1964, Veselov and Lyubimov, 1963) and screened microstrip line (Kovalenko, 1980). The properties of these waves were studied by Beliantsev and Gaponov (1964) and Krasnushkin (1974).

[17] This conclusion, generally speaking, cannot be drawn on the basis of an analysis of dispersion characteristics. For that we have to solve the problem of excitation of the waveguide by an external source. The solution of this problem (obtained by Beliantsev and Gaponov, 1964) proved this assumption.

The existence of complex waves usually results from the interaction of partial guided waves with differently directed energy flows. Complex waves are essential for correct formulation of excitation equations in complicated waveguide systems, rigorous analysis of waves scattering by various discontinuities etc.

In the presence of losses, the branch points slip from the co-ordinate axes in the plane of the complex variable γ; in this case the degeneration is possible only for modes of different types (backward waves corresponding to the replacement $\gamma \to -\gamma$ are shown in Figure 3.2*a* by broken lines). The degeneration points can be points of trivial, D- and J-multiplicity. In the last two cases the electromagnetic field in these points is of a complicated character: in the first it is a superposition of coupled degenerated eigenmodes, in the second a sum of the eigenmode and the associated wave (see Section 3.4).

3.6 Excitation of TM modes in a parallel-plate impedance waveguide

Excitation theory of waveguides with ohmic losses in the walls, described in Section 3.2, is quite general, though analysis of the relationships derived is a rather difficult task. The reason is that to arrive at an expansion in terms of normal modes of the waveguide under consideration it is necessary to analyse the dispersion equation given by eqn. 3.30 in the form of an infinite determinant.

There is, however, another approach to this problem which does not need an *a priori* modal expansion of the field sought: the integral-transform technique. The range of applicability of this method is considerably narrower than that of the approach examined in Section 3.2. However, the integral-transform method allows one to construct and analyse the solution in a much simpler way. We shall later examine it using as an example the problem of excitation of a plane waveguide with an impedance wall (Figure 3.3) by a 'magnetic' line source.

Mathematical formulation of the problem is as follows: it is necessary to find the solution of the Helmholtz equation for the y-component of the magnetic field $\psi(x, z)$

$$\nabla^2_{xz}\psi + k^2\psi = -j\omega\varepsilon\delta(x - x')\delta(z - z') \tag{3.46}$$

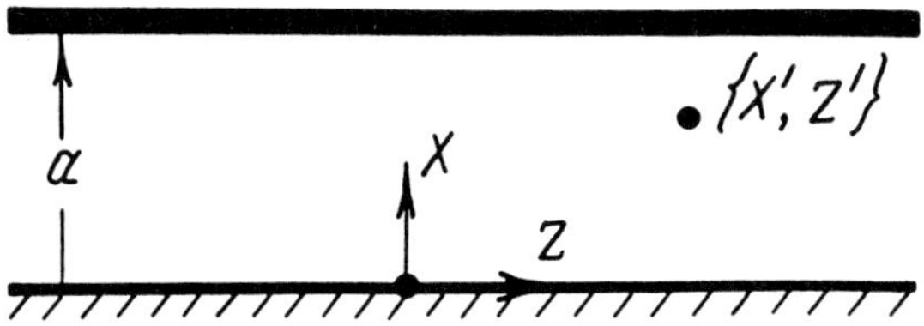

Figure 3.3 *Parallel-plate waveguide with impedance wall excited by 'magnetic' line source*

satisfying the boundary conditions

$$\left.\frac{\partial\psi}{\partial x}\right|_{x=0} = 0 \tag{3.47}$$

$$\left.\left(\frac{\partial\psi}{\partial x} - j\eta\psi\right)\right|_{x=a} = 0 \tag{3.48}$$

and the conditions for field reduction at $z \to \pm\infty$. The complex parameter η in eqn. 3.48 is expressed by the formula $\eta = kZ_S/W_0$.

This problem (set somewhat more generally) has been solved by means of the integral-transform method by Markov and Vasilyev (1970). We shall later briefly describe the procedure of solution and analyse the formulae obtained. A solution for eqn. 3.46 can be sought in the form

$$\psi = \psi_1 + \psi_2$$

where ψ_1 is the field excited by a 'magnetic' line source in the free space and ψ_2 is the so-called induced field satisfying the homogeneous 2-dimensional Helmholtz equation and chosen in such a way that the boundary conditions for the total field given by eqns. 3.47 and 3.48 should be fulfilled.

The field $\psi_1(x, z)$ is expressed by

$$\psi_1 = \frac{\omega\varepsilon}{4} H_0^{(1)}(kR) = \frac{\omega\varepsilon}{4\pi}\int_{-\infty}^{\infty} \frac{\exp\{j\gamma(z-z') + j\beta|x-x'|\}}{\beta}\,d\gamma \tag{3.49}$$

where $H_0^{(1)}(u)$ is a Hankel function of the first kind and zero order, $R = \{(x-x')^2 + (z-z')^2\}^{1/2}$, x' and z' are the co-ordinates of the line source and $\beta = (k^2 - \gamma^2)^{1/2}$.

The field $\psi_2(x, z)$ can be sought as Fourier's integral:

$$\psi_2 = j\omega\varepsilon\int_{-\infty}^{\infty} \{A(\gamma)\cos\beta x + B(\gamma)\sin\beta z\}\exp\{j\gamma(z-z')\}\,d\gamma \tag{3.50}$$

where $A(\gamma)$ and $B(\gamma)$ are unknown functions which should be determined from eqns. 3.47 and 3.48. Substituting eqns. 3.49 and 3.50 into the boundary conditions given by eqns. 3.47 and 3.48, we obtain a system of two equations for the unknown functions $A(\gamma)$ and $B(\gamma)$. Finding $A(\gamma)$ and $B(\gamma)$ from this system, we can write down the solution of our boundary-value problem as

$$\psi(x, z) = \frac{\omega\varepsilon}{2\pi j}\int_{-\infty}^{\infty} \frac{\cos(\beta x)[\beta\cos\{\beta(a-x')\} - j\eta\sin\{\beta(a-x')\}]}{\beta\{\beta\sin(\beta a) + j\eta\cos(\beta a)\}} \times \exp\{j\gamma(z-z')\}\,d\gamma \tag{3.51}$$

Since the line source is located right on the impedance wall ($x' = a$) and $z' = 0$, eqn. 3.51 can be simplified and will have the form

$$\psi(x, z) = \frac{\omega\varepsilon}{2\pi j}\int_{-\infty}^{\infty} \frac{\exp(j\gamma z)\cos(\beta x)}{\beta\sin(\beta a) + j\eta\cos(\beta a)}\,d\gamma \tag{3.52}$$

We shall use this formula as the basis for our further analysis. First note that the function under the integral is single-valued at every point of the complex plane γ (it has no branch points). Therefore $\psi(x, z)$ expressed by eqn. 3.52 can be completely determined by residues in the poles of the function under the integral. The poles, as can be seen from eqn. 3.52, are the roots of the transcendental equation

$$f(\beta) = \beta \sin(\beta a) + j\eta \cos(\beta a) = 0 \qquad (3.53)$$

i.e. a dispersion equation for TM modes.

First we assume that eqn. 3.53 has only simple roots. Then, calculating the integral in eqn. 3.52 through the Jordan lemma, we obtain

$$\psi(x, z) = \sum_{m=1}^{\infty} C_m \exp(j\gamma_m |z|) \cos(\beta_m x) \qquad (3.54)$$

where

$$C_m = -\frac{\omega\varepsilon\beta_m^2}{\{a(\beta_m^2 - \eta^2) - j\eta\}\gamma_m \cos(\beta_m a)}$$

and β_m are the transverse-wave numbers of TM modes of the plane impedance waveguide. Thus, eqn. 3.54 is an expansion of the field sought in terms of the TM modes of the waveguide under examination (TE modes are not excited). So the transition from eqn. 3.52 to eqn. 3.54 is only possible if there are no degenerated modes (multiple roots of eqn. 3.53).

Now suppose that eqn. 3.53 has multiple roots. This means that apart from eqn. 3.52 the following system of equities should be fulfilled:

$$\frac{d^p f(\beta)}{d\beta^p} = 0 \qquad (3.55)$$

$p = 1, 2, \ldots, N-1$, where N is the root multiplicity. Analysis shows that at $N > 2$ the chain of relations expressed by eqn. 3.55 for different p is incompatible. This explains why no more than doubly degenerated modes can exist in a plane impedance waveguide. It is evident that these waves will be of J-multiplicity. Equating the derivative $df(\beta)/d\beta$ to zero and using eqn. 3.53, we obtain

$$ja(\beta^2 - \eta^2) = -\eta \qquad (3.56)$$

which, together with the dispersion equation given by eqn. 3.53 determines conditions of J-multiplicity of the eigenmodes. It follows from eqn. 3.56 that no more than one pair of modes can be degenerate.

Expressing β from eqn. 3.56 and substituting the result obtained into eqn. 3.53, we have

$$F(\eta) = \left(\frac{j\eta}{a} + \eta^2\right)^{1/2} \tan\left\{\left(\frac{j\eta}{a} + \eta^2\right)^{1/2} a\right\} + j\eta = 0 \qquad (3.57)$$

which determines the values of the surface impedance at which there is a degeneration of eigenmodes. With purely imaginary η, solutions of eqn. 3.57 are nonexistent, i.e. points of J-multiplicity can only be observed in lossy waveguides.

At a sufficiently small $|\eta|$ the associated wave can only be initiated by a symmetrical eigenmode with small $|\beta| \sim O(|\eta|^{1/2})$. The function $F(\eta)$ is analytical in a certain circle with the centre $\eta = 0$, and at the same time $F(0) = 0$. This is why, according to the theorem of uniqueness for analytical functions (Evgrafov, 1965), $F(\eta) \neq 0$ in some neighbourhood of the point $\eta = 0$. The degeneration point $\eta = 0$ is evidently fictitious and does not lead to the excitation of an associated wave. Thus, for sufficiently small $|\eta|$ associated waves are absent, whatever the sign of η'' and relations between η' and η''.

The integral in eqn. 3.52, can be determined through the Jordan lemma as before, though it is necessary to take into consideration the fact that one of the poles of the expression under the integral (for example, the nth one) is a pole of the second order. After calculation, we obtain

$$\psi(x, z) = \sum_{\substack{m=1 \\ m \neq n}}^{\infty} C_m \exp(j\gamma_m|z|) \cos(\beta_m x)$$
$$+ \frac{\omega\varepsilon(2j\eta - 1)\gamma_n}{3\beta_n^2 a \cos(\beta_n a)} \cos(\beta_n x) \exp(j\gamma_n|z|)$$
$$+ \frac{\exp(j\gamma_n|z|)}{\beta_n a \cos(\beta_n a)} \left\{ j|z| \cos(\beta_n x) + \frac{\gamma_n}{\beta_n} x \sin(\beta_n x) \right\} \qquad (3.58)$$

The first two terms in eqn. 3.58 are a sum of eigenmodes; the third term, as can easily be proved by a direct check, is the associated wave.

Note that direct observation of the associated wave in the experiment is a rather difficult task. The reason is that the associated wave exists only at some discrete values of the surface-impedance parameter η determined from eqn. 3.57. For regular waveguides, owing to fluctuations of parameters, inaccuracies in manufacturing etc. we shall practically always be in the situation where only nondegenerated modes exist, though phase constants and field structures of two modes can turn out to be close enough. In this case an associated wave is a mathematical abstraction, convenient for description of wave-transformation processes when their phase constants and fields distributions draw nearer. Irregular waveguide transitions, such as, for example, impedance waveguides with a variable surface impedance, are another matter. If $\eta(z)$, when varying, passes through the point of J-multiplicity, new physical effects may occur due to the excitation of an associated wave. For a parallel-plate waveguide this problem was studied by Bichytskaya and Novikov (1979). The method of coupled waves was used as a basis; the solution of the system of differential equations has been carried out asymptotically in a zero-order approximation in the parameter of smallness $\varepsilon \sim |\partial\eta/\partial z|$. The main result is as follows: if the sth eigenmode is incident on the segment of a variable impedance and there is a point of J-multiplicity for the sth and pth modes, then the transformation of the sth mode into a pth mode takes place in a zero-order approximation in the parameter ε.[18] Note that at the point of J-multiplicity no energy is carried by eigenmodes; energy transmission through this cross-section is implemented by the associated wave.

This effect has some similarity to that of reflection of a mode from the cutoff cross-section in a smoothly irregular lossless waveguide. This parallel is understandable in the light of the fact that forward and backward modes of the same type in the cutoff cross-section are of J-multiplicity and the reflection means a transformation of the forward mode into the backward one. Here, in the cutoff cross-section an associated wave is also excited (Arkadaksky and Tsykin, 1976).

[18] Let us point out, for comparison, that when there are no points of degeneration the effects of transformation and reflection of modes in a smoothly irregular waveguide transition have the order $\varepsilon \ll 1$ (Katsenelenbaum, 1961).

It is important to note that the effects of associated wave excitations are of great interest and are examined intensively, for example, in connection with radiowave propagation along the 'earth–ionosphere' waveguide (Bichytskaya, 1990) and the theory of metallodielectric waveguides and wave processes in continuous media (Gureev, 1990*a*, *b*).

3.7 Attenuation coefficients of eigenmodes

Let us proceed now to the dispersion equation of a lossy waveguide. The conductivity of the walls is assumed to be sufficiently large ($\mathrm{Re}\, Z_S / W_0 \ll 1$). Neglecting intermodal coupling we can present eqn. 3.30 in the form

$$\prod_s^{\infty} (j\gamma - A_{ss}) = 0$$

from which we obtain

$$\gamma_p \simeq h_p + j \frac{Z_S}{N_p} \oint_C \boldsymbol{H}_{-p\tau} \boldsymbol{H}_{p\tau} \, \mathrm{d}C$$

or

$$\gamma'_p \simeq h_p - \frac{\mathrm{Im}\, Z_S \oint_C \boldsymbol{H}_{-p\tau} \boldsymbol{H}_{p\tau} \, \mathrm{d}C}{\int_{S_\perp} \{(\boldsymbol{E}_p \times \boldsymbol{H}_{-p}) - (\boldsymbol{E}_{-p} \times \boldsymbol{H}_p)\} \hat{i}_z \, \mathrm{d}S_\perp} \tag{3.59}$$

$$\gamma''_p \simeq \frac{\mathrm{Re}\, Z_S \oint_C \boldsymbol{H}_{-p\tau} \boldsymbol{H}_{p\tau} \, \mathrm{d}C}{\int_{S_\perp} \{(\boldsymbol{E}_p \times \boldsymbol{H}_{-p}) - (\boldsymbol{E}_{-p} \times \boldsymbol{H}_p)\} \hat{i}_z \, \mathrm{d}S_\perp} \tag{3.60}$$

Eqns. 3.59 and 3.60 are general relations, applicable to any types of mode including below-cutoff and complex modes if $|\omega - \omega_C|$ is not too small. For propagating modes, using the formulae $\boldsymbol{E}_{-s} = \pm \boldsymbol{E}^*_s$, $\boldsymbol{H}_{-s} = \mp \boldsymbol{H}^*_s$, we can transform eqns. 3.59 and 3.60 into

$$\gamma'_p \simeq h_p - \frac{\mathrm{Im}\, Z_S \oint_C \boldsymbol{H}_{-p\tau} \boldsymbol{H}_{p\tau} \, \mathrm{d}C}{2 \int_{S_\perp} (\boldsymbol{E}_p \times \boldsymbol{H}^*_p) \hat{i}_z \, \mathrm{d}S_\perp} \tag{3.61}$$

$$\gamma''_p \simeq \frac{\mathrm{Re}\, Z_S \oint_C \boldsymbol{H}_{-p\tau} \boldsymbol{H}_{p\tau} \, \mathrm{d}C}{2 \int_{S_\perp} (\boldsymbol{E}_p \times \boldsymbol{H}^*_p) \hat{i}_z \, \mathrm{d}S_\perp} \tag{3.62}$$

Eqn. 3.62 can be rewritten as $\gamma''_p \simeq P_1 / 2P_\perp$, where P_1 is the linear (per unit length) power of losses in the walls, and $P_\perp$ is the average over the period $T = 2\pi/\omega$ power carried through the waveguide cross-section $S_\perp$. Eqn. 3.62 reflects the energy method of loss calculation based on the concept of perturbations of the electromagnetic field due to finite conductivity of the walls (Weinstein, 1988, Ilyinsky and Slepyan, 1983). The quantity γ''_p is called the attenuation coefficient of the pth eigenmode.

Attenuation coefficient is a complete characteristic of the absorption process only for single-mode waveguides. For a multimode waveguide (i.e. a waveguide in which several eigenmodes can progagate) it is no longer so, since amplitude relations between different types of modes are of major importance.

It is seen from eqn. 3.61, that the propagation coefficient γ'_p due to finite conductivity of the walls undergoes a correction which can be neglected owing to the small size of the losses and sufficient remoteness from the cutoff frequency, i.e.

$$\gamma'_p \simeq h_p$$

For lossy waveguides the concept of a cutoff frequency does not have the same clear meaning as in lossless waveguides. In fact, there is no real frequency ω at which $\gamma_s(\omega) = \gamma'_s(\omega) + j\gamma''_p(\omega) = 0$. Nevertheless, when the losses are small the concept of a cutoff frequency can be introduced to divide two physically different regimes of waveguide operation. However, the border between these regimes is now not distinct and the cutoff frequency ω_C is not uniquely defined.[19] The easiest way to consider the cutoff frequency of a lossless waveguide is as that of an identical waveguide with losses. Then at $\omega \gtrsim \omega_C$, $\gamma'_s \gg \gamma''_s$ (travelling wave undergoes a weak active attenuation), at $\omega \lesssim \omega_C$, $\gamma'_s \ll \gamma''_s$ (the wave undergoes strong reactive attenuation, i.e. below-cutoff mode). The losses lead to a weak sine modulation of the exponentially decreasing field along the z-axis. Later in this book when discussing cutoff frequency of waveguides with losses we shall not make any specific reservations.

To discover the applicability limits of eqn. 3.62 let us obtain a more precise solution of the dispersion equation taking into account the coupling of two modes with the nearest propagation coefficients. Assuming that these are the sth and pth modes, we can write the dispersion equation in the form

$$\mathrm{Det}\begin{Vmatrix} A_{ss} - j\gamma & A_{sp} \\ A_{ps} & A_{pp} - j\gamma \end{Vmatrix} = 0$$

or

$$(A_{ss} - j\gamma)(A_{pp} - j\gamma) - A_{sp}A_{ps} = 0 \tag{3.63}$$

Solution of eqn. 3.63 gives the propagation coefficients

$$j\gamma_{1,2} = \frac{A_{ss} + A_{pp}}{2} \pm \left\{\frac{(A_{ss} - A_{pp})^2}{4} + A_{ps}A_{sp}\right\}^{1/2} \tag{3.64}$$

When there is no intermodal coupling ($A_{sp} = A_{ps} = 0$), we obtain, from eqn. 3.64,

$$j\gamma_1 = A_{ss} \qquad j\gamma_2 = A_{pp} \tag{3.65}$$

---

[19] To be more precise we mean not the cutoff frequency but the narrow cutoff-frequency range.

which leads to eqn. 3.62. Now let us assume that the pth and sth modes are coupled due to the losses in the walls but that inequality

$$|A_{ss} - A_{pp}| \gg 2|A_{ps}A_{sp}|^{1/2} \tag{3.66}$$

is correct. Ignoring the terms $O[(\mathrm{Re}\,\mathcal{Z}_S)^2]$ in eqn. 3.64 because of their extremely small size, we once again obtain eqn. 3.65. When the condition given by eqn. 3.66 is fulfilled, we can use eqn. 3.62 which results from the perturbation theory.

In the neighbourhood of degeneration points of coupled modes ($h_s - h_p \to 0$), the condition given by eqn. 3.66 is not satisfied. In this case corrections to eqn. 3.65 have an order of $\mathrm{Re}\,\mathcal{Z}_S$ and should be taken into consideration. Introducing the attenuation mutual coefficient γ_{sp} by the formula

$$\gamma_{sp} = \frac{\mathrm{Re}\,\mathcal{Z}_S \oint_C \boldsymbol{H}_{p\tau}\boldsymbol{H}^*_{s\tau}\,\mathrm{d}C}{2\left\{\int_{S_\perp}(\boldsymbol{E}_p \times \boldsymbol{H}^*_p)\hat{i}_z\,\mathrm{d}S_\perp \int_{S_\perp}(\boldsymbol{E}_s \times \boldsymbol{H}^*_s)\hat{i}_z\,\mathrm{d}S_\perp\right\}^{1/2}}$$

we can present the attenuation coefficients of D-degenerated eigenmodes as

$$\gamma''_{1,2} = \frac{\gamma''_s + \gamma''_p}{2} \pm \left\{\frac{(\gamma''_s - \gamma''_p)^2}{4} + \gamma_{ps}\gamma_{sp}\right\}^{1/2} \tag{3.67}$$

where $\gamma''_{s,p}$ are 'partial' attenuation coefficients for the sth and pth modes defined by eqn. 3.62. Eqn. 3.67 is valid when these modes are not very close to cutoff. Thus, for degenerated modes, coupled due to losses in the walls ($\gamma_{ps} \neq 0$), the conventional perturbation method is inapplicable (Kato, 1966) and attenuation coefficients should be determined according to eqn. 3.67.

When three or more modes in a lossless waveguide are degenerated, the analysis can be carried out in a similar manner though it becomes more cumbersome.

Let us consider, for example, degenerated modes in types of waveguide widely used in practice. In a circular waveguide the TM_{0n} and TE_{1n} modes are degenerated but because of different azimuthal dependency of the fields of these modes the attenuation mutual coefficient is zero (trivial multiplicity). Thus the perturbation method can be applied to calculate the attenuation coefficients of any modes in a circular waveguide at operational frequencies remote from the cutoff (see eqn. 3.62).

Expressions for fields in a circular lossless waveguide are well known and we shall not present them here. According to eqn. 3.62 we have

$$\gamma''_{in} = \frac{\mathrm{Re}\,\mathcal{Z}_S}{W_0}\frac{k}{b}\left\{\left(\frac{\mu_{in}}{kb}\right)^2 + \frac{i^2}{\mu^2_{in} - i^2}\right\}\left\{k^2 - \left(\frac{\mu_{in}}{b}\right)^2\right\}^{-1/2}$$

for TE_{in} modes and

$$\gamma''_{in} = \frac{\mathrm{Re}\,\mathcal{Z}_S}{W_0}\frac{k}{b}\left\{k^2 - \left(\frac{\nu_{in}}{b}\right)^2\right\}^{-1/2}$$

for TM_{in} modes (b is the waveguide radius).

In a rectangular waveguide the TE_{mn} and TM_{mn} modes ($m \geq 1, n \geq 1$) are degenerated at all operational frequencies and any values of a and b. Unlike the case of a circular waveguide, these modes are coupled owing to losses in the walls when $a \neq b$. Therefore the conventional perturbation method is inapplicable.[20] In fact, in a lossy rectangular waveguide at $m \geq 1$, $n \geq 1$ there are no separate TE and TM modes. The modes excited have all six components of the electromagnetic field and are linear combinations of the TE and TM modes. The amplitudes of the partial modes are comparable in the order of magnitude. In fact we have here a non-single-mode regime where, as already mentioned, attenuation coefficients of partial modes have only a conditional meaning and do not characterise completely the total losses.

Separate TE and TM modes with $m, n \geq 1$ in the presence of losses can be found only in a square waveguide. This is why the formulae for attenuation coefficients given by Chang (1989) are correct in this specific case.

The TE_{01} mode and TE_{10} mode in a rectangular waveguide are nondegenerated (they are degenerated only at $a = b$ but even in this case they are not coupled owing to the losses in the walls). This is why for these modes we can use the approach based on eqn. 3.62. In this case we obtain the following result for the TE_{10} mode:

$$\gamma''_{10} = \frac{\operatorname{Re} Z_S}{W_0}\left(1 + \frac{b\lambda^2}{2a^3}\right)\left\{1 - \left(\frac{\lambda}{2a}\right)^2\right\}^{-1/2}$$

For the TE_{01} mode a similar formula can be derived obtained from that given through replacement of a with b and b with a.

Eqn. 3.62 is applicable also to calculation of the attenuation coefficient of the TEM mode in doubly connected systems,[21] for example in a coaxial line. In the general case of an N-connected system eqn. 3.62 cannot be applied as the TEM mode in it is of the multiplicity $N - 1$ and, hence, described by more complicated laws (see above).

Attenuation in a coaxial line has been examined by Daywitt (1990). In this case the following relation is obtained for the attenuation coefficient of the TEM mode:

$$\gamma'' = \frac{\operatorname{Re} Z_S}{W_0} \frac{1 + (b/a)}{2b \ln (b/a)}$$

where a and b are the radii of the inner and outer co-axial conductors, respectively.

Now let us consider the attenuation in a cutoff waveguide. Here the forward and backward modes of the type under consideration are of J-multiplicity. Assuming that the operational frequency is not very close to the cutoff values of any other modes, let us use the formulae of a 'two-wave' approximation. Substituting $p = -s$ in eqn. 3.64, we obtain an expression for the complex

---

[20] For this reason the dependencies for the attenuation coefficients of the TE_{mn} and TM_{mn} modes ($m, n \geq 1$) in a rectangular waveguide, calculated using eqn. 3.62 and quoted sometimes in the literature (see, for example, Chang, 1989), have no meaning except for the case $a = b$. This fact has been pointed out repeatedly by Prof. L.A. Weinstein.

[21] In the presence of losses in the walls it is more correct to speak about a quasi-TEM mode as in this case the longitudinal field components are not exactly equal to zero.

propagation coefficients to the accuracy $\dot{O}[(\mathrm{Re}\,\zeta_S)^2]$:

$$\gamma_{1,2} \simeq \pm\left(h_s^2 + 2jh_s \frac{\zeta_S}{N_s} \oint_C |\boldsymbol{H}_{s\tau}^0|^2 \, \mathrm{d}C\right)^{1/2} \tag{3.68}$$

where the sign $\pm$ corresponds to forward and backward modes. Eqn. 3.68 is correct for all frequencies including $\omega \simeq \omega_C$. If the inequality

$$h_s \gg 2\frac{|\zeta_S|}{N_s} \oint_C |\boldsymbol{H}_{s\tau}^0|^2 \, \mathrm{d}C$$

is fulfilled, then on expanding the right-hand side of eqn. 3.68 into a power series in ζ_S and retaining the terms of the first order, we reach eqn. 3.62. Another limit case resulting from eqn. 3.68 is attenuation right at the cutoff frequency ($h_s \to 0$). The norm can be expressed as $N_s = \omega\varepsilon h_s M_s$, where M_s is some frequency-independent coefficient. Then at $h_s \to 0$

$$\gamma_{1,2} \simeq \pm\left(\frac{2j\zeta_S}{\omega\varepsilon M_s} \oint_C |\boldsymbol{H}_{s\tau}^0|^2 \, \mathrm{d}C\right)^{1/2} \tag{3.69}$$

At $\omega \to \omega_C$ the attenuation coefficients $\gamma''_{1,2} \sim (\mathrm{Re}\,\zeta_S)^{1/2}$, i.e. losses in a cutoff waveguide are considerably larger than those in a travelling waveguide of the same cross-section.

Finally, we shall discuss another case where the conventional perturbation method for calculating losses cannot be used, i.e. structures with unclosed infinitely thin metal screens, e.g. microstrip and slot lines. If the eigenmode under consideration has an electric-field component directed along the edge, then eqn. 3.62 gives an infinitely large value of attenuation coefficient at any frequencies, as the integral

$$\oint_C |\boldsymbol{H}_{s\tau}^0|^2 \, \mathrm{d}C$$

does not converge (Pregla, 1980, Heitkamper and Heinrich, 1991). This is because the magnetic field intensity $\boldsymbol{H}_s^0$ has a quadratically unintegrable singularity at the edge if $\zeta_S = 0$. For a finite impedance there is no edge singularity (see Section 2.4). Then the perturbation of the magnetic field due to an infinitesimal impedance cannot be taken as weak in some neighbourhoods of the edge.

In fact, the problem under study goes beyond the limits of perturbation theory. The problem is wider: to what extent are the impedance-boundary conditions themselves and the model of an infinitely thin strip applicable? Real structures, of course, have a finite thickness, exceeding sufficiently the skin depth, and an infinitely thin strip is simply a model giving an opportunity to apply highly effective computational methods. If this model is not used and we consider a strip of finite thickness with a real edge geometry, there will be no such difficulty, so that eqn. 3.62 will be valid. But in this case determination of $\boldsymbol{E}_p$, $\boldsymbol{H}_p$ as defined in eqn. 3.62 and its integration over $S_\perp$ and C may become very complicated.

Within the framework of the model of an infinitely thin strip for a mode having the current-density component directed along the edge, the Leontovich impedance boundary conditions are not valid (see Section 2.4); it is necessary to use the modified impedance expressed by eqns. 2.27 or 2.28. In this case $\gamma_p'' \sim \text{Re}\,\mathcal{Z}_S \ln|\mathcal{Z}_S|$, and the conventional perturbation method cannot be used. But the perturbation technique can be applied by considering $\text{Re}\,\mathcal{Z}_S$ (not $\mathcal{Z}_S$) as a small parameter; at the same time the terms $O(\text{Im}\,\mathcal{Z}_S)$ should be attributed to a zero-order approximation.

There is one more opportunity that allows us to avoid moving beyond the borders of the model of an infinitely thin strip: modification of the energy-perturbation method, which can be interpreted as follows: at the integration along the contour C in eqn. 3.62 some neighbourhoods of edges are excluded. The size Δ of a neighbourhood excluded should be expressed in terms of geometrical parameters of the edge (thickness of plates, edge curvature radius). Nosich and Shestopalov (1980) were the first to suggest this approach; correct determination of Δ for different edge configurations was carried out by Lewin (1984). Similar results were later also presented by Weinstein *et al.* (1986).

Such a modification of the energy-perturbation method is closely connected with that of the surface impedance described in Section 2.4. It can be seen from eqn. 2.22 that the quantity Δ is expressed in terms of $\mathcal{Z}_S$ by means of the equation $\Delta = 2|\mathcal{Z}_S|/\pi G^2 k W_0$. Substituting the impedances expressed by eqns. 2.27 and 2.28 into this formula and taking account of the fact that $|\mathcal{Z}_S| \simeq -\text{Im}\,\mathcal{Z}_S$, we obtain

$$\Delta = \frac{R}{8} \tag{3.70a}$$

for the edge of an elliptical shape, and

$$\Delta = \frac{D}{290} \tag{3.70b}$$

for the edge of a rectangular shape.

Eqns. 3.70*a* and 3.70*b* coincide with the corresponding formulae obtained by Lewin (1984). This fact indicates a principal equivalency of the energy-method modification considered and the concept of a surface impedance in Section 2.4. However, from the computational point of view these methods are not identical: they are complementary to one another and each has its own field of effective applicability.

Different cases of the perturbation-method application described above are shown in Table 3.1. Note that use of the energy method is not always expedient even if it is thoroughly substantiated: everything depends on the waveguide geometry and applied method of solving the boundary-value problem. For waveguides of simple configuration when separation of variables is possible, the energy method will in all probability be the shortest way of obtaining analytical formulae for attenuation coefficients. For waveguides of complicated shapes determination of eigenfunctions of corresponding lossless systems and subsequent integration in eqn. 3.62 may be rather difficult. Then it would be expedient to obtain a dispersion equation for the waveguide with losses using the numerical

Table 3.1 *Perturbation-method applications*

Mode	Asymptotics of γ_s'' at $Z_S \to 0$	Method of calculating γ_s''
Nondegenerated (ω is not near ω_C)	$O(\mathrm{Re}\, Z_S)$	energy perturbation technique; eqn. 3.62
D-degenerated (ω is not near ω_C)	$O(\mathrm{Re}\, Z_S)$	modified perturbation method taking into account the intermodal coupling; eqn. 3.67
Nondegenerated (ω is close to ω_C)	$O[(\mathrm{Re}\, Z_S)^{1/2}]$	perturbation method is inapplicable; eqn. 3.69
With the edge singularity of the current density	$O(\mathrm{Re}\, Z_S \ln \lvert Z_S \rvert)$	modified surface impedance technique; Z_S is expressed by eqns. 2.27 and 2.28

or numeric-analytical methods for impedance-boundary-value problems. Attenuation coefficients can be determined, solving the dispersion equation through the perturbation method considering Z_S or Re Z_S (for the case of modes with edge singularity of the current density) being small. Further on we shall be using both a direct solution of impedance problems and numerical integration of fields for lossless systems. Both approaches supplement one another and we shall come across cases when one is preferable to the other. We shall try to draw readers' attention to the reasons for this preference.

3.8 Attenuation in a generalised microstrip line; model of the infinitely thin strip

The cross-section of a generalised microstrip line is shown in Figure 3.4*a*. Let us consider the procedure for deriving the functional equations for determination of normal modes (Zarubanov and Ilyinsky, 1985). To simplify our analysis we assume that the relative magnetic permeability of all layers is 1. Let the strip be infinitely thin and the lower and the upper screens be perfectly conducting. Thus, we shall consider here the absorption in both the strip and the dielectric layers. The first component of losses is the most important to evaluate and the most difficult to determine. It is not difficult to take into account the losses in the screens in the framework of the technique described but, when comparing various methods of loss determination in a microstrip line, it would be more convenient to have separate components of the total losses.

In this section we shall describe an algorithm based on the impedance formulation of the problem. As we have accepted a model using an infinitely thin strip, the Leontovich boundary condition is not correct. In each of the regions S_i where the permittivity is constant, we are looking for the solution of a

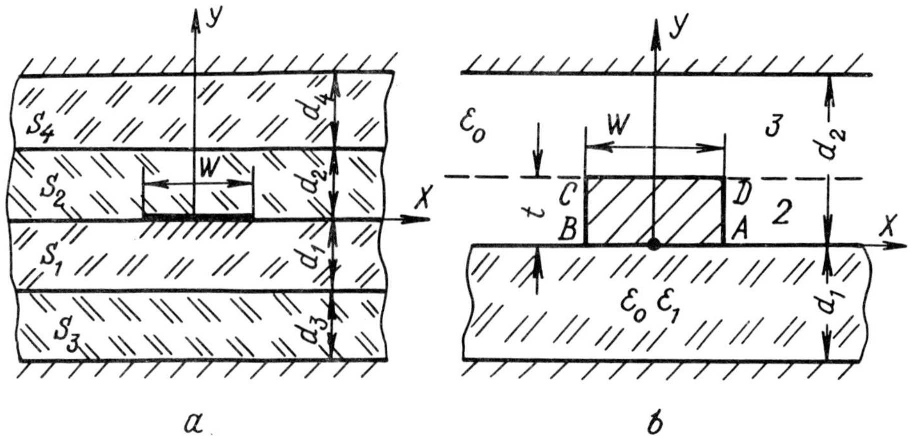

Figure 3.4 *Microstrip lines (sectional view)*

a With infinitely thin strip
b With strip of finite thickness

homogeneous system of Maxwell's equations in the form of normal modes propagating along the z-axis. Tangential components of fields should be continuous across the interfaces between the dielectric media and on the perfectly conducting surfaces of the screens the tangential component of the electric field should become zero. In addition, the edge condition should be satisfied.

Let us examine nonradiating modes of the line. This results in the additional condition that the field of the mode tends to zero when $(x^2+y^2)^{1/2}$ tends to infinity in every cross-section $z=$ constant. Introducing electric and magnetic Hertz vectors in the form of corresponding Fourier transforms with respect to x, expressing the field in every region in terms of the Hertz vectors, and matching the fields at the boundaries, we can obtain expressions for the components E_x and E_z at the interface between the regions S_1 and S_2 in the form of spectral expansions

$$\left.\begin{aligned} 2\pi j\omega\varepsilon_0 E_x &= \int_{-\infty}^{\infty} P(\alpha,\gamma)\hat{I}_x(\alpha)\exp(j\alpha x)\,d\alpha \\ &\quad + \int_{-\infty}^{\infty} G(\alpha,\gamma)\hat{I}_z(\alpha)\exp(j\alpha x)\,d\alpha \\ 2\pi j\omega\varepsilon_0 E_z &= \int_{-\infty}^{\infty} G(\alpha,\gamma)\hat{I}_x(\alpha)\exp(j\alpha x)\,d\alpha \\ &\quad + \int_{-\infty}^{\infty} Q(\alpha,\gamma)\hat{I}_z(\alpha)\exp(j\alpha x)\,d\alpha \end{aligned}\right\} \tag{3.71}$$

Here α is a spectral variable, $\hat{I}_{x,z}(\alpha)$ are Fourier transforms from jumps of z- and x-components of the magnetic field at the boundary,

$-\infty < x < \infty$,

$$P(\alpha, \gamma) = \frac{1}{\alpha^2 + \gamma^2}\left\{\frac{\alpha^2}{T_\varepsilon(\alpha)} - \frac{k^2\gamma^2}{T_\mu(\alpha)}\right\}$$

$$G(\alpha, \gamma) = \frac{\alpha\gamma}{\alpha^2 + \gamma^2}\left\{\frac{1}{T_\varepsilon(\alpha)} + \frac{k^2}{T_\mu(\alpha)}\right\}$$

$$Q(\alpha, \gamma) = \frac{1}{\alpha^2 + \gamma^2}\left\{\frac{\gamma^2}{T_\varepsilon(\alpha)} - \frac{k^2\alpha^2}{T_\mu(\alpha)}\right\}$$

$$T_\mu(\alpha) = \sum_{i=1}^{2} \beta_i \frac{\beta_i \tanh(\beta_i d_i) \tanh(\beta_{i+2} d_{i+2}) + \beta_{i+2}}{\beta_i \tanh(\beta_{i+2} d_{i+2}) + \beta_{i+2} \tanh(\beta_i d_i)}$$

$$T_\varepsilon(\alpha) = \sum_{i=1}^{2} \frac{\tilde{\varepsilon}_i\{\tilde{\varepsilon}_{i+2}\beta_i + \tilde{\varepsilon}_i\beta_{i+2} \tanh(\beta_i d_i) \tanh(\beta_{i+2} d_{i+2})\}}{\beta_i\{\tilde{\varepsilon}_{i+2}\beta_i \tanh(\beta_i d_i) + \tilde{\varepsilon}_i\beta_{i+2} \tanh(\beta_{i+2} d_{i+2})\}}$$

$\beta_i^2 = \alpha^2 + \gamma^2 - k^2\tilde{\varepsilon}_i$, $i = 1, 2, 3, 4$, and $\tilde{\varepsilon}_i$ is the relative permittivity of the layers.

Let us introduce now the main functional relations assuming that the impedance boundary conditions are satisfied on the surface of the strip. Then from eqn. 3.71 we obtain

$$\left.\begin{aligned}
&\int_{-\infty}^{\infty} P(\alpha, \gamma)\hat{I}_x(\alpha) \exp(j\alpha x)\, d\alpha \\
&\quad + \int_{-\infty}^{\infty} G(\alpha, \gamma)\hat{I}_z(\alpha) \exp(j\alpha x)\, d\alpha = \xi_1 I_x(x) \\
&\int_{-\infty}^{\infty} G(\alpha, \gamma)\hat{I}_x(\alpha) \exp(j\alpha x)\, d\alpha \\
&\quad + \int_{-\infty}^{\infty} Q(\alpha, \gamma)\hat{I}_z(\alpha) \exp(j\alpha x)\, d\alpha = \xi_2 I_z(x)
\end{aligned}\right\} \tag{3.72}$$

where $\xi_i = 2\pi j\omega\varepsilon_0 Z_{Si}$ $(i = 1, 2)$, Z_{S1} is the Leontovich surface impedance, Z_{S2} is the surface impedance expressed by eqn. 2.27 or eqn. 2.28, and $-W/2 < x < W/2$. Moreover, at $|x| > W/2$ the following functional relations are correct:

$$\int_{-\infty}^{\infty} \hat{I}_{x,z}(\alpha) \exp(j\alpha x)\, d\alpha = 0 \tag{3.73}$$

We stress that eqn. 3.73 is a spectral formulation of the fact that the electric current outside the strip is equal to zero and determines a permissible class of solutions together with the main functional relations given by eqn. 3.72.

To solve eqn. 3.72 let us use the moment method. We shall present the currents on the strip as expansions in terms of basis functions

$$\left.\begin{aligned}
I_z(x) &= \sum_{m=0}^{\infty} a_m \phi_m(x)\{1 - (2x/W)^2\}^{-1/2} \equiv \sum_{m=0}^{\infty} a_m \bar{\phi}_m(x) \\
I_x(x) &= \sum_{m=0}^{\infty} b_m \psi_m(x)\{1 - (2x/W)^2\}^{1/2} \equiv \sum_{m=0}^{\infty} b_m \bar{\psi}_m(x)
\end{aligned}\right\} \tag{3.74}$$

where a_m and b_m are complex coefficients. For even (symmetrical in x) modes $\phi_m(x) = T_{2m}(2x/W)$ and $\psi_n(x) = U_{2n+1}(2x/W)$; for odd modes $\phi_m(x) = T_{2m+1}(2x/W)$ and $\psi_n(x) = U_{2n}(2x/W)$, where $T_i(x)$ and $U_i(x)$ are Chebyshev polynomials of the first and second kinds, respectively. Substituting eqns. 3.74 into eqns. 3.72, we obtain

$$\int_{-\infty}^{\infty} P(\alpha, \gamma) \sum_{n=0}^{\infty} b_n \hat{\bar{\psi}}_n(\alpha) \exp(j\alpha x)\, d\alpha + \int_{-\infty}^{\infty} G(\alpha, \gamma) \sum_{m=0}^{\infty} a_m \hat{\bar{\phi}}_m(\alpha) \exp(j\alpha x)\, d\alpha = \xi_1 \sum_{n=0}^{\infty} b_n \bar{\psi}_n(x) \tag{3.75}$$

$$\int_{-\infty}^{\infty} G(\alpha, \gamma) \sum_{n=0}^{\infty} b_n \hat{\bar{\psi}}_n(\alpha) \exp(j\alpha x)\, d\alpha + \int_{-\infty}^{\infty} Q(\alpha, \gamma) \sum_{m=0}^{\infty} a_m \hat{\bar{\phi}}_m(\alpha) \exp(j\alpha x)\, d\alpha = \xi_2 \sum_{n=0}^{\infty} b_n \bar{\phi}_n(x) \tag{3.76}$$

We take the complete systems of functions $\{\bar{\psi}_n(x)\}$ and $\{\phi_m(x)\}$ as weighting functions for eqns. 3.75 and 3.76, respectively. After some transformation we obtain an infinite system of linear algebraic equations for the a_m and b_n coefficients. To be numerically solved this system has to be truncated:

$$\sum_{n=0}^{N} b_n \left\{ \int_{-\infty}^{\infty} P(\alpha, \gamma) \hat{\bar{\psi}}_n(\alpha) \hat{\bar{\psi}}_p^*(\alpha)\, d\alpha - \xi_1 \int_{-W/2}^{W/2} \bar{\psi}_n(x) \bar{\psi}_p(x)\, dx \right\} + \sum_{m=0}^{M} a_m \int_{-\infty}^{\infty} G(\alpha, \gamma) \hat{\bar{\phi}}_m(\alpha) \hat{\bar{\psi}}_p^*(\alpha)\, d\alpha = 0 \tag{3.77}$$

$$\sum_{n=0}^{N} b_n \int_{-\infty}^{\infty} P(\alpha, \gamma) \hat{\bar{\psi}}_n(\alpha) \hat{\phi}_r^*(\alpha)\, d\alpha + \sum_{m=0}^{M} a_m \left[\int_{-\infty}^{\infty} Q(\alpha, \gamma) \hat{\bar{\phi}}_m(\alpha) \hat{\phi}_r^*(\alpha)\, d\alpha - \xi_2 \int_{-W/2}^{W/2} \bar{\phi}_m(x) \phi_r(x)\, dx \right] = 0 \tag{3.78}$$

where $p = 0, 1, \ldots, N$; $r = 0, 1, \ldots, M$.

The integrals with respect to x in eqns. 3.77 and 3.78 after presenting U_n in a trigonometric form are easily found. It can be proved, using the orthogonality relations for Chebyshev polynomials, that these integrals in eqn. 3.78 at $m \neq r$ become zero. The improper integrals are absolutely and uniformly convergent. Equating the determinant of the system given by eqns. 3.77 and 3.78 to zero, we obtain a dispersion equation for complex propagation coefficients:

$$\text{Det}\, A(\gamma, \{\xi_j\}, \{\varepsilon_i\}) \equiv f(\gamma, \{\xi_j\}, \{\varepsilon_i\}) = 0 \tag{3.79}$$

where $j = 1, 2$; $i = 1, 2, 3, 4$.

Numerical solution of eqn. 3.79 should be treated with caution. That is because $f(\gamma, \xi_2)$ is an analytical function only for a finite matrix; while it approaches the limit $M, N \to \infty$ the analyticity at the point $\xi_2 = 0$ is broken;

hence the perturbation method considering $\mathcal{Z}_S$ as a small parameter cannot be applied: there is no convergence in this case. However, we can use a modification of the perturbation technique described in Section 3.7: $\mathrm{Re}\,\mathcal{Z}_{S2} = \mathrm{Re}\,\mathcal{Z}_{S1} = -\mathrm{Im}\,\mathcal{Z}_{S1} = \mathcal{Z}'_S$ is a small parameter whereas the terms $O(\mathrm{Im}\,\mathcal{Z}_{S2})$ are those of the zero-order approximation. When calculating the ohmic losses in the strip, the dielectric media are to be considered as being ideal ($\tan\delta_i = 0$).

Assuming $\mathcal{Z}'_S$ to be a small parameter, we expand the function $f(\gamma, \mathcal{Z}'_S)$ in eqn. 3.79 into a Taylor series in the neighbourhood of the point $\gamma = \gamma_0$, $\mathcal{Z}'_S = 0$, where γ_0 satisfies the equation $f(\gamma_0, 0) = 0$. Neglecting the terms above the first order, we obtain

$$f(\gamma_0, 0) + (\gamma - \gamma_0)\left.\frac{\partial f}{\partial \gamma}\right|_{\substack{\gamma=\gamma_0\\ \mathcal{Z}'_S=0}} + \mathcal{Z}'_S\left.\frac{\partial f}{\partial \mathcal{Z}'_S}\right|_{\substack{\gamma=\gamma_0\\ \mathcal{Z}'_S=0}} = 0 \tag{3.80}$$

The first summand is equal to zero by definition and as a result the following formula for the attenuation coefficient can be derived from eqn. 3.80:

$$\tilde{\gamma}''_\sigma = \mathrm{Im}\left\{\mathcal{Z}'_S\left.\frac{\partial f}{\partial \mathcal{Z}'_S}\right|_{\substack{\gamma=\gamma_0\\ \mathcal{Z}'_S=0}}\left(\left.\frac{\partial f}{\partial \gamma}\right|_{\substack{\gamma=\gamma_0\\ \mathcal{Z}'_S=0}}\right)^{-1}\right\} \tag{3.81}$$

The subscript σ is introduced to distinguish the attenuation in conducting surfaces of the line from that in dielectric layers (we shall calculate it later); the tilde sign denotes that only the losses in the strip are taken into account.

It is essential that while using eqn. 3.81 it is not necessary to carry out computations requiring complex arithmetic: derivatives of $f(\gamma, \mathcal{Z}'_S, \{\varepsilon_i\})$, are calculated with real arguments.

When creating an algorithm from eqn. 3.81 we face the problem of differentiation of the function given in the form of a determinant. Two methods are possible: the first is numerical differentiation (it needs caution in calculations). The other is analytical differentiation carried out according to the formula (Fadeev and Fadeeva, 1963)

$$\frac{\partial f}{\partial x} = \sum_{i=1}^{K} \mathrm{Det}\, A'_i \tag{3.82}$$

where $K = N + M + 2$. Here the matrix A'_i follows from A through replacing the ith row by that of the derivatives $\partial A_{ip}/\partial x$.

Which of these methods is better will depend on the labour required for analytical calculations of the derivatives in eqn. 3.82. The surface impedances $\mathcal{Z}_{S1}$ and $\mathcal{Z}_{S2}$ enter into eqn. 3.79 linearly; therefore differentiation with respect to $\mathcal{Z}'_S$ can be carried out on the basis of eqn. 3.82. The parameter γ is involved in the matrix elements A in a complicated transcendental way and analytical differentiation here implies rather cumbersome transformations. This is why the derivative with respect to γ was determined numerically—by the finite-difference method.

As the basis and weighting functions chosen are different (for the integrals to converge), so the matrix A will be asymmetrical. Only its top left-hand block of the order $N + 1$ will be symmetrical.

Let us proceed to determination of attenuation in dielectric layers. Now we assume $\mathcal{Z}_{S1} = \mathcal{Z}_{S2} = 0$ and the permittivities of layers to be complex:

$\varepsilon_i = \varepsilon_i' + j\varepsilon_i''$. Then ε_i'' are considered to be small parameters and the approach used is completely identical to calculation of γ_σ''. Omitting obvious intermediate calculations let us write down the final result:

$$\gamma_\varepsilon'' = -\sum_{i=1}^{4} \varepsilon_i'' \left.\frac{\partial f}{\partial \varepsilon_i}\right|_{\substack{\gamma=\gamma_0 \\ \varepsilon_i=\varepsilon_i'}} \left(\left.\frac{\partial f}{\partial \gamma}\right|_{\substack{\gamma=\gamma_0 \\ \varepsilon_i=\varepsilon_i'}} \right)^{-1} \tag{3.83}$$

Here γ_0 represents the solution of eqn. 3.79 at $\zeta_{S1} = \zeta_{S2} = \varepsilon_i'' = 0$ $(i = 1, 2, 3, 4)$. Differentiation of the function f with respect to ε_i in eqn. 3.83 is also performed by means of the finite-difference method.

The total losses within the framework of the perturbation method can be found by adding the independently calculated components. This means that the total attenuation coefficient is expressed by

$$\gamma'' = \tilde{\gamma}_\sigma'' + \gamma_\varepsilon'' + \gamma_{\sigma s}''$$

where $\tilde{\gamma}_\sigma''$ and γ_ε'' are determined by eqns. 3.81 and 3.83, respectively, and $\gamma_{\sigma s}''$ is the attenuation coefficient due to losses in the screen (its determination is not considered in this Section).

The algorithm presented allows us to calculate the attenuation in different types of planar lines. It is characterised by fast convergence: three basis functions are usually sufficient to approximate the longitudinal component of the current density and two functions to approximate the transverse component. The analysis and comparison of the numerical results obtained by means of various methods (including that described above) will be presented in Section 3.10.

3.9 Attenuation in a microstrip line; model of a strip of finite thickness

The cross-section of the transmission line under study is shown in Figure 3.4*b*. Let us consider the problem of determining the even normal modes (analysis of odd modes can be carried out in a similar way). For such a model we can use the conventional energy-perturbation method (see Section 3.7). Here it is necessary to calculate the field distribution in the cross-section of a line (in particular, the current density distribution on conducting surfaces), corresponding to the case of perfect conductivity. In this Section we shall describe the numerical algorithm for solving this problem suggested by Zarubanov and Ilyinsky (1990).

The mathematical formulation of the problem is similar to that in Section 3.8. Let us divide the cross-section of the microstrip line into three partial regions, as shown in Figure 3.4*b*, introducing into each the electric $\boldsymbol{\Pi}^e$ and magnetic $\boldsymbol{\Pi}^m$ Hertz vectors. The electromagnetic field is expressed in terms of Hertz vectors by the formulae

$$\left.\begin{aligned} \boldsymbol{E} &= \text{grad div } \boldsymbol{\Pi}^e + k^2 \boldsymbol{\Pi}^e + j\omega\mu_0 \text{ curl } \boldsymbol{\Pi}^m \\ \boldsymbol{H} &= \text{grad div } \boldsymbol{\Pi}^m + k^2 \boldsymbol{\Pi}^m - j\omega\tilde{\varepsilon}\varepsilon_0 \text{ curl } \boldsymbol{\Pi}^e \end{aligned}\right\} \tag{3.84}$$

As shown by Weinstein (1988), to represent an arbitrary electromagnetic field in a region without sources we can just use one component of each Hertz vector. One of the specific features of the approach described is as follows: because of the boundary conditions it is convenient to use different pairs of components of the Hertz vectors in various partial regions. According to Section 3.8, we look for solutions whose z-dependence is $\exp(jhz)$, where h is the propagation coefficient to be found. Thus, for regions 1 and 3 let us assume in eqn. 3.84 that $\mathbf{\Pi}^{e,m}(x, y, z) = \hat{i}_y \Pi_y^{e,m}(x, y) \exp(jhz)$, and for region 2 that $\mathbf{\Pi}^{e,m}(x, y, z) = \hat{i}_x \Pi_x^{e,m}(x, y) \exp(jhz)$ (in regions 2 and 3, $\tilde{\varepsilon} = 1$). For the functions $\Pi_{x,y}^{e,m}(x, y)$ in various partial regions the following spectral representations take place:

In region 1 $(-d_1 \leq y \leq 0)$

$$\Pi_{y1}^e(x, y) = \int_0^\infty C_1^e \frac{\cosh\{\beta_1(d_1 + y)\}}{\beta_1 \sinh(\beta_1 d_1)} \cos(\alpha x)\, d\alpha$$

$$\Pi_{y1}^m(x, y) = \int_0^\infty C_1^m \frac{\sinh\{\beta_1(d_1 + y)\}}{\omega\mu_0 \sinh(\beta_1 d_1)} \sin(\alpha x)\, d\alpha$$

in region 3 $(t \leq y \leq d_2)$

$$\Pi_{y3}^e(x, y) = -\int_0^\infty C_3^e \frac{\cosh\{\beta_3(d_2 - y)\}}{\beta_3 \sinh\{\beta_3(d_2 - t)\}} \cos(\alpha x)\, d\alpha$$

$$\Pi_{y3}^m(x, y) = \int_0^\infty C_3^m \frac{\sinh\{\beta_3(d_2 - y)\}}{\omega\mu_0 \sinh\{\beta_3(d_2 - t)\}} \sin(\alpha x)\, d\alpha$$

$$\beta_1^2 = \alpha^2 + h^2 - k^2\tilde{\varepsilon}, \qquad \beta_3^2 = \alpha^2 + h^2 - k^2, \qquad -\infty < x < \infty$$

in region 2 $(0 \leq y \leq t)$

$$\Pi_{y2}^e(x, y) = \int_0^\infty \{A \sinh(\beta_2 y) + B \cosh(\beta_2 y)\} \cos\{\xi(x - W/2)\}\, d\xi$$

$$\Pi_{y2}^m(x, y) = \frac{1}{\omega\mu_0} \int_0^\infty \{D \sinh(\beta_2 y) + F \cosh(\beta_2 y)\} \sin\{\xi(x - W/2)\}\, d\xi$$

$$\beta_2 = \beta_3 \qquad 0 \leq y \leq t \qquad W/2 \leq x \leq \infty$$

where $C_{1,3}^{e,m}$ are unknown spectral densities. Expressing the field tangential components in each region in terms of Hertz vectors we obtain their representations through the unknown spectral densities.

In this formulation the tangential components of the electric field are identically equal to zero on the screens and end faces of the strip. Now we have to impose the boundary conditions on the top and bottom faces and the edge conditions.

Let us introduce the following four functions

at $|x| < W/2$:

$$N_1(x) = H_x(x, t+0) - H_x(x, t-0) = j_z^t(x)$$

$$N_2(x) = j\{H_z(x, t+0) - H_z(x, t-0)\} = j_x^t(x)$$

$$N_3(x) = H_x(x, -0) - H_x(x, +0) = j_z^0(x)$$

$$N_4(x) = j\{H_z(x, -0) - H_z(x, +0)\} = j_x^0(x)$$

at $|x| > W/2$

$$\mathcal{N}_i(x) = 0 \qquad i = 1, 2, 3, 4$$

where $j_z^{0,t}(x)$ and $j_z^{0,t}(x)$ are unknown densities of the currents induced on the strip. According to the edge conditions for a right-angle wedge (Meixner, 1972), the current densities on the strip can be presented as

$$j_z^{0,t}(x) = \{1 - (2x/W)^2\}^{-1/3}\phi^{0,t}(x)$$

$$j_x^{0,t}(x) = \text{constant} + \{1 - (2x/W)^2\}^{2/3}\psi^{0,t}(x)$$

where $\phi(x)$ and $\psi(x)$ are smooth regular functions.

Let us substitute the integral representations of the fields into the continuity conditions for the tangential components of the electric fields at the interfaces between the partial regions, and into the expressions for the functions $\mathcal{N}_i$. Multiplying the equations for the components E_x and H_z by $\cos\{\alpha'(x - W/2)\}$ and those for the components E_z and H_x by $\sin\{\alpha'(x - W/2)\}$, and then integrating with respect to x along the positive semi-axis, we obtain, owing to the orthogonality of the corresponding trigonometric functions, a system of eight relationships allowing us to express the spectral densities of the field components in terms of Fourier transforms of the functions $\mathcal{N}_i$ (from here on indicated by $\hat{\mathcal{N}}_i$).

From the system obtained after simple but cumbersome transformations, we can obtain expressions for spectral densities in terms of the functions $\hat{\mathcal{N}}_i$. Substituting these expressions into integral representations for the tangential components of the electric field and using the boundary conditions on the upper and the lower sides of the strip, we obtain a system of four homogeneous functional equations of the first kind for the unknown functions $\hat{\mathcal{N}}_i$:

$$\left.\begin{aligned}
&\int_0^\infty (G_1\hat{\mathcal{N}}_1 + P_1\hat{\mathcal{N}}_2 - G\hat{\mathcal{N}}_3 - P\hat{\mathcal{N}}_4)\sin(\alpha x)\,d\alpha = 0\\
&\int_0^\infty (Q_1\hat{\mathcal{N}}_1 + G_1\hat{\mathcal{N}}_2 - Q\hat{\mathcal{N}}_3 - G\hat{\mathcal{N}}_4)\cos(\alpha x)\,d\alpha = 0\\
&\int_0^\infty (G_2\hat{\mathcal{N}}_1 + P_2\hat{\mathcal{N}}_2 + G_3\hat{\mathcal{N}}_3 + P_4\hat{\mathcal{N}}_4)\sin(\alpha x)\,d\alpha = 0\\
&\int_0^\infty (Q_2\hat{\mathcal{N}}_1 + G_2\hat{\mathcal{N}}_2 + Q_3\hat{\mathcal{N}}_3 - G_3\hat{\mathcal{N}}_4)\cos(\alpha x)\,d\alpha = 0
\end{aligned}\right\} \tag{3.85}$$

where $|x| \leq W/2$, and

$$P = \frac{1}{\alpha^2 + h^2}\left(\frac{\alpha^2}{T_\varepsilon} - \frac{k^2h^2}{T_\mu}\right)$$

$$G = \frac{\alpha h}{\alpha^2 + h^2}\left(\frac{1}{T_\varepsilon} + \frac{k^2}{T_\mu}\right)$$

$$Q = \frac{1}{\alpha^2 + h^2}\left(\frac{h^2}{T_\varepsilon} - \frac{k^2\alpha^2}{T_\mu}\right)$$

$$\begin{Bmatrix} P_1 \\ G_1 \\ Q_1 \end{Bmatrix} = \frac{1}{T_t \cosh(\beta_2 t)} \begin{Bmatrix} P \\ G \\ Q \end{Bmatrix}$$

$$\begin{Bmatrix} P_2 \\ G_2 \\ Q_2 \end{Bmatrix} = \frac{1}{T_t} \left[\tanh(\beta_2 t) \coth(\beta_1 d_1) \begin{Bmatrix} P_\varepsilon \\ G_\varepsilon \\ Q_\varepsilon \end{Bmatrix} + \begin{Bmatrix} P \\ G \\ Q \end{Bmatrix} \right]$$

$$\begin{Bmatrix} P_3 \\ G_3 \\ Q_3 \end{Bmatrix} = \cosh(\beta_2 t)\{\tanh(\beta_2 t) \coth(\beta_2 d_2) - 1\} \begin{Bmatrix} P \\ G \\ Q \end{Bmatrix}$$

$$P_\varepsilon = \frac{1}{\alpha^2 + h^2} \left(\frac{\tilde{\varepsilon}\beta_2 \alpha^2}{\beta_1 T_\varepsilon} - \frac{k^2 \beta_1 h^2}{\beta_2 T_\mu} \right)$$

$$G_\varepsilon = \frac{\alpha h}{\alpha^2 + h^2} \left(\frac{\tilde{\varepsilon}\beta_2}{\beta_1 T_\varepsilon} + \frac{k\beta_1^2}{\beta_2 T_\mu} \right)$$

$$Q_\varepsilon = \frac{1}{\alpha^2 + h^2} \left(\frac{\tilde{\varepsilon}\beta_2 h^2}{\beta_1 T_\varepsilon} - \frac{k^2 \beta_1 \alpha^2}{\beta_2 T_\mu} \right)$$

$$T_t = \frac{1 - \tanh^2(\beta_2 t)}{1 - \tanh(\beta_2 t) \tanh(\beta_2 d_2)}$$

$$T_\mu = \beta_1 \coth(\beta_1 d_1) + \beta_2 \coth(\beta_2 d_2)$$

$$T_\varepsilon = \frac{\tilde{\varepsilon} \coth(\beta_1 d_1)}{\beta_1} - \frac{\coth(\beta_2 d_2)}{\beta_2}$$

The functions T_ε and T_μ have roots corresponding to the propagation coefficients of the LM and LE modes of a 2-layer plane dielectric waveguide.

Taking into account the behaviour of current density near the edges of the strip, we expand the current densities in the series in terms of the complete systems of Gegenbauer polynomials with the corresponding weight multipliers:

$$\left.\begin{aligned} \begin{Bmatrix} j_z^t(x) \\ j_z^0(x) \end{Bmatrix} &= \sum_{n=0}^{\infty} \begin{Bmatrix} a_n^t \\ a_n^0 \end{Bmatrix} \{1 - (2x/W)^2\}^{-1/3} C_{2n}^{1/6}(2x/W) \\ \begin{Bmatrix} j_x^t(x) \\ j_x^0(x) \end{Bmatrix} &= \begin{Bmatrix} b_0^t \\ b_0^0 \end{Bmatrix} 2x/W + \sum_{m=1}^{\infty} \begin{Bmatrix} b_m^t \\ b_m^0 \end{Bmatrix} \{1 - (2x/W)^2\}^{2/3} C_{2m-1}^{7/6}(2x/W) \end{aligned}\right\} \quad (3.86)$$

The coefficients $a_n^{t,0}$, $b_n^{t,0}$ have to satisfy the conditions

$$\sum_{n=0}^{\infty} |a_n^{t,0}|^2 < \infty \qquad \sum_{n=0}^{\infty} |b_n^{t,0}|^2 < \infty$$

The terms containing $b_n^{t,0}$ are added into eqn. 3.86 to satisfy the matching conditions for the longitudinal component of the magnetic field at $x = \pm W/2$.

The Fourier transforms of the basis functions are calculated in the explicit form

$$\hat{\phi}_n(\alpha) = (-1)^n \pi W \frac{\Gamma(2n+1/3)}{2(2n)!\,\Gamma(1/6)2^{1/6}} \frac{J_{2n+1/6}(\alpha W/2)}{(\alpha W/2)^{1/6}}$$

$$\hat{\Psi}_m(\alpha) = (-1)^{m-1} \pi W \frac{\Gamma(2m+1/3)}{2(2m+1)!\,\Gamma(7/6)2^{7/6}} \frac{J_{2m+1/6}(\alpha W/2)}{(\alpha W/2)^{7/6}}$$

$$(n = 0, 1, \ldots; m = 1, 2, \ldots)$$

Let us substitute the expansions for current densities given by eqn. 3.86 into eqn. 3.85. Applying the moment method with the weighting functions ϕ_r $(r = 0, 1, \ldots)$, ψ_s $(s = 1, 2, \ldots)$ and changing the order of the operations of integration and summation [which is possible in this case (Ilyinsky, 1981)], we obtain an infinite system of homogeneous linear algebraic equations for the coefficients $a_n^{t,0}$ and $b_n^{t,0}$. Note that at this stage the number of equations is two less than the number of unknowns. Matching the expansions for the transverse currents given by eqn. 3.86 with the integral representations for the tangential magnetic field H_z in region 2 at $y = 0$ and $y = t$, we obtain the required equations.

Using the Wiener–Paley theorem in a similar way to Ilyinsky (1981), we can show that the infinite system of linear algebraic equations and the original problem of normal-mode determination are equivalent in the sense that the field of a normal mode gives rise to a nontrivial solution for the homogeneous system of linear equations and, vice versa, a nontrivial solution of the latter enables us to calculate the field of the normal mode.

Equating the determinant of the system to zero, we obtain a dispersion equation to find the propagation coefficient. The $a_n^{t,0}$ and $b_n^{t,0}$ coefficients, and hence the functions $\hat{N}_i(\alpha)$, are determined from the corresponding homogeneous linear system. The attenuation coefficient due to losses in the metal surfaces was determined by the formula

$$\gamma''_\sigma = \frac{\operatorname{Re} Z_S \int_C |\boldsymbol{J}|^2 \, dl}{2P} \tag{3.87}$$

where

$$P = \int_{S_\perp} (E_x H_y^* - E_y H_x^*) \, dx \, dy$$

The integral in the numerator of eqn. 3.87 is taken along the contour around the metal conductors in the cross-section of the line. The components of the current density in the expression under the integral are presented by eqn. 3.86 (see also Appendix 2). The integral in the denominator of eqn. 3.87 is taken over the cross-section of the line. Integration with respect to y was performed analytically; later, using Parseval's theorem, we transfer from integration with respect to x to that with respect to the spectral variable α:

$$\int_{S_\perp} (E_x H_y^* - E_y H_x^*)\, dx\, dy = \frac{1}{2\pi} \int_{-\infty}^{\infty} P_z(\alpha)\, d\alpha. \tag{3.88}$$

The expression for the function $P_z(\alpha)$ is presented in Appendix 2.

In practice, for numerical computations the infinite system of linear equations has been truncated. Inner convergence of the method is fairly fast: in practice it always is enough to consider three terms in the expansions of the longitudinal and transverse current densities. Some matrix blocks are symmetrical. This fact allows us to reduce the number of matrix elements to be calculated. It is important from the computational point of view, because every element is expressed by an improper integral of a rapidly varying nonmonotonic function. The numerical integration was carried out to the upper limit $25\pi/W$ by means of the Gaussian quadrature formula. The relative error in calculating the integrals did not exceed 10^{-7}. The search for a root of the dispersion equation was carried out in the interval $(h_{LM}, k\sqrt{\tilde{\varepsilon}})$, where h_{LM} is the propagation coefficient of the lower LM mode and only the root closest to $k\sqrt{\tilde{\varepsilon}}$ was sought. Once the root had been found a nontrivial solution for the homogeneous system of linear algebraic equations was determined by means of the method of back iterations (Ilyinsky and Zarubanov, 1980). In practice, to achieve a high accuracy five iterations are sufficient.

3.10 Attenuation in a microstrip line; numerical results

In this section we analyse the numerical results obtained on the basis of algorithms described in Sections 3.8 and 3.9. Here we compare them with other data presented so far in the literature. This clarifies which of the physical models and methods are most effective for evaluating the attenuation in various planar transmission lines: slot, coplanar etc. To determine $\mathcal{Z}_S$, the value $\sigma_0 = 5.9 \times 10^7$ S/m (copper) was used.

Tables 3.2 and 3.3 present the results of calculation of attenuation for a line with $W = 0.96$ mm, $d_1 = 1$ mm, $d_2 = \infty$, $\tilde{\varepsilon} = 9.6$ at $f = 2$ GHz and $f = 12$ GHz for strip conductors of various thicknesses. The second, third and fourth columns correspond to the contributions (in percent) to the ohmic losses made by the currents on the upper, lower and side surfaces of the strip, respectively; the fifth column gives the contribution of the current on the screen. The sixth and the seventh columns present the absolute value of losses (in dB/m) in the screen and

Table 3.2 *Attenuation coefficients of microstrip lines: contribution of different components in total losses at $f = 2$ GHz*

t/W	$\gamma''_{\sigma t}$ (%)	$\gamma''_{\sigma 0}$ (%)	$\gamma''_{\sigma 1}$ (%)	$\gamma''_{\sigma s}$ (%)	$\gamma''_{\sigma s}$ (dB/m)	γ''_{σ} (dB/m)
10^{-3}	7.34	80.81	0.39	11.46	0.156	1.36
10^{-2}	7.38	79.08	2.64	11.00	0.157	1.42
10^{-1}	6.22	78.84	5.25	9.69	0.163	1.69
1	0.31	89.43	2.75	7.51	0.191	2.54

Table 3.3 *Attenuation coefficients of microstrip lines: contribution of different components in total losses at $f = 12$ GHz*

t/W	$\gamma''_{\sigma t}$ (%)	$\gamma''_{\sigma 0}$ (%)	$\gamma''_{\sigma 1}$ (%)	$\gamma''_{\sigma s}$ (%)	$\gamma''_{\sigma s}$ (dB/m)	γ''_{σ} (dB/m)
10^{-3}	6.78	78.00	0.37	14.85	0.487	3.28
10^{-2}	6.73	76.47	2.51	14.29	0.489	3.43
10^{-1}	5.65	76.61	5.00	12.74	0.506	3.97
1	0.18	87.36	2.55	9.91	0.544	5.49

the total ohmic losses in the line ($\gamma''_{\sigma} = \gamma''_{\sigma t} + \gamma''_{\sigma 0} + \gamma''_{\sigma 1} + \gamma''_{\sigma s} = \tilde{\gamma}''_{\sigma} + \gamma''_{\sigma s}$). In spite of the statement by Vymorokov *et al.* (1987) that losses on the end faces of the strip for a 50 Ω microstrip line make up 30% of the total losses, according to our calculations these losses are one fifth the size. Note that Vymorokov *et al.* calculated the losses through the static approximation.

Figure 3.5 shows the dependencies of total attenuation coefficient in the microstrip line with $W = 1.2446$ mm, $d_1 = 0.508$ mm, $d_2 = \infty$, $\tilde{\varepsilon} = 9.35$, $\tan\delta = 2.5 \times 10^{-4}$ and $t/W = 0.001$. Curve 1 is taken from the work by Gopinath *et al.* (1970). It was obtained for a model using an infinitely thin strip in TEM-approximation. Curve 3 (broken line) corresponds to the results of experiments by the same authors. Curve 4 shows data by Nikolsky and Kozlov (1987*b*) and curve 5 gives results according to the method in Section 3.9. Curve 2 is a result of using the technique in Section 3.8, where Z_S indicates the Leontovich impedance (the losses in the screen are taken into account). Such a model, as was mentioned above, is not physically correct: by curve 2 we mean to show the errors implicated in the use of the conventional model of an infinitely thin strip. It is seen that curves 4 and 5 demonstrate best agreement with the experiment. Curve 2 and the experimental data differ noticeably. Zarubanov and Ilyinsky (1990) studied transmission lines with wide strips ($W = 9.15$ mm, $d_1 = 0.64$ mm). In this case the model, corresponding to curve 2, leads to sufficiently accurate results. This can be easily explained physically: inapplica-

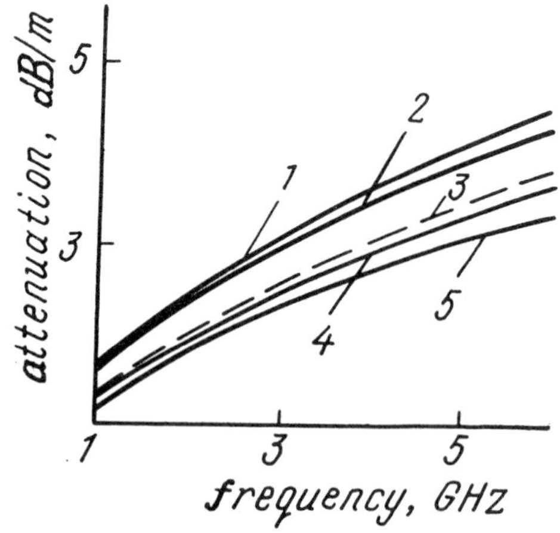

Figure 3.5 *Total losses in microstrip line calculated by various methods*

bility of the Leontovich-impedance condition in combination with a model using an infinitely thin strip results from incorrect description of edge effects. However, for such wide strips the contribution of edge effects is so small that errors in their description are not significant. Let us underline that the model corresponding to curve 2 invariably leads to an overestimated value of the attenuation coefficient. This can be explained by a more rapid increase of the longitudinal current density in the neighbourhood of the edge than for a conductor of finite thickness.

In Figure 3.6 are shown the results of calculation of the attenuation caused by the losses in a strip (the parameters of the line are the same as for Figure 3.5). Curve 1 corresponds to the algorithm from Section 3.8 with the Leontovich impedance ($\mathcal{Z}_{S2} \rightarrow \mathcal{Z}_{S1}$), curve 2 to the algorithm from Section 3.9, curve 3 to the data by Nikolsky and Kozlov (1987*b*), and curve 4 to the energy-perturbation-method modification for an infinitely thin strip (Section 3.7) and to the algorithm from Section 3.8 with the impedance $\mathcal{Z}_S$ expressed by eqn. 2.28 (calculation results for these methods coincide within the accuracy of presentation, and therefore in Figure 3.6 they are presented by one curve). This information allows us to conclude that the model of an infinitely thin strip modified according to Sections 2.4 or 3.7 enables us to describe the edge effects correctly and to obtain accurate results.

When analysing the reliability of numerical data on losses, a qualitative investigation of current distribution on the conducting surfaces is important. Figure 3.7 shows the current-density distributions for a line with $W = 0.96$ mm, $d_1 = 1$ mm, $d_2 = \infty$, $\tilde{\varepsilon} = 9.6$ and different values of the strip thickness at $f =$ 12 GHz. Curves 1 correspond to $t/W = 0.01$, curves 2 correspond to $t/W = 0.1$, and broken lines correspond to an infinitely thin strip. In the currents designations the superscript t corresponds to the upper side of the plate, 0 to the lower side, 1 to the lateral side and s to the currents on the lower screen. When the strip thickness decreases, the total currents densities $j_z^0 + j_z^t$ and $j_x^0 + j_x^t$ tend to the corresponding current distributions for a line with an infinitely thin strip.

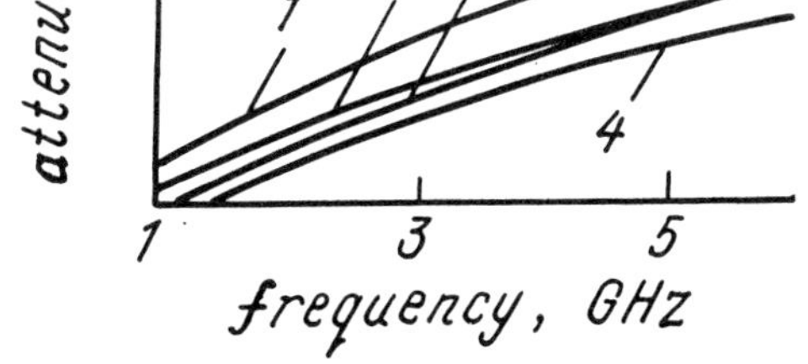

Figure 3.6 *Partial losses in the strip calculated by various methods*
Parameters of microstrip line as in Figure 3.5

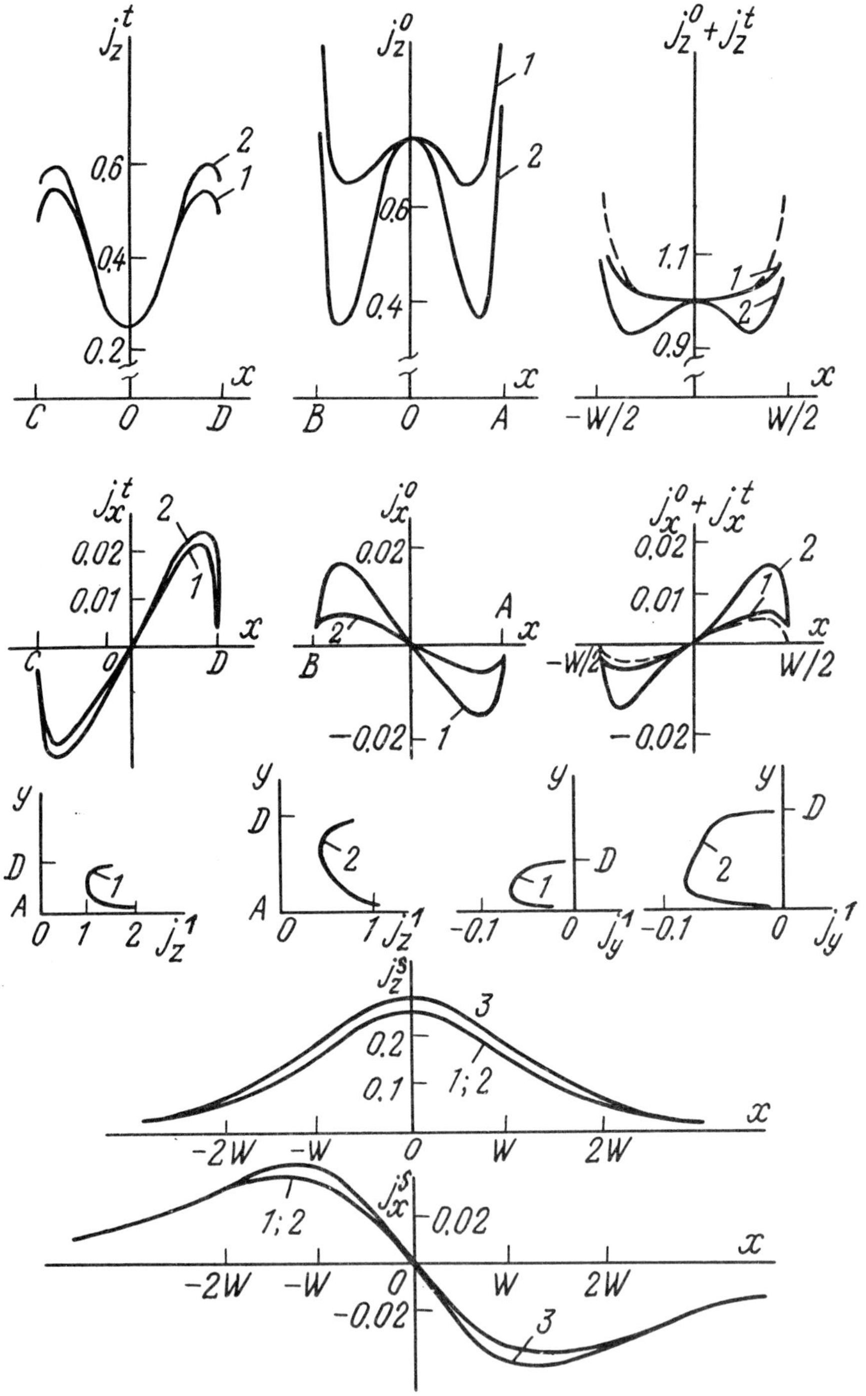

0.6
0.4
0.2
C
0
D
x
1
2
0.6
0.4
B
A
1.1
0.9
−W/2
W/2
0.02
0.01
−0.02
y
A
0
1
2
−0.1
0.2
0.1
1; 2
3
−2W
−W
W
2W

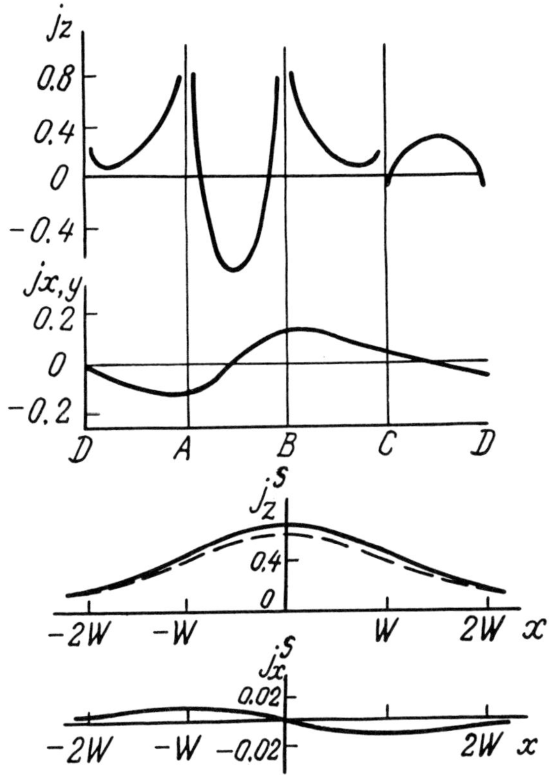

Figure 3.8 *As for Figure 3.7 for microstrip line with $t/W = 1$*

The strip thickness slightly influences the currents induced on the lower plate. The distributions of these currents are close to each other even at $t/W = 1$ (curves 3) and $t/W = 0.01$ (curves 2). In Figure 3.8 current distributions at $f =$ 12 GHz of the same line with a square section strip are given. The broken line presents a distribution of the longitudinal current on the lower plate for an infinitely thin strip. Compared with the case of thin strips, the character of longitudinal currents for the upper and lower sides of the plate has been changed.

It is known that for a line with a dielectric substratum a concept of characteristic impedance cannot uniquely be introduced. For microstrip lines the following concept for characteristic impedance is mainly used: $Z_2 = P/I^2$, where I is the total longitudinal current of the central conductor and P is defined by eqn. 3.88. For an open microstrip line with an infinitely thin strip we can define the

---

Figure 3.7 *Surface-current-density distributions for conductors of a microstrip line*

1 $t/W = 0.01$, 2 $t/W = 0.1$
– – – – – $t = 0$

impedance in the form (Brews, 1987):

$$Z_1 = \frac{Z_T k}{h} \qquad Z_3 = \frac{V^2}{P} \qquad Z_4 = \frac{V}{I}$$

where Z_T is the impedance of the line in the TEM approximation, and V can be regarded as tension in the line:

$$V = \int_{-d_1}^{0} E_y(0, y)\, dy$$

Table 3.4 shows the values of impedances of the line with parameters corresponding to Figures 3.5, 3.6, determined using various formulae. The magnitude of Z_T was calculated according to the method suggested by Lerer and Mikhalevsky (1983). At $f \to 0$, all the formulae produce very similar results. With increases in frequency, Z_1 decreases and Z_3 increases faster than Z_2 and Z_4. For a strip of finite thickness at $t/W = 0.01$, the behaviour of Z_2 remains the same and its numerical value slightly diminishes (27 Ω at $f = 6$ GHz). The small spread in the impedances shows a good approximation of the field by the TEM mode in this case.

Finally, let us dwell upon several general points about which models would be acceptable for use in evaluation of losses in planar structures. First, let us point out that direct analysis of structures with strips of finite thickness (both with impedance and approximating to perfect conductivity) is rather complicated: to prove this we can compare algorithms described in Sections 3.8 and 3.9. Moreover, in Section 3.9 a strip of rectangular geometry is examined (see also Rawal and Jackson, 1991). In fact, the shape of the strip is determined by technological factors and may not be rectangular: rounded, trapezoidal etc. In this case opportunities for solving the problem become even more limited: use of the partial-regions method is out of the question (or at least, problematic). It would then seem natural to use the integral-equation technique, but for thin strips ($t \ll W, d_{1,2}$) it may lead to ill-conditioned systems of linear algebraic equations (an example of this type as applied to a different class of problems is discussed in Section 5.7).

This is why it would seem very attractive to modify the model of an infinitely thin conductor, and attempts of this kind have been discussed in the literature more than once. In all probability, one of the first such attempts was made by Mirshekar-Syahkai and Davies (1982). They suggested excluding the edge neighbourhood in the quadratic integration of the current density when using the

Table 3.4 *Comparison of the values of Z_i*

f (GHz)	Z_1 (Ω)	Z_2 (Ω)	Z_3 (Ω)	Z_4 (Ω)
1	30.15	30.03	30.39	30.21
2	30.10	30.02	30.58	30.30
3	30.04	30.01	30.81	30.41
4	29.96	30.00	31.09	30.54
5	29.88	30.00	31.09	30.70
6	29.80	30.02	31.75	30.87

energy-perturbation method. The size of the neighbourhood Δ excluded was determined by the number of basis functions $\mathcal{N}$ in the current-density expansion: $\Delta = W/\mathcal{N}$. It was asserted that the numerical results for different $\mathcal{N}$ confirm a convergence of the method. It can be seen from the results of Sections 2.4 and 3.7 (see also Pregla, 1980, Heitkamper and Heinrich, 1991) that this approach at $\mathcal{N} \to \infty$ diverges according to $O(\ln \mathcal{N})$. Obviously, this slow divergence gave rise to the illusion of convergence at $\mathcal{N} \to \infty$. Another example is the work by Rozzi *et al.* (1990), where attenuation in a slot line was calculated. The current density on the surface of the plates was approximated with 'corrections' to the distribution near the edge: instead of the singularity $x^{-1/2}$ corresponding to the infinitely thin screen, Rozzi *et al.* used $x^{-1/3}$ corresponding to a perfectly conducting wedge with a right angle. In this case integration in accordance with the conventional energy-perturbation method results in a finite value of γ''_σ. However, this approach cannot be considered as physically justified, or the results obtained reliable, because in the above current-density approximation there are no parameters characterising the edge geometry (thickness, radius of curvature etc.).

There is one more class of model which does not imply use of the impedance-boundary conditions. Klimenko and Fialkovsky (1990), Vakanas *et al.* (1990), Nikolsky and Kozlov (1986, 1987*a*, 1987*b*) and Kiang (1991) presented the strip as a dielectric body with permittivity $\varepsilon = j\sigma_0/\omega$. It can be assumed that the development of such models is partially stimulated by the difficulties arising when using standard approaches (see Section 3.7). From the physical point of view, these models are faultless, of course, but their practical application is very complicated. In the paper by Klimenko and Fialkovsky (1990) only a most simple key problem is considered—plane-wave excitation of a finitely conducting halfplane in free space. Analysis has been carried out with the aid of the modified Wiener–Hopf method: through factorisation the problem has been reduced to a rapidly converging system of linear algebraic equations. However, the algorithm is so complicated that its generalisation for a more realistic model of the elements of microwave and millimetre-wave circuits is rather problematic. In the numerical solution we shall certainly face the problem of field presentation inside the conducting body, for example its expansion in terms of some basis functions. However, if the skin effect is strong this field is very inhomogeneous (it is large in a thin near-boundary layer and sharply damped beyond its limits). Therefore, when a conventional system of basis functions is applied, the convergence of the expansions obtained is very slow. The situation can be improved to some extent, but only for very thin strips with thickness comparable with the field-penetration depth (Vakanas *et al.*, 1990, Kiang, 1991).

Owing to mathematical difficulties Nikolsky and Kozlov (1987*b*) have modified this model, considering a rectangular strip with face ends impermeable by the field (they are covered with ideally conducting films). A heuristic hypothesis is used here which needs to be checked and proved. First the losses in the face ends are not taken into account. Though their dimensions include usually a small part of the total strip perimeter, the current density near the face ends is high and their contribution to losses can be significant. Secondly, the changes in the edge condition caused by the arbitrary introduction of 'impermeable' face ends can distort the currents on the side surfaces of the strip near the edge.

Analysis and comparison of various models show that the most promising are the models of an infinitely thin strip modified in accordance with the real edge geometry as shown in Sections 2.4 and 3.7. These models provide sufficient accuracy in the loss calculation and, at the same time, do not lead to major mathematical difficulties (see also Barsotti *et al.*, 1991). In addition, they are very flexible as regards the edge geometry: its change results in that of the numerical coefficient in the expressions for Δ or $\mathcal{Z}_S$.

Chapter 4

Normal oscillations in resonators with losses

This Chapter deals with the theory of normal oscillations in resonators with losses in the walls, and also cavities filled with inhomogeneous dissipative dispersive media. Concepts of eigenoscillations and free oscillations and corresponding Q-factors are introduced. General formulae for the Q-factors of eigenoscillations and free oscillations are derived and the difference between them, resulting from the dispersion properties of the media, is considered.

The Q-factors of various resonators (cylindrical, spherical, conical, biconical and cylindrical with coaxial metal plug) are calculated. Different approaches are used to calculate Q-factor (energy-perturbation method, impedance-perturbation technique for solving the characteristic equation), and the advantages and disadvantages of each method are discussed.

Comparisons are made of resonators of different shapes with respect to their Q-factor. Ways of increasing the Q-factor through optimisation of the resonator shape and filling are analysed.

4.1 Expansion of eigenoscillations of a resonator with losses in the walls in terms of resonant modes of an identical lossless resonator

Let S be a closed surface with the interior V (Figure 4.1). The solutions of Maxwell's homogeneous equations

$$\left.\begin{aligned} \operatorname{curl} \boldsymbol{E} &= j\omega\mu \boldsymbol{H} \\ \operatorname{curl} \boldsymbol{H} &= -j\omega\varepsilon \boldsymbol{E} \end{aligned}\right\} \tag{4.1}$$

satisfying the impedance boundary condition

$$(\boldsymbol{n}_0 \times \boldsymbol{E}) + Z_S\{\boldsymbol{n}_0 \times (\boldsymbol{n}_0 \times \boldsymbol{H})\} = 0 \tag{4.2}$$

on S are called eigenoscillations of the cavity V.

As the eigenvalue the frequency ω is used. The permittivity ε and permeability

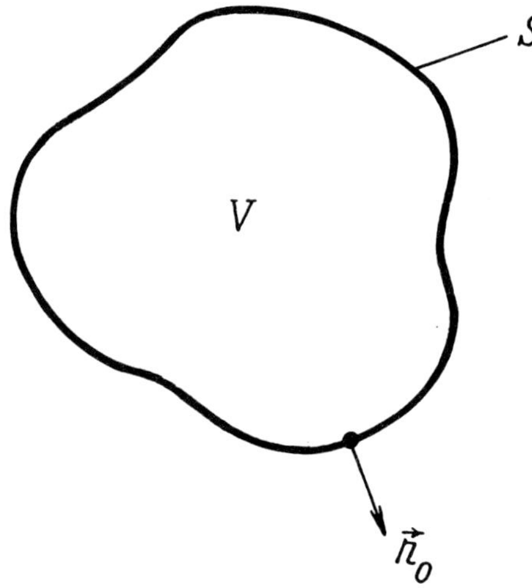

Figure 4.1 *Geometry of cavity resonator*

μ are assumed to be frequency-independent, purely real and constant throughout the cavity (a more general case will be discussed in Section 4.3).

For complex Z_S the problem considered is not self-adjoint and the eigenvalues are complex $\omega_{(r)} = \omega'_{(r)} - j\omega''_{(r)}$, r being the eigenoscillation number. The parenthetical subscripts are used to denote the eigenfrequencies of the resonator with losses in the walls, whereas ω_r are the eigenfrequencies of the resonator with a perfectly conducting surface.

Let us dwell briefly upon the physical meaning of the problem formulated. Eqns. 4.1 and 4.2 describe oscillations in the resonator after the external sources have been switched off. For the resonator with losses such oscillations are attenuated, and to analyse them we therefore need a nonstationary statement of the problem. In formulating the problem using eqns. 4.1 and 4.2 we in fact postulate the harmonic time dependence $\exp\{-j\omega_{(r)}t\}$ or, in other words, the exponential character of the oscillation attenuation $\exp\{-\omega''_{(r)}t\}$. Generally speaking, the existence of such solutions has to be proved as well as whether all solutions are characterised by this form of time dependence. Such a proof can be performed using the more rigorous approach described in Section 4.3. The energy losses complicate considerably the procedure of determining the eigenfrequencies and eigenfunctions even for regions with simple shapes. The relevant technique (compare with Section 3.2) for finding the eigenoscillations of a resonator with losses is to expand them in the series in terms of resonant modes of an identical lossless resonator (Mashkovtsev *et al.*, 1966, Weinstein, 1988). Then for the coefficients in such expansions we obtain a homogeneous infinite system of linear algebraic equations. The approximate solution of this system, obtained with the condition that the losses are small, results in simple formulae for the electrodynamic parameters of eigenoscillations of the resonator with losses.

The surface S will be considered as arbitrary and not containing infinitely thin edges. Such an analysis can also be carried out for a surface with edges of this shape but the modified impedance described in Section 2.4 should be used in eqn. 4.2 instead of the Leontovich impedance.

Let us assume that the eigenoscillations of a resonator without losses satisfying the system of equations

$$\left.\begin{aligned} \operatorname{curl} \hat{\boldsymbol{E}}_s &= j\omega_s \mu \hat{\boldsymbol{H}}_s \\ \operatorname{curl} \hat{\boldsymbol{H}}_s &= -j\omega_s \varepsilon \hat{\boldsymbol{E}}_s \end{aligned}\right\}$$

and the boundary condition

$$(\boldsymbol{n}_0 \times \hat{\boldsymbol{E}}_s)|_S = 0$$

are known. Here $\{\omega_s\}$ ($s = 1, 2, \ldots$) is the sequence of eigenfrequencies of the resonator without losses in increasing order of magnitude

$$0 < \omega_1 \leq \omega_2 \cdots \leq \omega_s \leq \cdots$$

Let us consider also the resonator-potential functions expressed as

$$\begin{Bmatrix} \boldsymbol{e}_s \\ \boldsymbol{h}_s \end{Bmatrix} = \operatorname{grad} \begin{Bmatrix} \Phi_s^E \\ \Phi_s^M \end{Bmatrix}$$

$$\nabla^2 \Phi_s^{E,M} = 0 \qquad \Phi_s^E|_S = 0 \qquad \left.\frac{\partial \Phi_s^M}{\partial \boldsymbol{n}}\right|_S = 0$$

The potential functions can be treated conventionally as degenerated eigenoscillations of infinite multiplicity with the frequency $\omega_s = 0$. According to Weinstein (1988) the systems of functions complete in V are formed through the unification $\{\boldsymbol{E}_s\} = \{\hat{\boldsymbol{E}}_s\} \oplus \{\boldsymbol{e}_s\}$, $\{\boldsymbol{H}_s\} = \{\hat{\boldsymbol{H}}_s\} \oplus \{\boldsymbol{h}_s\}$. The families $\{\boldsymbol{E}_s\}$ and $\{\boldsymbol{H}_s\}$ are each orthogonal:

$$\varepsilon \int_V \boldsymbol{E}_p \boldsymbol{E}_s \, \mathrm{d}V = -\mu \int_V \boldsymbol{H}_p \boldsymbol{H}_s \, \mathrm{d}V = \delta_{sp} m_p \qquad (m_p > 0)$$

Thus, the sets $\{\boldsymbol{E}_s\}$ and $\{\boldsymbol{H}_s\}$ are bases in V. Then the vectors $\boldsymbol{E}$ and $\boldsymbol{H}$ to be found can be expressed in the form of the expansions

$$\boldsymbol{E} = \sum_s A_s \boldsymbol{E}_s \qquad \boldsymbol{H} = \sum_s B_s \boldsymbol{H}_s \tag{4.3}$$

A_s and B_s being unknown constant coefficients. In accordance with a general idea of the method of moments let us replace eqn. 4.1 by an equivalent system of projection relations:

$$\left.\begin{aligned} \int_V (\operatorname{curl} \boldsymbol{E} - j\omega\mu \boldsymbol{H}) \boldsymbol{H}_p \, \mathrm{d}V = 0 \\ \int_V (\operatorname{curl} \boldsymbol{H} + j\omega\varepsilon \boldsymbol{E}) \boldsymbol{E}_p \, \mathrm{d}V = 0 \end{aligned}\right\} \tag{4.4}$$

where $p = 1, 2, \ldots$.

To substitute eqn. 4.3 into eqn. 4.4 we exclude the differential operations on $\boldsymbol{E}$ and $\boldsymbol{H}$, as the expressions for curl $\boldsymbol{E}$ and curl $\boldsymbol{H}$ cannot be obtained by performing term-by-term differentiation[22] of the series for $\boldsymbol{E}$ and $\boldsymbol{H}$. Using the

[22] This is because the series in eqn. 4.3 are not uniformly convergent on the boundary S, as the fields and the basis functions satisfy different boundary conditions. This kind of peculiarity of such series is also discussed in Section 3.3.

standard vector identities and eqn. 4.2, we rewrite eqn. 4.4 as

$$\left.\begin{aligned}&\int_V \operatorname{curl} \boldsymbol{H}_p \boldsymbol{E}\, \mathrm{d}V - \mathcal{Z}_S \oint_S \boldsymbol{H}_p\{\boldsymbol{n}_0 \times (\boldsymbol{n}_0 \times \boldsymbol{H})\}\, \mathrm{d}S \\ &\quad - j\omega\mu \int_V \boldsymbol{H}\boldsymbol{H}_p\, \mathrm{d}V = 0 \\ &\int_V \boldsymbol{H} \operatorname{curl} \boldsymbol{E}_p\, \mathrm{d}V + j\omega\varepsilon \int_V \boldsymbol{E}\boldsymbol{E}_p\, \mathrm{d}V = 0\end{aligned}\right\} \tag{4.5}$$

Substituting eqn. 4.3 into eqn. 4.5 and using the orthogonality properties of the functions $\hat{\boldsymbol{E}}_s$ and $\hat{\boldsymbol{H}}_s$, we obtain the following system of equations for A_s, B_s:

$$\omega_p B_p = \omega A_p \tag{4.6}$$

$$jm_p(\omega_p A_p - \omega B_p) + \mathcal{Z}_S \sum_s B_s \Lambda_{ps} = 0 \tag{4.7}$$

where

$$\Lambda_{ps} = \oint_S \boldsymbol{H}_p\{\boldsymbol{n}_0 \times (\boldsymbol{n}_0 \times \boldsymbol{H}_s)\}\, \mathrm{d}S = -\oint_S \boldsymbol{H}_{p\tau}\boldsymbol{H}_{s\tau}\, \mathrm{d}S$$

(the subscript τ stands for tangential to S component of the vector $\boldsymbol{H}$).

From eqn. 4.6 it follows that coefficients A_p with potential functions $\boldsymbol{e}_p$ are zero ($\omega_p \equiv 0$). However, the potential functions $\boldsymbol{h}_p$ are preserved in the expansion for the magnetic-field intensity.

Excluding A_p from eqn. 4.7 we obtain the homogeneous system of linear equations for B_p

$$jm_p \frac{\omega_p^2 - \omega^2}{\omega} B_p + \mathcal{Z}_S \sum_s^{\infty} B_s \Lambda_{sp} = 0 \tag{4.8}$$

Nontrivial solutions of eqn. 4.8 determine eigenoscillations of the resonator with finitely conducting walls. Complex eigenfrequencies are calculated as roots of the determinantal equation

$$\operatorname{Det} \left\| jm_p \frac{\omega_p^2 - \omega^2}{\omega} \delta_{sp} + \mathcal{Z}_S \Lambda_{sp} \right\| = 0 \tag{4.9}$$

4.2 Resonance frequencies and Q-factors of eigenoscillations

Let us suppose that among the modes of the lossless resonator there are no degenerate modes ($\omega_{p'} \neq \omega_p$) and the intervals between eigenfrequencies are rather large. Then, owing to the small size of $\mathcal{Z}_S$ ($|\mathcal{Z}_S| \ll W_0$) for the pth eigenoscillation it is sufficient to take into account only the pth equation in the system given by eqn. 4.8. Then the characteristic equation can be written as

$$jm_p \frac{\omega_p^2 - \omega^2}{\omega} + \mathcal{Z}_S \Lambda_{pp} = 0 \tag{4.10}$$

Considering that the difference $|\omega_p - \omega|$ is small, we can replace $\mathcal{Z}_S(\omega)$ by $\mathcal{Z}_S(\omega_p)$ in eqn. 4.10. Then we obtain the expression for the eigenfrequency of the pth eigenoscillation

$$\omega_{(p)} \simeq \omega_p + j\mathcal{Z}_S \frac{\Lambda_{pp}}{2m_p} \tag{4.11}$$

The real part of the eigenfrequency $\omega'_{(p)}$ determines the conditions of resonance; let us call it a resonance frequency of the resonator. According to eqn. 4.11

$$\omega'_{(p)} \simeq \omega_p - \operatorname{Im} \mathcal{Z}_S \frac{\Lambda_{pp}}{2m_p} \tag{4.12}$$

Later, when discussing resonators, we shall be considering the impedance $\mathcal{Z}_S$ as frequency-independent, using its value at the eigenfrequency ω_p. Usually, because of the small size of $\mathcal{Z}_S$, the second term in eqn. 4.12 is negligibly small, and $\omega'_{(p)} \simeq \omega_p$, i.e. the resonance frequencies can be determined assuming the walls to be perfectly conducting. The rate of attenuation of the eigenoscillation due to losses in the walls is characterised by the imaginary part of the eigenfrequency. The value

$$Q_p^C = \frac{\omega'_{(p)}}{2\omega''_{(p)}} \simeq \frac{\omega_p}{2\omega''_{(p)}} \tag{4.13}$$

is called the ohmic Q-factor of a free oscillation (the superscript c indicates that the losses in the metal conducting walls are considered). Substituting the expressions for ω_p and $\omega''_{(p)}$ into eqn. 4.13, we obtain

$$Q_p^C = \frac{\omega_p m_p}{\operatorname{Re} \mathcal{Z}_S \Lambda_{pp}} = \frac{\omega_p}{\operatorname{Re} \mathcal{Z}_S} \frac{\int_V \mu \boldsymbol{H}_p^2 \, \mathrm{d}V}{\oint_S \boldsymbol{H}_{p\tau}^2 \, \mathrm{d}S} \tag{4.14}$$

Eqn. 4.14 reflects the energy-perturbation method for resonators (Jones, 1986, Weinstein, 1988). The following interpretation of eqn. 4.14 is possible: let $\mathscr{P}$ be the power dissipated in the walls, then $\mathscr{P} = -\mathrm{d}\bar{W}/\mathrm{d}t$, where $\bar{W}$ is the average energy stored in the resonator. As $\bar{W}(t) = \bar{W}(0)\exp(-2\omega'' t)$, we have

$$\frac{\omega'}{2\omega''} = Q_p^C = \frac{\omega' \bar{W}}{\mathscr{P}} = \frac{\omega'}{\operatorname{Re} \mathcal{Z}_S} \frac{\int_V \mu \boldsymbol{H}^2 \, \mathrm{d}V}{\oint_S \boldsymbol{H}_\tau^2 \, \mathrm{d}S} \tag{4.15}$$

where $\boldsymbol{H}$ is magnetic-field distribution in the resonator with losses. If the disturbance of the resonant-mode field due to losses is small, then assuming that $\boldsymbol{H} \simeq \boldsymbol{H}_p$ in eqn. 4.15, we obtain eqn. 4.14. This assumption and, consequently, eqn. 4.14 are not always valid, e.g. for the modes with close resonance frequencies. Neither can this formula be used in the presence of infinitely thin diaphragms in the resonator, because of the singular behaviour of the field at the edges. The above considerations are similar to those in Section 3.7 which deals with the waveguides, so we do not present any detailed discussion of these here. Eqn. 4.15, generally speaking, is free from the limitations described but precise calculation of the magnetic field $\boldsymbol{H}$ is a rather complicated problem.

Within the limits of applicability eqn. 4.14 allows us to determine very easily (by calculating the integrals) the ohmic Q-factor if the field distribution of the resonant mode in the lossless resonator is known.

Let us consider now a case of two oscillations with close resonance frequencies ($\omega_p \simeq \omega_s$). Then the two equations in the system given by eqn. 4.8

$$jm_p \frac{\omega_p^2 - \omega^2}{\omega} B_p + Z_S(\Lambda_{pp} B_p + \Lambda_{ps} B_s) = 0$$

$$jm_s \frac{\omega_s^2 - \omega^2}{\omega} B_s + Z_S(\Lambda_{ss} B_s + \Lambda_{sp} B_p) = 0$$

have to be retained.

Assuming the determinant of this system to be equal to zero we obtain the characteristic equation

$$\left(jm_p \frac{\omega_p^2 - \omega^2}{\omega} + Z_S \Lambda_{pp}\right)\left(jm_s \frac{\omega_s^2 - \omega^2}{\omega} + Z_S \Lambda_{ss}\right) - Z_S^2 \Lambda_{sp} \Lambda_{ps} = 0 \tag{4.16}$$

To simplify the analysis consider $\omega_p = \omega_s = \omega_0$ assuming that $\Delta\omega = \omega_0 - \omega$ is small owing to the small size of the losses. Then eqn. 4.16 can be reduced to a quadratic equation for $\Delta\omega$:

$$(\Delta\omega)^2 - \Delta\omega \frac{jZ_S}{2}\left(\frac{\Lambda_{pp}}{m_p} + \frac{\Lambda_{ss}}{m_s}\right) - Z_S^2 \frac{(\Lambda_{pp}\Lambda_{ss} - \Lambda_{ps}^2)}{4m_s m_p} = 0 \tag{4.17}$$

When deriving eqn. 4.17, we take into consideration the relationship $\Lambda_{sp} = \Lambda_{ps}$. The roots of eqn. 4.17 are expressed by the formula

$$\Delta\omega_{1,2} = -\frac{jZ_S}{4}\left[\left(\frac{\Lambda_{pp}}{m_p} + \frac{\Lambda_{ss}}{m_s}\right) \pm \left\{\left(\frac{\Lambda_{pp}}{m_p} + \frac{\Lambda_{ss}}{m_s}\right)^2 - \frac{4(\Lambda_{pp}\Lambda_{ss} - \Lambda_{ps}^2)}{m_s m_p}\right\}^{1/2}\right]$$

The real part of $\Delta\omega$ determines a small shift of the resonant frequency under the influence of finite conductivity of the walls. Such shifts are different for the two eigenoscillations, i.e. because there is finite conductivity the degeneration is removed.

The ohmic Q-factor of coupled *sp*-oscillations is determined by the imaginary part of $\Delta\omega_{1,2}$

$$Q_{1,2}^c = \frac{(\omega_p + \Delta\omega'_{1,2})}{2\Delta\omega''_{1,2}} \simeq \frac{\omega_p}{2\Delta\omega''_{1,2}}$$

and can be expressed through the ohmic Q-factors of the pth and sth eigenoscillations calculated according to eqn. 4.14 and the parameter

$$\tilde{Q}_{sp} = -\frac{\omega_p (m_s m_p)^{1/2}}{\text{Re}\, Z_S \Lambda_{sp}}$$

The latter can reasonably be called a mutual Q-factor of the pth and sth eigenoscillations. The final result is as follows:

$$\frac{1}{Q^C_{1,2}} \simeq \frac{1}{2}\left[\frac{1}{Q^C_p} + \frac{1}{Q^C_s} \pm \left\{\left(\frac{1}{Q^C_p} - \frac{1}{Q^C_s}\right)^2 + \frac{4}{\tilde{Q}^2_{sp}}\right\}^{1/2}\right] \tag{4.18}$$

If $\Lambda_{sp} = 0$ ($\tilde{Q}_{sp} = \infty$) eqn. 4.18 for each eigenoscillation transforms into eqn. 4.14. Thus, the approximate formula for the Q-factor given by eqn. 4.14 should be corrected only when the resonant frequencies of any two eigenoscillations coincide (are close), and these eigenoscillations are coupled due to losses in the walls (the mutual Q-factor is not infinitely large). A similar (but more cumbersome) consideration with identical conclusions can be carried out also for three or more modes with close resonance frequencies. The analysis presented is closely similar to that for waveguides, given in Section 3.7.

Let us consider, as examples, two types of commonly used resonators: rectangular (in the shape of a parallelepiped) and cylindrical (segments of a circular cylinder).

Let a parallelepiped have the dimensions $a \times b \times l$. The transverse electric and transverse magnetic types of modes will be determined relative to the face $a \times b$. The TE_{rnm} and TM_{rnm} modes with $r, n, m \neq 0$ are degenerate for any relationships between the dimensions. Let the multi-indices s and p correspond to the TM and TE modes, respectively; then for the Q-factors in eqn. 4.18 we obtain

$$Q^C_s = \frac{W_0 2\pi abl}{\mathrm{Re}\, Z_S \lambda^3_p \left\{(a+l)\dfrac{m^2 b}{l^2} + l(a+b)\left(\dfrac{r^2}{a^2} + \dfrac{n^2}{b^2}\right)\right\}} \tag{4.19}$$

$$Q^C_p = \frac{W_0 \pi (ab)^2 l\{(r^2/a^2) + (n^2/b^2)\}}{\mathrm{Re}\, Z_S 2\lambda_p \left\{(na)^2\left(1 + \dfrac{l}{b}\right) + (rb)^2\left(1 + \dfrac{l}{a}\right)\right\}} \tag{4.20}$$

$$\tilde{Q}_{sp} = \frac{W_0 (ab)^2 l}{\mathrm{Re}\, Z_S \lambda^2_p mnr(b-a)}\left(\frac{r^2}{a^2} + \frac{n^2}{b^2}\right) \tag{4.21}$$

where $\lambda_p = 2\{(m/l)^2 + (r/a)^2 + (n/b)^2\}^{-1/2}$ is the resonance wavelength.

Thus the Q-factor of the mode in a rectangular resonator, when all indices are nonzero, cannot be determined by means of eqn. 4.14, resulting from the energy-perturbation method. Eqns. 4.18–4.21 are to be used instead, except for the case $a = b$, when, according to eqn. 4.21, $\tilde{Q}_{sp} = \infty$.

In a cylindrical resonator, according to the calculations, the degenerate eigenoscillations are not coupled despite the losses in the walls and the modes for which $\Lambda_{sp} \neq 0$ cannot be degenerate at any relationships between the diameter and length. That is why the energy-perturbation method is valid for any types of eigenoscillations in a cylindrical resonator. The ohmic Q-factor can always be determined according to eqn. 4.14.[23] The corresponding formulae are given in Section 4.4.

[23] This is correct for resonators with ideally smooth walls. Even small deformations or roughness of the walls may result in the coupling of degenerate modes. In this case, the actual ohmic Q-factor for degenerate modes can differ greatly from the Q-factor calculated according to eqn. 4.14 (see also Section 4.4).

In a resonator with losses in the walls, apart from the eigenmodes, the associated oscillations may occur. To analyse them we can use the results of Section 3.4. Associated oscillations are mathematically described by the multiple roots of eqn. 4.9 (compare with Section 3.4).

4.3 Eigenoscillations and free oscillations in a resonator with a magnetodielectric absorbing body

In this Section we shall examine the oscillations in a resonator with perfectly conducting walls but containing an isotropic absorbing magnetodielectric body. We shall take into consideration also the frequency dependence of the permittivity ε and permeability μ of the body material.

Let us consider the problem strictly formulated as nonstationary, because the attenuation of oscillations in the resonator without sources is, in fact, a transient process. So we have to solve Maxwell's equations

$$\left.\begin{aligned} \operatorname{curl} \boldsymbol{H} &= \hat{\varepsilon}\frac{\partial \boldsymbol{E}}{\partial t} \\ \operatorname{curl} \boldsymbol{E} &= -\hat{\mu}\frac{\partial \boldsymbol{H}}{\partial t} \end{aligned}\right\} \tag{4.22}$$

where $\hat{\varepsilon}$ and $\hat{\mu}$ are integral operators of the form

$$\begin{Bmatrix} \hat{\varepsilon} \\ \hat{\mu} \end{Bmatrix} = \int_{-\infty}^{t} \begin{Bmatrix} \varepsilon(t-\tau, \boldsymbol{r}) \\ \mu(t-\tau, \boldsymbol{r}) \end{Bmatrix} \cdots \mathrm{d}\tau$$

describing the frequency dispersion of the body material (kernels $\varepsilon(t-\tau, \boldsymbol{r})$, $\mu(t-\tau, \boldsymbol{r})$ are scalar functions).

We shall consider the moment of switching off the sources as the initial instant ($t=0$) of our observation. Then the initial conditions for eqn. 4.22 have the following form

$$\int_{-\infty}^{0} \varepsilon(-\tau, \boldsymbol{r})\boldsymbol{E}(\tau, \boldsymbol{r})\,\mathrm{d}\tau = \mathfrak{E}(\boldsymbol{r})$$

$$\int_{-\infty}^{0} \mu(-\tau, \boldsymbol{r})\boldsymbol{H}(\tau, \boldsymbol{r})\,\mathrm{d}\tau = \mathfrak{H}(\boldsymbol{r})$$

where $\mathfrak{E}(\boldsymbol{r})$ and $\mathfrak{H}(\boldsymbol{r})$ are known functions.

The boundary condition for eqn. 4.22 is expressed by

$$(\boldsymbol{n}_0 \times \boldsymbol{E})|_S = 0 \qquad (t > 0)$$

In addition, the tangential components of $\boldsymbol{E}$ and $\boldsymbol{H}$ must be continuous across the surface of the magnetodielectric body.

In a similar way to that used by Weinstein and Vackman (1983), we represent

$\boldsymbol{E}(\boldsymbol{r}, t)$ and $\boldsymbol{H}(\boldsymbol{r}, t)$ in the form of Fourier integrals

$$\left.\begin{aligned} \boldsymbol{E}(\boldsymbol{r}, t) &= \frac{1}{2\pi}\int_{-\infty}^{\infty} \boldsymbol{E}(\boldsymbol{r}, \omega)\exp(-j\omega t)\,d\omega \\ \boldsymbol{H}(\boldsymbol{r}, t) &= \frac{1}{2\pi}\int_{-\infty}^{\infty} \boldsymbol{H}(\boldsymbol{r}, \omega)\exp(-j\omega t)\,d\omega \end{aligned}\right\} \tag{4.23}$$

These representations are valid, as it is physically obvious that

$$\lim_{t\to\infty}\begin{Bmatrix}\boldsymbol{E}(t)\\ \boldsymbol{H}(t)\end{Bmatrix} = 0$$

Substituting eqn. 4.23 into eqn. 4.22 we obtain the equations for $\boldsymbol{E}(\boldsymbol{r}, \omega)$ and $\boldsymbol{H}(\boldsymbol{r}, \omega)$

$$\left.\begin{aligned} \text{curl}\,\boldsymbol{H} &= -j\omega\varepsilon(\omega)\boldsymbol{E} + \mathfrak{E} \\ \text{curl}\,\boldsymbol{E} &= j\omega\mu(\omega)\boldsymbol{H} - \mathfrak{H} \end{aligned}\right\} \tag{4.24}$$

and the boundary condition

$$(\boldsymbol{n}_0 \times \boldsymbol{E})|_S = 0 \tag{4.25}$$

Thus we obtain for the Fourier amplitudes the problem of the resonator excitation by the currents given. The functions $\varepsilon(\boldsymbol{r}, \omega)$ and $\mu(\boldsymbol{r}, \omega)$ are Fourier transforms of kernels $\varepsilon(t-\tau, \boldsymbol{r})$ and $\mu(t-\tau, \boldsymbol{r})$, $\varepsilon(\omega)$ and $\mu(\omega)$ being complex-valued functions:

$$\varepsilon(\omega) = \varepsilon'(\omega) + j\varepsilon''(\omega), \quad \mu(\omega) = \mu'(\omega) + j\mu''(\omega), \quad \varepsilon''(\omega), \quad \mu''(\omega) > 0$$

Assume that we know the eigenfunctions $\boldsymbol{E}_s$ and $\boldsymbol{H}_s$ of the resonator under examination, satisfying the equations

$$\left.\begin{aligned} \text{curl}\,\boldsymbol{H}_s &= -j\omega_s\varepsilon(\omega)\boldsymbol{E}_s \\ \text{curl}\,\boldsymbol{E}_s &= j\omega_s\mu(\omega)\boldsymbol{H}_s \end{aligned}\right\} \tag{4.26}$$

and eqn. 4.25. The eigenvalue is ω_s, and ω is some free parameter. It is obvious that $\boldsymbol{E}_s = \boldsymbol{E}_s(\boldsymbol{r}, \omega)$, $\boldsymbol{H}_s = \boldsymbol{H}_s(\boldsymbol{r}, \omega)$ and $\omega_s = \omega_s(\omega)$. Let us present the solution of eqn. 4.24 in the form

$$\left.\begin{aligned} \boldsymbol{E}(\boldsymbol{r}, \omega) &= \sum_s^{\infty} A_s(\omega)\boldsymbol{E}_s(\boldsymbol{r}, \omega) \\ \boldsymbol{H}(\boldsymbol{r}, \omega) &= \sum_s^{\infty} B_s(\omega)\boldsymbol{H}_s(\boldsymbol{r}, \omega) \end{aligned}\right\} \tag{4.27}$$

where $A_s(\omega)$ and $B_s(\omega)$ are the coefficients sought. Quasistatic terms in eqn. 4.27 are rejected as

$$\text{div}\,\mathfrak{E} = \text{div}\,\mathfrak{H} = 0$$

Following Weinstein (1988), the coefficients $A_s(\omega)$ and $B_s(\omega)$ are expressed in the form

$$A_s(\omega) = -\frac{j}{(\omega^2 - \omega_s^2)m_s}\int_V \{\omega\mathfrak{E}(\boldsymbol{r}')\boldsymbol{E}_s(\boldsymbol{r}', \omega) - \omega_s\mathfrak{H}(\boldsymbol{r}')\boldsymbol{H}_s(\boldsymbol{r}', \omega)\}\,dV \tag{4.28}$$

$$B_s(\omega) = -\frac{j}{(\omega^2 - \omega_s^2)m_s}\int_V \{\omega_s \mathfrak{E}(\boldsymbol{r}')\boldsymbol{E}_s(\boldsymbol{r}', \omega) - \omega\mathfrak{H}(\boldsymbol{r}')\boldsymbol{H}_s(\boldsymbol{r}', \omega)\}\,dV \tag{4.29}$$

The field sought is determined using eqns. 4.27–4.29 by means of the Fourier transform according to eqn. 4.23. The result is

$$\boldsymbol{E}(\boldsymbol{r}, t) = -\frac{1}{2\pi j}\sum_s^{\infty}\int_{-\infty}^{\infty}\frac{D_s(\omega)\boldsymbol{E}_s(\boldsymbol{r}, \omega)}{(\omega^2 - \omega_s^2)m_s(\omega)}\exp(-j\omega t)\,d\omega \tag{4.30}$$

where

$$D_s(\omega) = \int_V \{\omega\mathfrak{E}(\boldsymbol{r}')\boldsymbol{E}_s(\boldsymbol{r}', \omega) - \omega_s\mathfrak{H}(\boldsymbol{r}')\boldsymbol{H}_s(\boldsymbol{r}', \omega)\}\,dV$$

A similar formula can be written for $\boldsymbol{H}(\boldsymbol{r}, t)$.

Let us analyse eqn. 4.30 in more detail. According to the principle of causality (see, for example, Agranovich and Ginzburg, 1979), the functions $\varepsilon(\omega)$ and $\mu(\omega)$ are analytical in ω in the lower halfplane;[24] hence $\boldsymbol{E}_s(\boldsymbol{r}, \omega)$ and $D_s(\omega)$ are also analytical. Taking into account the fact that the expression under the integral has only simple poles [$m_s(\omega) \neq 0$ in the lower halfplane], we obtain, according to the Jordan lemma, the expression for $\boldsymbol{E}(\boldsymbol{r}, t)$ at $t > 0$

$$\boldsymbol{E}(\boldsymbol{r}, t) = \mathrm{Re}\left\{\sum_s^{\infty}\frac{D_s(\Omega_s)\boldsymbol{E}_s(\boldsymbol{r}, \Omega_s)\exp(-j\Omega_s t)}{\Omega_s m_s(\Omega_s)\left(1 - \left.\dfrac{d\omega_s}{d\omega}\right|_{\omega=\Omega_s}\right)}\right\} \tag{4.31}$$

where Ω_s satisfy the equation

$$\Omega_s = \omega_s(\Omega_s) \tag{4.32}$$

As can be seen from eqns. 4.31 and 4.32 the electromagnetic field in the resonator after the sources are switched off can be presented as a sum of the eigenoscillations $\boldsymbol{E}_s(\boldsymbol{r}, \omega)$ and $\boldsymbol{H}_s(\boldsymbol{r}, \omega)$ at $\omega = \Omega_s$.

The functions $\boldsymbol{E}_s(\boldsymbol{r}, \Omega_s)$, $\boldsymbol{H}_s(\boldsymbol{r}, \Omega_s)$ will be called free oscillations of the resonator, the values Ω_s being their complex frequencies (Weinstein and Solntsev, 1973).

The free oscillations satisfy Maxwell's homogeneous equations of the type

$$\left.\begin{aligned}\mathrm{curl}\ \boldsymbol{H}^f &= -j\Omega\varepsilon(\Omega)\boldsymbol{E}^f\\ \mathrm{curl}\ \boldsymbol{E}^f &= j\Omega\mu(\Omega)\boldsymbol{H}^f\end{aligned}\right\} \tag{4.33}$$

These equations present the eigenvalue problem nonlinear in the spectral parameter Ω (the boundary condition for free oscillations in the case of perfect conductivity of the walls is the same as for eigenoscillations).[25]

---

[24] Only nonconducting media are considered.

[25] This is correct also in the presence of losses in the walls when the surface impedance is frequency-independent.

Free oscillations are harmonic and have the complex frequencies Ω_s ($\Omega_s = \Omega_s' - j\Omega_s''$, $\Omega_s'' > 0$) and are therefore nonmonochromatic. They are amplitude-modulated oscillations with the amplitude decreasing exponentially as $\exp(-\Omega_s'' t)$. Thus the validity of the formal statement of the problem given by eqns. 4.1 and 4.2 in Section 4.1 is obvious. Now we introduce the quantity

$$Q_s^f = \frac{\Omega_s'}{2\Omega_s''}$$

called the Q-factor of the free oscillation in the resonator. This value characterises the rate of oscillation attenuation (after the sources are switched off) in the resonator due to dielectric and magnetic losses.

Let us calculate the Q-factor of the free oscillations. From eqn. 4.33 we can obtain the following 'energy' identity for the sth free oscillation:

$$\int_V (\Omega_s^* \varepsilon^*(\Omega_s)|\boldsymbol{E}_s^f|^2 - \Omega_s \mu(\Omega_s)|\boldsymbol{H}_s^f|^2)\,\mathrm{d}V = 0 \tag{4.34}$$

Assuming that $\Omega_s'' \ll \Omega_s'$, let us expand the functions $\varepsilon(\Omega_s)$ and $\mu(\Omega_s)$ in Taylor's series in the neighbourhood of the point Ω_s' (thanks to analyticity of $\varepsilon(\omega)$ and $\mu(\omega)$ in the lower halfplane such expansions are valid). Neglecting terms above the first order in Ω_s'', we have

$$\Omega_s \varepsilon(\Omega_s) \simeq \Omega_s' \varepsilon'(\Omega_s') - j\Omega_s'' \frac{\mathrm{d}}{\mathrm{d}\Omega}\{\Omega_s' \varepsilon'(\Omega_s')\} + j\Omega_s' \varepsilon''(\Omega_s')$$

A similar expansion can also be written for the function $\mu(\Omega)$. Substituting the expansions for $\varepsilon(\Omega_s)$ and $\mu(\Omega_s)$ into eqn. 4.34, we obtain

$$\Omega_s' \int_{V_0} \{\varepsilon''(\Omega_s')|\boldsymbol{E}_s^f|^2 + \mu''(\Omega_s')|\boldsymbol{H}_s^f|^2\}\,\mathrm{d}V$$
$$= \Omega_s'' \int_V \left[\frac{\mathrm{d}}{\mathrm{d}\Omega}\{\Omega_s' \varepsilon'(\Omega_s')\}|\boldsymbol{E}_s^f|^2 + \frac{\mathrm{d}}{\mathrm{d}\Omega}\{\Omega_s' \mu'(\Omega_s')\}|\boldsymbol{H}_s^f|^2\right]\mathrm{d}V$$

Now we can easily derive the following formula for Q_s^f:

$$Q_s^f = \frac{\displaystyle\int_V \left[\frac{\mathrm{d}}{\mathrm{d}\Omega}\{\Omega_s' \varepsilon'(\Omega_s')\}|\boldsymbol{E}_s^f|^2 + \frac{\mathrm{d}}{\mathrm{d}\Omega}\{\Omega_s' \mu'(\Omega_s')\}|\boldsymbol{H}_s^f|^2\right]\mathrm{d}V}{\displaystyle 2\int_{V_0} \{\varepsilon''(\Omega_s')|\boldsymbol{E}_s^f|^2 + \mu''(\Omega_s')|\boldsymbol{H}_s^f|^2\}\,\mathrm{d}V} \tag{4.35}$$

Eqn. 4.35 is sometimes used for active systems (in a linear operation) considering the losses as formally negative ($\varepsilon'' < 0$, $\mu'' < 0$). In this case $Q \leq 0$, which corresponds to the exponentially increasing amplitude of the oscillations [according to the factor $\exp(\Omega_s' t/2|Q|)$]. We would like to stress here that such an interpretation is purely formal and should be treated with caution, because the difference between passive and active systems even in linear operation is not merely reflected by that in the sign of ε'' and μ''. With active systems some additional difficulties arise. The first is purely mathematical: in active systems the fields grow with time and Fourier's ordinary integrals to represent them do not converge. This difficulty can be bypassed by shifting the contour of integration to the upper halfplane (Weinstein and Vakman, 1983). However, this gives rise to another important problem: the integration path in eqn. 4.30 is in the upper halfplane where the functions $\varepsilon(\omega)$ and $\mu(\omega)$ are not necessarily analytical. Their singularities when deriving eqn. 4.31 from eqn. 4.30 should be taken into account. In this case the structure of the spectrum of free oscillations may change considerably: new components which are not typical of passive systems may also emerge, in particular oscillations of a continuous spectrum.

Now we consider excitation of the resonator under consideration by the time-harmonic current $\boldsymbol{J}(\boldsymbol{r}, t) = \mathrm{Re}\{\boldsymbol{J}^E(\boldsymbol{r})\exp(-j\omega t)\}$. Such a problem, even in the presence of losses, can be stationary: losses in the resonator in this case are compensated for by the exciting current. Thus, it is necessary to find the solution of Maxwell's equations

$$\left.\begin{aligned} \mathrm{curl}\,\boldsymbol{H} &= -j\omega\varepsilon(\omega)\boldsymbol{E} + \boldsymbol{J}^E \\ \mathrm{curl}\,\boldsymbol{E} &= j\omega\mu(\omega)\boldsymbol{H} \end{aligned}\right\} \tag{4.36}$$

satisfying the boundary condition given by eqn. 4.25.

According to Weinstein (1988), when $|\omega - \omega_s| \ll \omega$, we can obtain the approximate solution of eqn. 4.36

$$\boldsymbol{E} \simeq A_s \boldsymbol{E}_s \qquad \boldsymbol{H} \simeq B_s \boldsymbol{H}_s \qquad B_s(\omega) = \frac{\omega_s}{\omega} A_s(\omega)$$

$$A_s(\omega) = -\frac{j\omega}{(\omega^2 - \omega_s^2)m_s}\int_V \boldsymbol{J}^E \boldsymbol{E}_s \,\mathrm{d}V$$

These formulae describe the phenomenon of resonance: a sharp increase in the intensity of the field excited when the resonant frequency of an eigenoscillation and the frequency of the source coincide. In this case the spatial structure of the field of forced oscillations is practically the same as that of the resonance mode. In the presence of losses the eigenfrequencies $\omega_s(\omega)$ are complex ones: $\omega_s(\omega) = \omega_s'(\omega) - j\omega_s''(\omega)$. The eigenfrequency ω_s can be presented in the form

$$\omega_s(\omega) = \omega_s'(\omega)\left(1 - j\frac{1}{2Q_s^e}\right)$$

where $Q_s^e(\omega) = \omega_s'/2\omega_s''$ is the eigen Q-factor of the resonator. This value directly determines the field amplitude at the resonance:

$$\max_{\omega} |A_s(\omega)| \simeq \frac{1}{m_s Q_s^e(\omega)\omega_s'}\left|\int_V \boldsymbol{J}^E \boldsymbol{E}_s \,\mathrm{d}V\right|$$

Let us dwell on the derivation of the formula for unloaded Q-factor of the cavity considered. We introduce a system of eigenfunctions $\boldsymbol{e}_q^0, \boldsymbol{h}_q^0$ of the identical resonator without losses ($\varepsilon'' = \mu'' = 0$). Having added corresponding potential functions to these systems, we obtain the complete systems $\{\boldsymbol{e}_q\}$, $\{\boldsymbol{h}_q\}$ in terms of which the fields sought can be expanded. Let us present the eigenoscillations in the form

$$\boldsymbol{E}_s = \sum_q^{\infty} a_q^{(s)} \boldsymbol{e}_q \qquad \boldsymbol{H}_s = \sum_q^{\infty} b_q^{(s)} \boldsymbol{h}_q \tag{4.37}$$

The system of projection relations is

$$\left.\begin{aligned}\omega_p^0 \int_V \varepsilon'(\boldsymbol{r})\boldsymbol{e}_p\boldsymbol{E}_s\,\mathrm{d}V + \omega_s \int_V \mu(\boldsymbol{r})\boldsymbol{h}_p\boldsymbol{H}_s\,\mathrm{d}V = 0\\ \omega_p^0 \int_V \mu'(\boldsymbol{r})\boldsymbol{h}_p\boldsymbol{H}_s\,\mathrm{d}V + \omega_s \int_V \varepsilon(\boldsymbol{r})\boldsymbol{e}_p\boldsymbol{E}_s\,\mathrm{d}V = 0\end{aligned}\right\} \tag{4.38}$$

where ω_p^0 are eigenfrequencies of the corresponding lossless resonator.

Substituting eqn. 4.37 into eqn. 4.38 we obtain a system of equations for $a_p^{(s)}$ and $b_p^{(s)}$:

$$\left.\begin{aligned}\omega_p^0 a_p^{(s)} - \omega_s b_p^{(s)} + j\omega_s \sum_q b_q^{(s)} \frac{\Lambda_{pq}^{\mu}}{m_p} = 0\\ -\omega_p^0 b_p^{(s)} + \omega_s a_p^{(s)} + j\omega_s \sum_q a_q^{(s)} \frac{\Lambda_{pq}^{\varepsilon}}{m_p} = 0\end{aligned}\right\} \tag{4.39}$$

where

$$\Lambda_{pq}^{\mu} = \int_V \mu''(\boldsymbol{r})\boldsymbol{h}_p\boldsymbol{h}_q\,\mathrm{d}V$$

$$\Lambda_{pq}^{\varepsilon} = \int_V \varepsilon''(\boldsymbol{r})\boldsymbol{e}_p\boldsymbol{e}_q\,\mathrm{d}V$$

$$m_p = \int_V \varepsilon'(\boldsymbol{r})\boldsymbol{e}_p^2\,\mathrm{d}V = -\int_V \mu'(\boldsymbol{r})\boldsymbol{h}_p^2\,\mathrm{d}V$$

The nontrivial solutions of the system represented by eqn. 4.39 determine the eigenoscillations in the resonator with a dissipative filling. The complex eigenfrequencies can be calculated as roots of the corresponding determinantal equation. When the losses are small and the frequency ω_s is not degenerated and is not near degenerate, we can obtain the following characteristic equation from eqn. 4.39:

$$(\omega_s^0)^2 - \omega_s^2\left(1 - \frac{j\Lambda_{ss}^{\mu}}{m_s}\right)\left(1 + \frac{j\Lambda_{ss}^{\varepsilon}}{m_s}\right) = 0 \tag{4.40}$$

Assuming that $\Lambda_{ss}^{\varepsilon,\mu}/m_s \ll 1$ (the condition when the losses are small) we obtain from eqn. 4.40 the approximate formula for ω_s

$$\omega_s \simeq \omega_s^0\left\{1 - j\frac{(\Lambda_{ss}^{\varepsilon} - \Lambda_{ss}^{\mu})}{2m_s}\right\} \tag{4.41}$$

The real part of ω_s determines the resonant frequency:

$$\mathrm{Re}\,\omega_s = \omega_s^0$$

The imaginary part of ω_s determines the unloaded Q-factor of the resonator; using eqn. 4.41 it can be presented as

$$Q_s^u(\omega) = \frac{\int_V \varepsilon'(\boldsymbol{r}, \omega)\{\boldsymbol{e}_s^0(\omega)\}^2 \, dV}{\int_V [\varepsilon''(\boldsymbol{r}, \omega)\{\boldsymbol{e}_s^0(\omega)\}^2 + \mu''(\boldsymbol{r}, \omega)\{\boldsymbol{h}_s^0(\omega)\}^2] \, dV} \tag{4.42}$$

Eqn. 4.42, like eqn. 4.15, is a result of an interpretation of the losses as perturbations (see Section 4.2). For degenerated modes eqns. 4.15 and 4.42 are not applicable. The corresponding generalisation can be carried out in a similar way to that in Section 4.2.

Without dispersion eqns. 4.35 and 4.42 coincide; they determine the same value ($Q_s^u = Q_s^f$) that can be presented as

$$Q = \omega_0 \frac{\text{stored energy}}{\text{loss power}} \tag{4.43}$$

With a dispersive filling of the resonator $Q_s^u \neq Q_s^f$. These quantities represent the different characteristics of the losses in the resonator: Q_s^u specifies the transmission band in the regime of forced-oscillation mode, and Q_s^f the intensity of attenuation of oscillations in the resonator after the sources are switched off. Neither Q_s^u nor Q_s^f can be presented by eqn. 4.43 if dispersion is taken into account. Moreover, for dispersive dissipative media the numerator and denominator in eqn. 4.43, as pointed out by Agranovich and Ginzburg (1979), cannot be determined correctly at all within the framework of phenomenological electrodynamics. When the losses are small, we can substitute $\boldsymbol{E}_{s0}^f$ and $\boldsymbol{H}_{s0}^f$, corresponding to a given resonator without losses [at $\varepsilon''(\omega) = \mu''(\omega) = 0$], for $\boldsymbol{E}_s^f$ and $\boldsymbol{H}_s^f$ in eqn. 4.35. In this case the numerator in eqn. 4.35 can be interpreted as the energy, stored by the sth free oscillation, in a similar resonator without losses.

When the losses occur in both the walls and the magnetodielectric filling, it is again possible to make an analysis similar to that described above. We do not present it here and only point out that within the framework of applicability of the perturbation method the complete losses can be found as a sum of partial losses, so that the complete Q-factor is expressed as

$$\frac{1}{Q^{\Sigma}} = \frac{1}{Q^c} + \frac{1}{Q^m} \tag{4.44}$$

where Q^m is the 'magnetodielectric' Q-factor.

Thus the partial Q-factors—ohmic and 'magnetodielectric'—can be determined independently of each other, assuming that the losses of each type are the only losses under consideration, and the complete Q-factor is determined through the partial Q-factors according to eqn. 4.44.

When there are losses of a different physical nature—due to radiation into free space, coupling to a load etc.—each of these types of losses can also be characterised by a partial Q-factor determined independently from other partial Q-factors and included in the complete Q-factor according to eqn. 4.44. Note that the condition of additivity of losses is not reduced to the smallness of all partial losses. The additivity takes place only when there is no coupling of degenerate modes due to losses.

Eqn. 4.44 is not valid when losses of different physical nature are interdependent. This is the case, for example, when we have a cavity coupled with regular semi-infinite below-cutoff waveguides. When ohmic losses in waveguides are absent then there are no radiation losses either but the finite conductivity of the waveguide walls also accounts for radiation losses.

4.4 *Q*-factor of a cylindrical cavity

In this Section we begin a description of dissipative properties of simply shaped cavity resonators: cylindrical, spherical and conical. Let us underline that we deal with calculation of the *Q*-factor characterising the losses in conducting walls (the cavities have no filling inside).

As in the determination of attenuation in waveguides (Section 3.7) it is necessary to make a choice between two approaches: the energy-perturbation method based on eqn. 4.14 and the impedance-perturbation technique for solving the characteristic equation of the resonator with finitely conducting walls. In principle, both methods are equivalent, but we have to choose the one which implies the simpler calculations and more illustrative representations of the final results. Note that for resonators of simple shapes the eigenfields are determined by the method of separation of variables and are expressed in terms of well known special functions (in particular, the integrals in eqn. 4.14 are calculated analytically). Eqn. 4.14 for the *Q*-factor can be presented in the form $Q = q/k_0\Delta_0$, where q is a dimensionless coefficient which does not depend on the material of the resonator and is determined by its shape and the type of oscillations. Later we shall analyse expressions for this coefficient.

In the case of perfect conductivity of the walls there are two classes of eigenoscillations in a cylindrical resonator: TM_{mnp} and TE_{mnp}. For the TM_{mnp} oscillations

$$k_0 = \{(\nu_{mn}/b)^2 + (p\pi/l)^2\}^{1/2} \qquad (m, p = 0, 1, \ldots;\ n = 1, 2, \ldots)$$

and

$$\left.\begin{aligned} H_r &= \frac{jmk_0}{r\nu_{mn}^2 W_0}\mathcal{J}_m\left(\frac{\nu_{mn}}{b}r\right)\begin{Bmatrix} -\sin(m\phi) \\ \cos(m\phi)\end{Bmatrix}\cos\left(\frac{p\pi z}{l}\right) \\ H_\phi &= -\frac{jk_0}{\nu_{nm}W_0 b}\mathcal{J}'_m\left(\frac{\nu_{mn}}{b}r\right)\begin{Bmatrix}\cos(m\phi) \\ \sin(m\phi)\end{Bmatrix}\cos\left(\frac{p\pi z}{l}\right) \\ H_z &= 0 \end{aligned}\right\} \qquad (4.45)$$

where b and l are the radius and the length of the cylinder, respectively. For the TE_{mnp} oscillations

$$k_0 = \{(\mu_{mn}/b)^2 + (p\pi/l)^2\}^{1/2} \qquad (m = 0, 1, \ldots;\ n, p = 1, 2, \ldots)$$

and

$$\left.\begin{aligned}
H_r &= \frac{p\pi b}{l\mu_{mn} W_0} \mathcal{J}'_m\left(\frac{\mu_{mn}}{b} r\right)\left\{\begin{matrix}\cos(m\phi)\\ \sin(m\phi)\end{matrix}\right\}\cos\left(\frac{p\pi z}{l}\right)\\
H_\phi &= \frac{p\pi m b^2}{l\mu_{nm}^2 W_0 r} \mathcal{J}_m\left(\frac{\mu_{mn}}{b} r\right)\left\{\begin{matrix}-\sin(m\phi)\\ \cos(m\phi)\end{matrix}\right\}\cos\left(\frac{p\pi z}{l}\right)\\
H_z &= \frac{1}{W_0} \mathcal{J}_m\left(\frac{\mu_{mn}}{b} r\right)\left\{\begin{matrix}\cos(m\phi)\\ \sin(m\phi)\end{matrix}\right\}\sin\left(\frac{p\pi z}{l}\right)
\end{aligned}\right\} \qquad (4.46)$$

We will not write out the electric components of eigenfields here as we do not need them for the Q-factor determination.

The TE_{0np} and TM_{1np} oscillations are degenerate with any relationships between the radius and the length of the resonator. Moreover, for some ratios b/l other modes can also be degenerate. However, it is possible to show that all degeneration points in a cylindrical resonator are of trivial multiplicity (the mutual Q-factor of degenerate oscillations is infinitely large). This means that the Q-factor of a cylindrical resonator for any mode can be determined by means of the energy-perturbation method according to eqn. 4.14. Using eqns. 4.45, 4.46, and 4.14 we obtain

$$q = \frac{\{v_{mn}^2 + (p\pi b/l)^2\}^{1/2}}{1 + b(2 - \delta_{0p})/l} \qquad (4.47)$$

for the TM_{mnp} oscillations and

$$q = \frac{\{\mu_{mn}^2 + (p\pi b/l)^2\}^{3/2}(1 - m^2/\mu_{mn}^2)}{\mu_{mn}^2 + (2b/l)(p\pi b/l)^2 + \{1 - (2b/l)\}(mp\pi b/\mu_{mn} l)^2} \qquad (4.48)$$

for the TE_{mnp} oscillation.

This does not mean, however, that the points of trivial multiplicity should not be taken into consideration at all. Though degenerate oscillations in these points are not coupled due to ohmic losses they can be coupled on account of the skewness and deformation of the walls, adjustment elements etc. These irregularities are usually rather small but they may result in a noticeable interaction of degenerate oscillations. This is why the actual values of the Q-factors for all degenerate oscillations can differ considerably from those calculated by eqn. 4.14. A detailed study of the problem of mode interaction in cavity resonators was carried out by Shteinshleyger (1955).

In Figures 4.2–4.4 the dependencies of q on $2b/l$ for different modes are shown. As expected, the TE_{0np} oscillations are characterised by the largest values of q. It is interesting that for these oscillations the Q-factor is maximal when $2b = l$ for any n and p. The magnitude of this maximum grows slowly with n and p increasing.

Eqns. 4.47 and 4.48 can also be obtained using the method of separation of variables for solving the interior-impedance problem for a cylindrical domain. In this case the original eigenvalue problem is reduced to the transcendental characteristic equations for complex eigenfrequencies. The solution of the equations obtained by means of the impedance-perturbation method results in eqns. 4.47 and 4.48. However, it is necessary to take into account here the fact

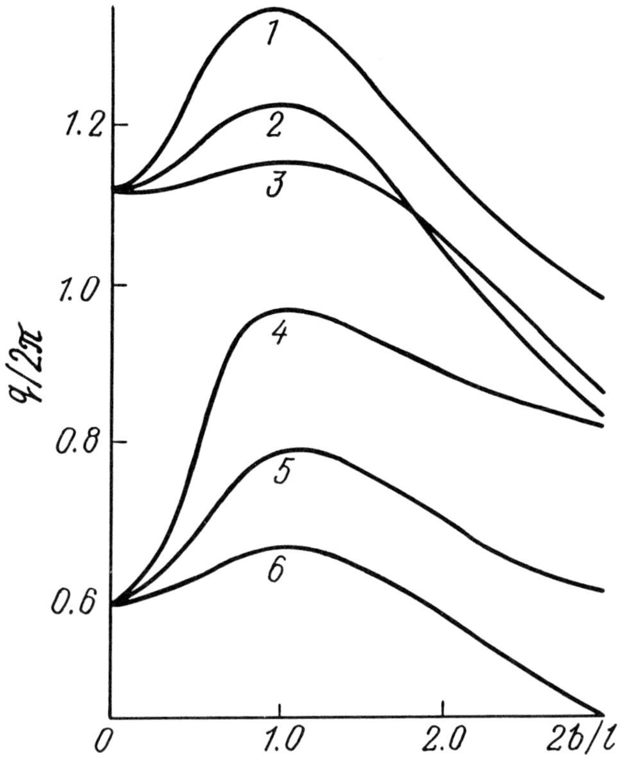

Figure 4.2 *Ohmic Q-factor of cylindrical cavity for various modes*

1 TE_{023} 2 TE_{022} 3 TE_{021} 4 TE_{013}
5 TE_{012} 6 TE_{011}

that in an impedance cylindrical resonator the oscillations with $m \neq 0$ are not purely transverse magnetic and transverse electric (all components of the electric and magnetic fields are nonzero).

4.5 *Q*-factor of spherical and conical cavities

Apart from the cylindrical resonators examined in the previous Section, other types of cavity resonator can be treated by the method of separation of variables: rectangular resonators (see Section 4.2), resonators in the shape of a sphere, spherical cone, oblate and prolate spheroids (angular symmetric modes) etc. Below we present some of the results obtained using this method.

Spherical resonator: TM oscillations: Separation of variables is carried out in the spherical polar co-ordinates ρ, ϕ, ϑ; the origin of the co-ordinate system is in the centre of the sphere. The class of eigenoscillations under examination is characterised by the condition $H_\rho = 0$. If the Debye potential is introduced, the

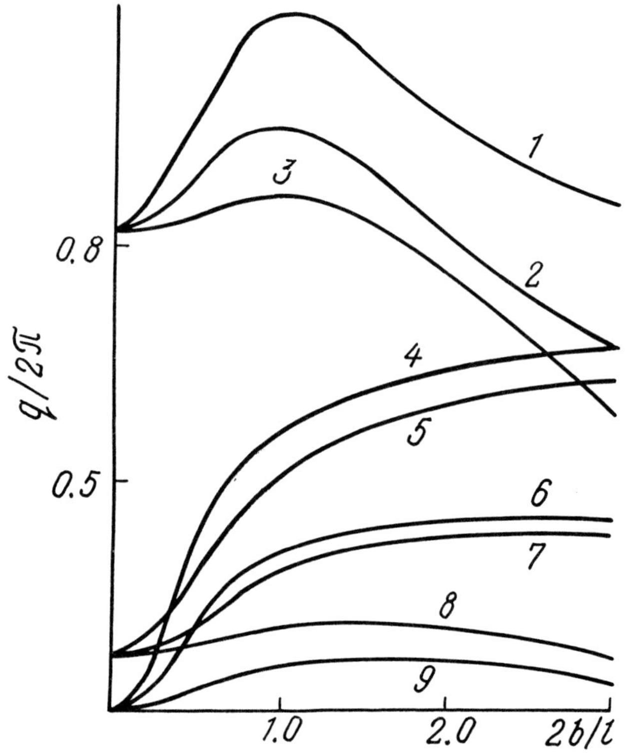

Figure 4.3 *Ohmic Q-factor of cylindrical cavity for various modes*

1 TE_{123}	2 TE_{122}	3 TE_{121}	4 TE_{113}	5 TE_{213}
6 TE_{112}	7 TE_{212}	8 TE_{211}	9 TE_{111}	

problem is reduced to a scalar wave equation which can be easily solved by the method mentioned above. The magnetic components of the TM_{mnp} oscillation in the approximation of perfect conductivity of the walls are

$$\left.\begin{aligned} H_\rho &= 0 \\ H_\vartheta &= \frac{jkm(k\rho)^{1/2}}{\rho \sin \vartheta W_0} J_{n+1/2}(k\rho) P_n^m(\cos \vartheta) \begin{Bmatrix} \cos(m\phi) \\ -\sin(m\phi) \end{Bmatrix} \\ H_\phi &= -\frac{jk(k\rho)^{1/2}}{\rho W_0} J_{n+1/2}(k\rho) \frac{d}{d\vartheta} P_n^m(\cos \vartheta) \begin{Bmatrix} \sin(m\phi) \\ \cos(m\phi) \end{Bmatrix} \end{aligned}\right\} \tag{4.49}$$

where $P_n^m(x)$ are associated Legendre functions of the first kind.

The resonant wave number k_0 of the TM_{mnp} oscillation is determined as the pth root of the characteristic equation

$$\frac{J_{n+1/2}(k_0 R)}{J_{n-1/2}(k_0 R)} = \frac{k_0 R}{n} \tag{4.50}$$

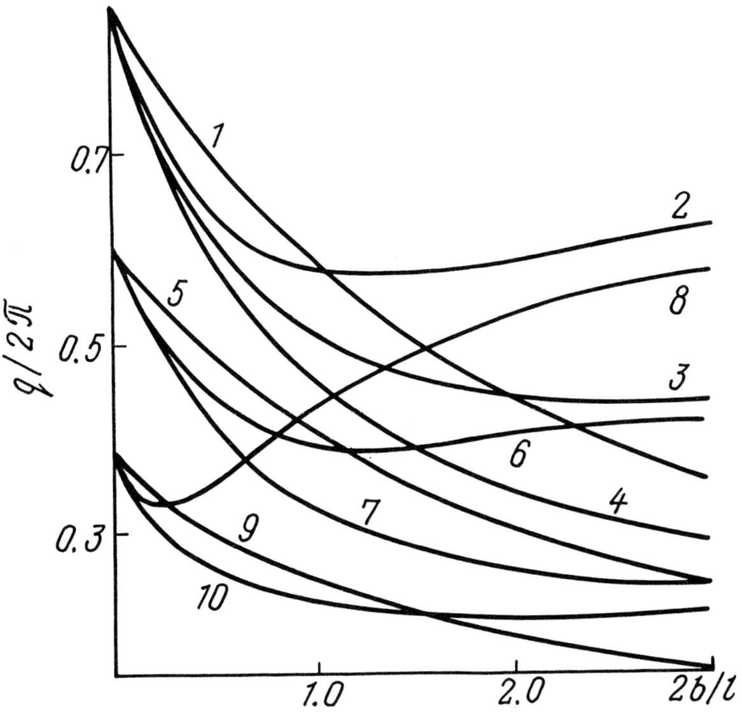

Figure 4.4 *Ohmic Q-factor of cylindrical cavity for various modes*

1	TM_{020}	2	TM_{023}	3	TM_{022}	4	TM_{021}	5	TM_{110}
6	TM_{112}	7	TM_{111}	8	TM_{013}	9	TM_{011}	10	TM_{010}

where R is the radius of the sphere. Eqn. 4.50 follows from the boundary condition $E_\phi = 0$ at $r = R$.

The characteristic equation given by eqn. 4.50 does not contain the azimuthal index m. Thus every eigenvalue is of infinite multiplicity. As a result of some disturbance the degeneration can be removed; in this case the eigenfrequencies split. These questions were studied using methods of group theory (Sapogova and Kontorovich, 1971, Gaplevskii and Kontorovich, 1971). It is easy to prove that in this case the degeneration has a trivial character.

The Q-factor for this type of oscillation can be found by the energy-perturbation method according to eqn. 4.14. Substituting eqn. 4.49 into eqn. 4.14 we obtain after integration

$$q = \left\{1 - \frac{n(n+1)}{(k_0 R)^2}\right\} k_0 R$$

This formula coincides with that obtained by Fel *et al.* (1962). They solved the interior problem for an impedance sphere by the method of separation of variables and applied the impedance-perturbation method for solving the characteristic equation obtained. Fel *et al.* (1962) pointed out, in particular, that the corresponding formula in the well known book by Smythe (1950) was incorrect.

Spherical resonator: TE oscillations: The solution procedure is completely similar to the previous case. The distributions of magnetic components of the TE_{mnp} eigenoscillation in the case of perfectly conducting walls are

$$\left.\begin{aligned} H_\rho &= \frac{n(n+1)(k\rho)^{1/2}}{\rho^2 W_0} \mathcal{J}_{n+1/2}(k\rho) P_n^m(\cos\vartheta) \begin{Bmatrix} \sin(m\phi) \\ \cos(m\phi) \end{Bmatrix} \\ H_\vartheta &= \frac{1}{\rho W_0} \frac{d}{d\rho} [\{\sqrt{(k\rho)}\}\mathcal{J}_{n+1/2}(k\rho)] \frac{d}{d\vartheta} P_n^m(\cos\vartheta) \begin{Bmatrix} \sin(m\phi) \\ \cos(m\phi) \end{Bmatrix} \\ H_\phi &= \frac{m}{\rho W_0 \sin\vartheta} \frac{d}{d\rho} [\{\sqrt{(k\rho)}\}\mathcal{J}_{n+1/2}(k\rho)] P_n^m(\cos\vartheta) \begin{Bmatrix} \cos(m\phi) \\ -\sin(m\phi) \end{Bmatrix} \end{aligned}\right\} \tag{4.51}$$

The resonance wave number k_0 of the TE_{mnp} oscillation is determined as the pth root of the characteristic equation

$$\mathcal{J}_{n+1/2}(k_0 R) = 0 \tag{4.52}$$

Using eqns. 4.14, 4.51 and 4.52, we obtain the expression for the coefficient q

$$q = k_0 R \tag{4.53}$$

Thus, for example, for the TE_{011} oscillation $k_0 R = 4.4934$, i.e. the Q-factor of a spherical resonator is the largest among those of resonators of simple configurations. For comparison, for a cylindrical resonator with the same type of oscillations $q_{max} = 4.14$ which is achieved when the length and diameter of the cylinder are equal.

In some works it is suggested that spherical resonators have the highest Q-factor possible. This statement is supposed to be based on the approximate formula

$$Q \sim \frac{V}{\Delta_0 S} \tag{4.54}$$

where V and S are the volume of the resonator and its inner surface area (at a given V the sphere has a minimal S). We would like to stress, however, that such considerations are not very convincing. Eqn. 4.54 evaluates only the order of magnitude of Q and can be used to draw a rough comparison between cavity resonators and other classes of resonance systems (for example with quasistationary or quasioptical resonators) as regards their Q-factors. The nonuniformity and the specific features of the induced current-density distribution over S are not reflected in eqn. 4.54. This is why it is impossible to compare Q-factors of cavity resonators of different shapes using this formula.

Conical resonator: TE_{0np} oscillations: For this type of oscillation only the components E_ϕ, H_ρ, H_ϑ differ from zero. The separation of variables is carried out in the spherical polar co-ordinates with the origin in the cone vertex. The expressions for fields are

$$\left.\begin{aligned} E_\phi &= j\frac{k}{\rho}\{\sqrt{(k\rho)}\}J_\nu(k\rho)\frac{d}{d\vartheta}P_{\nu-1/2}(\cos\vartheta) \\ H_\rho &= \frac{1}{\rho^2 W_0}\left(\nu^2-\frac{1}{4}\right)\{\sqrt{(k\rho)}\}J_\nu(k\rho)P_{\nu-1/2}(\cos\vartheta) \\ H_\vartheta &= \frac{1}{\rho W_0}\frac{d}{d\rho}[\{\sqrt{(k\rho)}\}J_\nu(k\rho)]\frac{d}{d\vartheta}P_{\nu-1/2}(\cos\vartheta) \end{aligned}\right\} \tag{4.55}$$

where $P_{\nu-1/2}(\cos\vartheta) = P^0_{\nu-1/2}(\cos\vartheta)$, and ν is determined from the boundary condition $E_\phi = 0$ at $\vartheta = \vartheta_0$. The latter can be written in the form

$$\frac{d}{d\vartheta}P_{\nu-1/2}(\cos\vartheta) = 0$$

The resonant wave numbers satisfy the following equation

$$J_\nu(k_0 R) = 0 \tag{4.56}$$

For simplicity let us consider only small flare angles: $\vartheta_0^2/2 \ll 1$. Then

$$\nu = (\mu_{0n}^2/\vartheta_0^2 + \tfrac{1}{4})^{1/2}, \qquad P_{\nu-1/2}(\cos\vartheta) \simeq J_0\{\vartheta(\nu^2-\tfrac{1}{4})^{1/2}\}$$

and the expressions for the field components given by eqn. 4.55 can be rewritten as

$$\left.\begin{aligned} E_\phi &= \frac{jk}{\rho}\frac{\mu_{0n}}{\vartheta_0}(k\rho)^{1/2}J_\nu(k\rho)J_1\left(\mu_{0n}\frac{\vartheta}{\vartheta_0}\right) \\ H_\rho &= \frac{\mu_{0n}^2}{\vartheta_0^2\rho^2 W_0}(k\rho)^{1/2}J_\nu(k\rho)J_0\left(\mu_{0n}\frac{\vartheta}{\vartheta_0}\right) \\ H_\vartheta &= -\frac{\mu_{0n}}{W_0\vartheta_0}\frac{d}{\rho d\rho}\{(k\rho)^{1/2}J_\nu(k\rho)\}J_1\left(\mu_{0n}\frac{\vartheta}{\vartheta_0}\right) \end{aligned}\right\} \tag{4.57}$$

Using eqns. 4.14 and 4.57 and the equality

$$\begin{aligned} \int_0^R \frac{J_\nu^2(k\rho)}{\rho^2}d\rho = k\Bigg\{&\frac{kR\{J_{\nu+1}(kR)J_{\nu-1}(kR) - J_\nu^2(kR)\}}{2\nu+1} \\ &+ \frac{J_\nu(kR)J_{\nu+1}(kR)}{2\nu+1} - \frac{J_\nu(kR)J_{\nu-1}(kR)}{2\nu-1} \\ &+ \frac{kR\{J_{\nu-1}^2(kR) - J_\nu(kR)J_{\nu-2}(kR)\}}{2\nu-1}\Bigg\} \\ &- \frac{J_\nu^2(kR)}{R} \end{aligned}$$

we obtain the formula

$$q = \frac{k_0 R\vartheta_0}{1+\vartheta_0} \tag{4.58}$$

for the coefficient q.

For negligible ϑ_0 ($\vartheta_0 \ll 1$) we obtain, from eqn. 4.58,

$$q \simeq k_0 R \vartheta_0$$

Physically the transition from eqn. 4.58 to this formula corresponds to neglecting the losses on the cone end wall ($\rho = R$).

For the resonance wave number k_0, from eqn. 4.56 we obtain the relation

$$k_0 R \simeq \nu + 1.856\nu^{1/3} + 1.033\nu^{-1/3} + \cdots \tag{4.59}$$

the accuracy of which increases for decreasing ϑ_0.

The calculations according to eqns. 4.58 and 4.59 show that in the angle range $5^0 < \vartheta_0 < 15^0$ $q \simeq 4$ and depends weakly on ϑ_0.

It is necessary to make the following comment. Unlike the case of a cylinder and sphere, for a cone with impedance boundary conditions the method of separation of variables cannot be applied. This is why the energy-perturbation method based on eqn. 4.14 is practically the only possible simple method for determining the Q-factor of a conical resonator.

Of other resonators to which the method of separation of variables can be applied, we should also mention prolate and oblate spheroidal cavities. The separation of variables is carried out in prolate and oblate spheroidal coordinates, respectively, for angular symmetric modes only. The eigenfields are expressed in terms of wave spheroidal functions. Unfortunately, the integrals for these functions in eqn. 4.14 are not evaluated in analytical form. Numerical calculation of these integrals is not simple to perform, making numerical methods relevant for resonators of more complicated shapes than those examined above. One such method will be described in Section 4.6.

Another kind of resonator of practical importance is the quasi-optical open resonator. The theory of such resonators is based on the parabolic-equation method (Weinstein, 1969*a*). The ohmic Q-factor can be determined by the energy-perturbation method according to eqn. 4.14. The volume V in eqn. 4.14 is understood as the region of the field localisation with the caustic surfaces as boundaries, the surface S being the illuminated part of the mirrors.

Calculations of this nature were carried out by Hargreaves *et al.* (1991). According to them, for a resonator with spherical mirrors

$$q = \tfrac{1}{2} k_0 d \tag{4.60}$$

where d is the distance between the mirrors. It should be underlined that eqn. 4.60 does not allow us to calculate the complete Q-factor because it ignores radiation losses. The question of the radiation Q determination has been described at length (see, for example, Weinstein, 1969*a*) and we shall not dwell upon it here.

Generally speaking, resonators of this type are characterised by high unloaded Q-factors. According to the experimental data obtained by Kuraev *et al.* (1991) the complete unloaded Q at the band $f \simeq 83$–117 GHz is of the order of 10^5.

4.6 Galerkin's method for calculation of a complex-shaped cavity resonator in the form of a body of revolution

Let us consider a cavity resonator in the form of a body of revolution with an arbitrary continuous single-valued generatrix $b(z)$ (Figure 4.5). Let us examine the TE_{0np} oscillations characterised by the field components E_ϕ, H_r, H_z:

$$\boldsymbol{E} = \hat{i}_\phi \Phi \tag{4.61}$$

$$\boldsymbol{H} = \frac{1}{j\omega\mu_0} \operatorname{curl}(\hat{i}_\phi \Phi) \tag{4.62}$$

The analysis presented is based on a variant of Galerkin's method proposed by Slepyan (1977). The objective of the analysis, unlike that of Sections 4.4 and 4.5, is to investigate resonators of essentially different shapes within the framework of a single approach. A special set of basis functions is used, each of which satisfies precisely the boundary condition for arbitrary $b(z)$. Such functions can easily be constructed for the boundary condition of perfect conductivity of the cavity material. However, such a construction is much more difficult to perform for impedance boundary conditions, so the Q-factor evaluation here will be carried out by means of the energy-perturbation method, on the basis of eqn. 4.14.

Formulation of the eigenvalue problem is as follows: it is necessary to find nontrivial solutions of the equation

$$\hat{L}\Phi = k^2 \Phi \tag{4.63}$$

where $\hat{L} = -\nabla_{rz}^2 + r^{-2}$. The boundary condition for Φ is

$$\Phi(r, z)|_S = 0 \tag{4.64}$$

where S is the inner surface of the resonator.

Let us introduce a system of functions

$$\Phi_\lambda(r, z) = \zeta(z)\varepsilon_i^{-1} J_1\left(\frac{\mu_{0i}}{b(z)} r\right) \sin\left(\frac{\pi m z}{l}\right) \tag{4.65}$$

where $\lambda = \{i, m\}$, $i, m = 1, 2, \ldots$, $\zeta(z)$ is a function, different from zero at the interval $0 \le z \le l$ with the exception of a finite number of isolated points, and ε_i are constant multipliers.

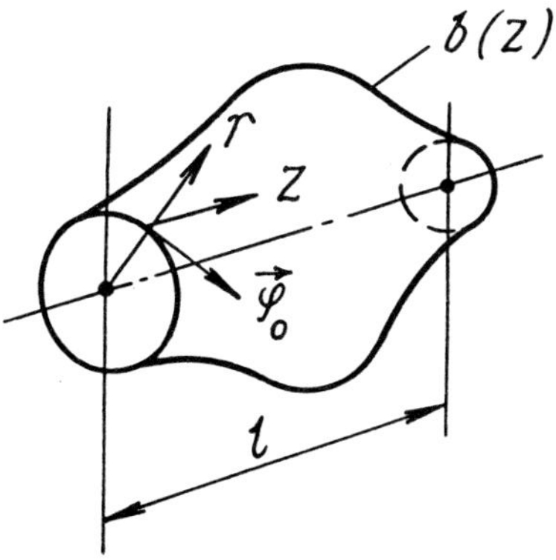

Figure 4.5 *Cavity resonator in the form of a body of revolution with an arbitrary generatrix*

It is easy to see that the functions Φ_λ satisfy the boundary condition given by eqn. 4.64. Following the general idea of Galerkin's method, we replace eqn. 4.63 by projection relationships

$$\int_0^l \int_0^{b(z)} (-\hat{L}\Phi + k^2\Phi)\Phi_{\lambda'} r \, dr \, dz = 0 \tag{4.66}$$

representing Φ as

$$\Phi = \Phi^N = \sum_{\lambda}^{N} C_\lambda \Phi_\lambda \tag{4.67}$$

where $\lambda' = \{i', m'\}$, $N = \{I, M\}$, and C_λ are unknown coefficients. Substituting eqn. 4.67 into eqn. 4.66 we obtain the following matrix eigenvalue problem

$$\mathfrak{B}\boldsymbol{C} = (k^N)^2 \mathfrak{M}\boldsymbol{C} \tag{4.68}$$

where $\boldsymbol{C}$ is an N-dimensional vector of the coefficients C_λ to be found, and $\mathfrak{B}$ and $\mathfrak{M}$ are matrices of the order N. The structure of $\boldsymbol{C}$ and $\mathfrak{B}$ is as follows:

$$\boldsymbol{C} = \left\| \begin{matrix} C_{11} \\ \cdots \\ C_{1M} \\ \hline \cdots \\ \hline C_{I1} \\ \cdots \\ C_{IM} \end{matrix} \right\|, \quad \mathfrak{B} = \left\| \begin{matrix} \boxed{\mathfrak{B}_1^1} & \cdots & \boxed{\mathfrak{B}_{i'}^1} & \cdots & \boxed{\mathfrak{B}_I^1} \\ \cdots & \cdots & \cdots & \cdots & \cdots \\ \boxed{\mathfrak{B}_1^i} & \cdots & \boxed{\mathfrak{B}_{i'}^i} & \cdots & \boxed{\mathfrak{B}_I^i} \\ \cdots & \cdots & \cdots & \cdots & \cdots \\ \boxed{\mathfrak{B}_1^I} & \cdots & \boxed{\mathfrak{B}_{i'}^I} & \cdots & \boxed{\mathfrak{B}_I^I} \end{matrix} \right\|$$

$$\mathfrak{B}_{i'}^i = \left\| \begin{matrix} \mathfrak{B}_{i'1}^{i1} & \cdots & \mathfrak{B}_{i'1}^{iM} \\ \cdots & \mathfrak{B}_{i'm'}^{im} & \cdots \\ \mathfrak{B}_{i'M}^{i1} & \cdots & \mathfrak{B}_{i'M}^{iM} \end{matrix} \right\|$$

The structure of the matrix $\mathfrak{M}$ is similar to that of the matrix $\mathfrak{B}$. Matrix elements are expressed by

$$\mathfrak{B}_{\lambda\lambda'} = 2\pi \int_0^l \int_0^{b(z)} \{\nabla\Phi_\lambda \nabla\Phi_{\lambda'} + r^{-2}\Phi_\lambda \Phi_{\lambda'}\} r \, dr \, dz \tag{4.69}$$

$$\mathfrak{M}_{\lambda\lambda'} = 2\pi \int_0^l \int_0^{b(z)} \Phi_\lambda \Phi_{\lambda'} \, r \, dr \, dz \tag{4.70}$$

The inner integrals in eqns. 4.69 and 4.70 can be determined in a closed form. After some transformations we have

$$\mathfrak{B}_{im'}^{im} = \pi \frac{J_0^2(\mu_{0i})}{\varepsilon_i^2} \int_0^l \left(\left[\zeta^2 \left\{ \mu_{0i}^2 + \left(\frac{\mu_{0i}^2}{3} + 1 \right) \left(\frac{db}{dz} \right)^2 \right\} \right. \right.$$
$$\left. + b^2 \left(\frac{d\zeta}{dz} \right)^2 + b \left(\frac{d\zeta^2}{dz} \right) \frac{db}{dz} \right] \sin(\bar{m}z) \sin(\bar{m}'z)$$

$$+ b\zeta\left(\frac{db}{dz}\zeta + b\frac{d\zeta}{dz}\right)\{\bar{m}'\cos(\bar{m}'z)\sin(\bar{m}z) + \bar{m}\cos(\bar{m}z)\sin(\bar{m}'z)\}$$

$$+ \zeta^2 b^2 \bar{m}\bar{m}'\cos(\bar{m}z)\cos(\bar{m}'z)\Big)\,dz$$

$$\mathfrak{B}^{im}_{i'm'} = \frac{2\pi\mu_{0i}\mu_{0i'}J_0(\mu_{0i})J_0(\mu_{0i'})}{(\mu_{0i}^2 - \mu_{0i'}^2)\varepsilon_i\varepsilon_{i'}}\int_0^l \left[2\frac{\mu_{0i}^2 + \mu_{0i'}^2}{\mu_{0i}^2 - \mu_{0i'}^2}\right.$$

$$\times \sin(\bar{m}z)\sin(\bar{m}'z) - \frac{db}{dz}b\zeta^2\{\bar{m}\cos(\bar{m}'z)\sin(\bar{m}z)$$

$$\left. - \bar{m}'\cos(\bar{m}z)\sin(\bar{m}'z)\}\right]dz \qquad i \neq i'$$

$$\mathfrak{M}^{im}_{im'} = \pi\frac{J_0^2(\mu_{0i})}{\varepsilon_i^2}\int_0^l b^2\zeta^2 \sin(\bar{m}z)\sin(\bar{m}'z)\,dz$$

$$\mathfrak{M}^{im}_{i'm'} = 0 \qquad i \neq i'$$

Here $\bar{m} = m\pi/l$ and $\bar{m}' = m'\pi/l$.

To derive the expressions for $\mathfrak{B}^{im}_{i'm}$ and $\mathfrak{M}^{im}_{i'm'}$, we use the formulae

$$\int_0^1 x^2 J_0(\mu_{0i}x)J_1(\mu_{0i'}x)\,dx = \begin{cases} \dfrac{\mu_{0i'}J_0(\mu_{0i})J_0(\mu_{0i'})}{\mu_{0i}^2 - \mu_{0i'}^2} & (i \neq i') \\ 0 & (i = i') \end{cases}$$

$$\int_0^1 x^3 J_0(\mu_{0i}x)J_0(\mu_{0i'}x)\,dx = \begin{cases} \dfrac{2(\mu_{0i'}^2 + \mu_{0i}^2)J_0(\mu_{0i})J_0(\mu_{0i'})}{(\mu_{0i}^2 - \mu_{0i'}^2)^2} & (i \neq i') \\ \frac{1}{6}J_0^2(\mu_{0i}) & (i = i') \end{cases}$$

The matrices $\mathfrak{B}$ and $\mathfrak{M}$ are symmetrical:

$$\mathfrak{B}^{im}_{i'm'} = \mathfrak{B}^{i'm'}_{im} \qquad \mathfrak{M}^{im}_{i'm'} = \mathfrak{M}^{i'm'}_{im}$$

The eigenvalues of the problem given by eqn. 4.68 determine approximately (as N is finite) the resonator eigenfrequencies ω_α^N ($\omega_\alpha^N = k_\alpha^N c$, $\alpha = \{n, p\}$ being the two-dimensional mode index), and the eigenvectors determine electric-field distributions of eigenmodes (in accordance with eqns. 4.61 and 4.67). The magnetic-field components of eigenoscillations are expressed in terms of Φ by eqn. 4.62.

The ohmic Q-factor of the resonator, as already mentioned, is calculated by means of the energy-perturbation method. The final result is

$$q_\alpha = \frac{2k_\alpha^3\chi^{3/2}\displaystyle\int_0^1\int_0^1 [\Phi_{(\alpha)}(x, T)]^2 g^2(T)x\,dx\,dT}{\displaystyle\int_0^1 \left\{f_\alpha(T)\left\{1 + \left(\chi\frac{dg}{dT}\right)^2\right\}^{3/2} + \chi^3\{g_0^2 F_0^\alpha(T) + g_1^2 F_1^\alpha(T)\}\right\}dT}$$

where

$$q_\alpha = k_\alpha \Delta_0 Q_\alpha$$

$$f_\alpha(T) = \left\{ \sum_{i=1}^{I} \sum_{m=1}^{M} C_{im}^{(\alpha)} \frac{\mu_{0i} J_0(\mu_{0i})}{\varepsilon_i \sqrt{g(T)}} \zeta(T) \sin(m\pi T) \right\}^2$$

$$F_{0,1}^{\alpha}(T) = \left\{ \sum_{i=1}^{I} \sum_{m=1}^{M} C_{im}^{(\alpha)} \frac{J_1(\mu_{0i} T)}{\varepsilon_i} m\pi(\sqrt{T}) v_m^{0,1} \right\}^2$$

$$g(T) = b(T)/b_{max} \qquad T = z/l \qquad \chi = b_{max}/l \qquad x = r/b(z)$$

$$g_0 = g(0) \qquad g_1 = g(1) \qquad v_m^0 = \zeta(0) \qquad v_m^1 = (-1)^m \zeta(1)$$

The function $\zeta(z)$ and the coefficients ε_i in eqn. 4.65 can be chosen specifically to improve the efficiency of the computational algorithm, making it possible to enhance the convergence rate or to simplify the formulae for the matrix elements $\mathfrak{B}_{\lambda\lambda'}$ and $\mathfrak{M}_{\lambda\lambda'}$ by presetting the function $\zeta(z)$ in an appropriate manner.

To ensure that the algorithm is correct it is necessary to carry out a mathematical substantiation of all formal operations performed above. We shall not present any detailed description here (see, if necessary, the work by Slepyan, 1977). We merely note that it contains the proof of convergence of approximate values of k^N and q^N to the accurate values at $I, M \to \infty$ and also the condition of algorithm stability against the errors in the matrix-element calculations (the concept of stability is discussed in Section 1.4).

The proof of convergence is based on the properties of linear independence and completeness of the function systems $\{\Phi_\lambda\}$ and $\{\text{curl}(\hat{i}_\phi \Phi_\lambda)\}$ in corresponding functional spaces. To prove the convergence of the Q-factor it is necessary also to substantiate the possibility of a term-by-term differentiation of the series in eqn. 4.67 at $N \to \infty$ (when the magnetic field is determined from eqn. 4.62). It is essential that the matrices $\mathfrak{B}$ and $\mathfrak{M}$ are positive definite. Finally, the necessary and sufficient condition of the algorithm stability, according to Mikhlin (1966), is a strong minimality of the set of functions $\{\Phi_\lambda\}$ in the energy space of the operator $\tilde{L}$. It takes place when the asymptotic evaluation $\varepsilon_i = O(\mu_{0i}^{1/2})$ is fulfilled [it is convenient, in particular, to set $\varepsilon_i = J_0(\mu_{0i})$].

As a test example, the eigenfrequency and the Q-factor of the TE_{011} mode in the spherical resonator have been determined. When the matrix elements $\mathfrak{B}_{\lambda\lambda'}$, $\mathfrak{M}_{\lambda\lambda'}$ for a sphere and similar configurations (spheroid etc.) are being calculated, the following difficulty occurs: at the points $z = 0, l$ $b(z) \to 0$ and $db/dz \to \infty$, which leads to indeterminate forms (the expressions under the integral are finite at these points). The method of evaluation of these indeterminate forms depends additionally upon the choice of the function $\zeta(z)$. In our calculations $\zeta(z) = 1$ and the function under the integral equals zero at $z = 0$ and $z = l$. This is why the end sections with a length of $\Delta T = \Delta l/l$ are ignored and the integration is carried out between the limits Δl and $l - \Delta l$. The calculation results for a sphere at $\Delta T = 0.01$ and different values of I, M are presented in Table 4.1. The values of q and the normalised resonant wave number obtained from eqns. 4.52 and 4.53 are assumed to be accurate. Excellent agreement is observed for even six basis functions ($I = 2$, $M = 3$). Thus, the convergence rate of the algorithm is fairly high.

The eigenfrequencies of the TE_{011}, TE_{012}, TE_{013} modes of prolate and oblate

Table 4.1 *Convergence of normalised resonant wave number and Q-factor for spherical cavity*

Parameter	$I=1$ $M=3$	$I=1$ $M=7$	$I=2$ $M=3$	Accurate value
kl	9.0330	9.0081	8.9885	8.9868
q	4.0091	4.3100	4.4883	4.4934

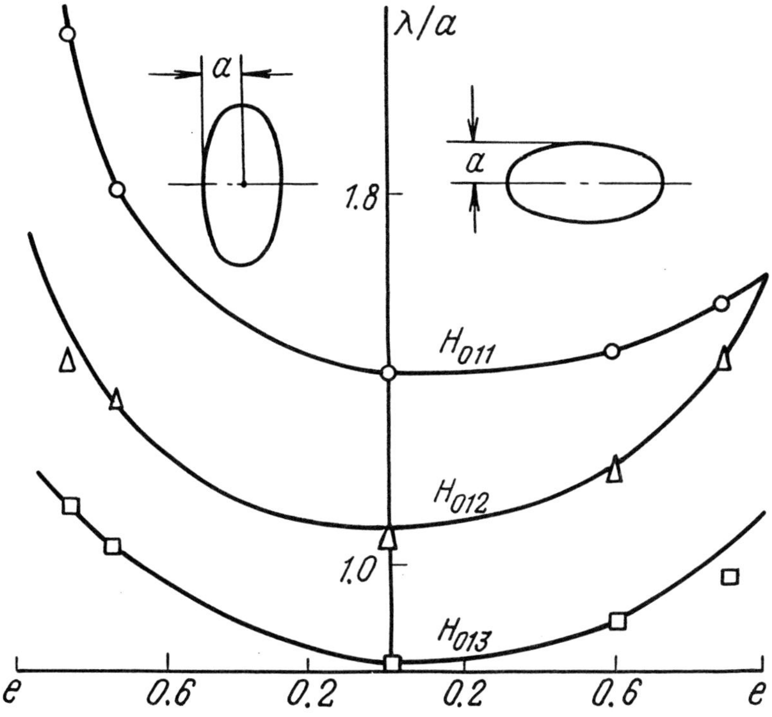

Figure 4.6 *Normalised resonant wavelength of prolate and oblate spheroidal resonators as a function of eccentricity*

spheroidal resonators have also been calculated. In Figure 4.6 the values of $2\pi/ka$ (a being the minor semiaxis) are shown as functions of the eccentricity e. The calculations have been carried out at $I=2$, $M=3$, $\Delta T=0.01$. The solid lines correspond to the data obtained by Semakov and Tereschenko (1975) through the method of separation of variables in spheroidal co-ordinates. Circles, triangles and squares represent the results of our calculations according to the algorithm described in this Section. As in the previous example, there is again very good agreement between the results obtained by the two approaches. It is necessary to stress that we have taken only six basis functions for a wide range of eccentricities of both prolate and oblate spheroids. This allows us to conclude that our algorithm is effective for resonators of various configurations.

Using this algorithm we have determined the electrodynamic characteristics of the TE_{011} mode in a symmetrical truncated biconical resonator. The computational results and the sectional view of this resonator are shown in Figure 4.7 (the solid line being the coefficient q, and the broken line the dimensionless parameter $\alpha = 2\mu_{01}(1 - g_0)/k_0 l \tan \vartheta_0$). The calculations have been carried out for $g_0 = 0.1$, $I = 1$, $M = 7$. Numerical experiments have shown that the increase in I and M does not in practice influence the results described.

Of some interest is the comparison of cavity resonators of simple shapes (cylindrical, spherical, conical, biconical) examined in this Chapter with respect to the Q-factor of the TE_{011} mode. This is because the TE_{0np} modes in cavities in the form of a body of revolution are characterised by the largest values of Q and the TE_{011} mode has the lowest resonant frequency among the TE_{0np} modes. For convenience, we shall compare the values of the coefficient q directly: this corresponds to resonators with the same resonant frequency and the same material of the walls. Spherical resonators have the highest Q of all the types of resonator under study ($q = 4.4934$). For cylindrical cavities the coefficient q equals 4.14. The values of q for conical and biconical resonators are close to that for a cylindrical resonator. The differences in Q, though, are not significant for all types of configuration, i.e. spherical resonators cannot be recommended for use, taking into account the difficulties in their manufacture and frequency retuning. For this reason we cannot yet gain much by way of optimising the generatrix shape of a body of revolution. The cylindrical cavity appears to be promising for applications, but in this case higher modes in combination with

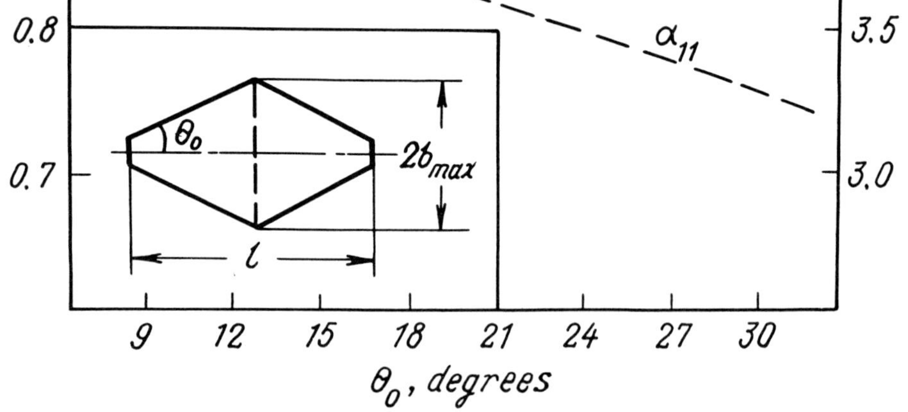

Figure 4.7 *Normalised eigenfrequency and Q-factor of TE_{011} mode in truncated biconical resonator*

some means of spectrum rarefaction (for example, narrow azimuthal slits) should be used.

4.7 Cylindrical resonator with dielectric slabs

In this Section we shall consider one specific case of a cylindrical resonator with a partial dielectric filling. We shall demonstrate that the insertion of dielectric bodies can increase the complete unloaded *Q*-factor for certain types of eigenoscillations (Ilyinsky and Slepyan, 1983).

Consider a cavity resonator of cylindrical shape with two identical dielectric discs made of a material with the permittivity $\varepsilon = \varepsilon' + j\varepsilon''$ ($\varepsilon' \gg \varepsilon''$) and permeability $\mu = \mu_0$. The discs are positioned symmetrically as regards the $z = l/2$ plane (l being the length of the resonator). The geometry of the problem is shown in Figure 4.8(i). Taking account of the fact that the $z = l/2$ plane is a plane of symmetry for the resonator under consideration, the field distributions can be either symmetrical or antisymmetrical with respect to this plane. This allows us to analyse the field in only half the resonator. Let us consider the TE_{0np} oscillations which are described by the azimuthal component of the electric field E_ϕ according to eqns. 4.61 and 4.62. To determine the *Q*-factor we use the energy-perturbation method, based on eqns. 4.14 and 4.42. For this we have to calculate the eigenfields in a lossless resonator of the same shape. Using the method of partial regions [they are denoted by Roman numerals in Figure 4.8(i)], we represent the fields in the form

$$E_\phi^{(\nu)} = C^{(\nu)}(z)\, \mathcal{J}_1\left(\frac{\mu_{0n}}{b} r\right)$$

$$H_z^{(\nu)} = \frac{1}{j\omega\mu_0} C^{(\nu)}(z) \frac{1}{r}\frac{\mathrm{d}}{\mathrm{d}r}\left\{r \mathcal{J}_1\left(\frac{\mu_{0n}}{b} r\right)\right\}$$

$$H_r^{(\nu)} = -\frac{1}{j\omega\mu_0}\frac{\mathrm{d}C^{(\nu)}}{\mathrm{d}z} \mathcal{J}_1\left(\frac{\mu_{0n}}{b} r\right)$$

where ν ($\nu = 1, 2, 3$) is the number of a partial region. The functions $C^{(\nu)}(z)$ are solutions of ordinary differential equations of the form

$$\frac{\mathrm{d}^2 C^{(\nu)}}{\mathrm{d}z^2} + \{k^2 \varepsilon'_{(\nu)} - \mu_{0n}^2\} C^{(\nu)} = 0 \tag{4.71}$$

where $\varepsilon'_{(1)} = \varepsilon'_{(3)} = \varepsilon_0$, $\varepsilon'_{(2)} = \varepsilon'$.

The boundary conditions for p even are

$$\left.\begin{aligned} C^{(1)}(0) &= 0 \\ C^{(1)}(a) - C^{(2)}(a) &= 0 \\ C^{(2)}(a+d) - C^{(3)}(a+d) &= 0 \\ \left(\frac{\mathrm{d}C^{(1)}}{\mathrm{d}z} - \frac{\mathrm{d}C^{(2)}}{\mathrm{d}z}\right)\Bigg|_{z=a} &= 0 \end{aligned}\right\} \tag{4.72}$$

$$\left.\begin{aligned} \left(\frac{\mathrm{d}C^{(2)}}{\mathrm{d}z} - \frac{\mathrm{d}C^{(3)}}{\mathrm{d}z}\right)\bigg|_{z=a+d} &= 0 \\ C^{(3)}(l/2) &= 0 \end{aligned}\right\} \tag{4.73}$$

The last relationship in eqns. 4.73 should be replaced when p is odd by

$$\left.\frac{\mathrm{d}C^{(3)}}{\mathrm{d}z}\right|_{z=l/2} = 0$$

Solution of the eigenvalue problem given by eqns. 4.71, 4.72 and 4.73 leads to the transcendental characteristic equation for resonant-wave numbers which allows us to determine the eigenfrequencies and eigenfields of the resonator. We shall not present this analysis in a general form here. Our objective is to illustrate the typical physical effects in one specific case. Let $d = a$ and

$$ha = \pi/2 \tag{4.74}$$

$$\tilde{h}d = \pi/2 \tag{4.75}$$

where h and $\tilde{h}$ are propagation coefficients of the TE_{0n} mode in circular waveguides of radius b, filled, respectively, with air and dielectric with the permittivity ε':

$$h = (k^2 - \mu_{0n}^2/b^2)^{1/2}$$

$$\tilde{h} = (k^2\tilde{\varepsilon} - \mu_{0n}^2/b^2)^{1/2}$$

where $\tilde{\varepsilon}$ is relative dielectric permittivity.

Omitting simple intermediate calculations, let us write out the final expressions for the electric field in partial regions:

$$\left.\begin{aligned} E_\phi^{(1)} &= B\sin(hz)\,J_1\left(\frac{\mu_{0n}}{b}r\right) \\ E_\phi^{(2)} &= B\cos\{\tilde{h}(z-a)\}\,J_1\left(\frac{\mu_{0n}}{b}r\right) \\ E_\phi^{(3)} &= -B\frac{\tilde{h}}{h}\cos\left\{h\left(z-\frac{l}{2}\right)\right\}J_1\left(\frac{\mu_{0n}}{b}r\right) \end{aligned}\right\} \tag{4.76}$$

where B is an arbitrary constant.

One of eqns. 4.74 and 4.75 can be considered as a characteristic equation [determining the resonant frequency for given a (or d)]; then from the other we find d (or a).

The complete unloaded Q-factor can be obtained (see Section 4.3) from eqn. 4.44

$$Q_\Sigma = \frac{Q^c Q^\varepsilon}{Q^c + Q^\varepsilon} \tag{4.77}$$

where Q^c is expressed by eqn. 4.14, and Q^ε by eqn. 4.42 at $\mu'' = 0$ (dispersive

Table 4.2 *Calculational results for copper cylindrical resonator with polycore slabs*

b (mm)	a (mm)	d (mm)	Q^{ε}	Q^{c}	Q_{Σ}
12	2.180	0.638	5.34×10^4	7.72×10^4	3.16×10^4
14	2.136	0.636	5.11×10^4	8.14×10^4	3.14×10^4
8	2.520	0.645	7.32×10^4	7.32×10^4	3.00×10^4

media are not considered). In this case eqn. 4.42 can be rewritten in the form:

$$Q^{\varepsilon} = \frac{\int_V \varepsilon' E_{\phi}^2 \, dV}{\int_{V_d} \varepsilon'' E_{\phi}^2 \, dV} \tag{4.78}$$

where V_d is the volume of the dielectric filling. In accordance with the energy-perturbation method we shall substitute the expressions of eigenfields in the corresponding lossless resonator (they are given by eqn. 4.76) into eqns. 4.14 and 4.78. Evaluating the integrals in eqn. 4.78, we obtain

$$Q^{\varepsilon} = \frac{1}{\tilde{\varepsilon} \tan \delta} \left[\frac{\tilde{h}}{h} \left\{ 1 + \left(\frac{\tilde{h}}{h} \right)^2 \right\} + \tilde{\varepsilon} \right] \tag{4.79}$$

where δ is the angle of dielectric losses,

$$Q^{c} = \frac{(kb)^2 d \left[\frac{\tilde{h}}{h} \left\{ 1 + \left(\frac{\tilde{h}}{h} \right)^2 \right\} + \tilde{\varepsilon} \right]}{\Delta_0 \left\{ (hb)^2 + \frac{\pi \mu_{0n}^2}{2hb} \left(1 + \frac{h}{\tilde{h}} + \frac{\tilde{h}^2}{h^2} \right) \right\}} \tag{4.80}$$

Table 4.2 presents the results of numerical calculations by eqns. 4.77, 4.79 and 4.80 for copper resonators with polycore slabs ($\tilde{\varepsilon} = 10$, $\tan \delta = 10^{-4}$, $\lambda_0 = 8$ mm, TE_{013} operational mode). For comparison we point out that, for a cylindrical resonator made of copper without dielectric slabs, $Q_{max} \simeq 2 \times 10^4$ ($\lambda_0 = 8$ mm). Thus, the Q-factor of the resonator with dielectric discs is 1.5 times as high as that of the optimal unfilled resonator. When the length of the resonator is significantly larger than its radius, there is no gain in Q.

Physically, this result can be explained by the fact that the main part of the stored electromagnetic energy is concentrated in the central region of the resonator; the electromagnetic field in the peripheral regions and inside the dielectric slabs is comparatively weak [see Figure 4.8(ii)]. That is why the surface currents in the end walls are small, and so are the ohmic losses. So, at $l < b$ Q^c increases, in this case essentially because the losses in the cylindrical wall are smaller. Then the losses in the end walls are considerably reduced owing to the influence of dielectric slabs. Even though the losses in the dielectric filling lead to a decrease in the complete Q-factor, it remains fairly high—higher than the maximum Q obtainable in unfilled resonators at the same operational frequency.

The increase in Q is still higher when using dielectrics with $\tan \delta$ smaller than

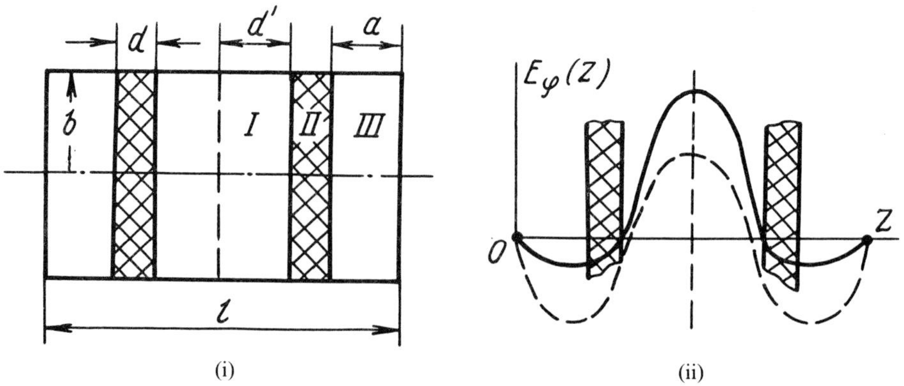

Figure 4.8 *Cylindrical resonator with dielectric slabs*

(i) General view and main designations
(ii) TM_{013}-mode field distribution along the z-axis
——— with dielectric slabs
– – – – without dielectric slabs

those of polycore. Similarly, the Q-factor of the TE_{0np} modes with $p \geq 3$ can be increased. Note that the greatest increase in Q occurs when the conditions given by eqns. 4.74 and 4.75 are fulfilled.

A similar physical mechanism is used by Kazantsev *et al.* (1974) to reduce attenuation of the TE_{0n} mode in a circular waveguide. They studied a circular waveguide with a hollow coaxial dielectric pipe. A reduction in losses is possible compared with those of ordinary circular waveguides.

4.8 *Q*-factor of a cylindrical resonator with a coaxial insert

The configuration of the resonator under examination and the main notations are shown in Figure 4.9. Such resonators are widely used in microwave electronic devices, measuring equipment etc. Many publications have been dedicated to determination of their electrodynamic characteristics. Two types of coaxial inserts were studied, one in the shape of a hollow cylinder with an infinitely thin wall, and the other in the shape of a cylinder with a flat metallised end face. In most published work the walls of the resonator were supposed to be perfectly conducting; the Q-factor was not determined.

A rough (as to the order) evaluation of Q for the fundamental mode has been made by Kleev and Manenkov (1982). The problem of calculating the Q-factor of the TM_{0np} oscillations was considered in papers by Gubsky *et al.* (1982), Kleev and Manenkov (1984) and Rodionova and Slepyan (1986). In the first two papers the energy-perturbation method was used, while the third dealt with solving the impedance problem for a given resonator. Some results on the Q-factor of cylindrical resonators with coaxial inserts, and explanatory notes to them, can be found in the survey by Ilyinsky and Slepyan (1990).

When using the methods of integral equations and of partial regions, calcu-

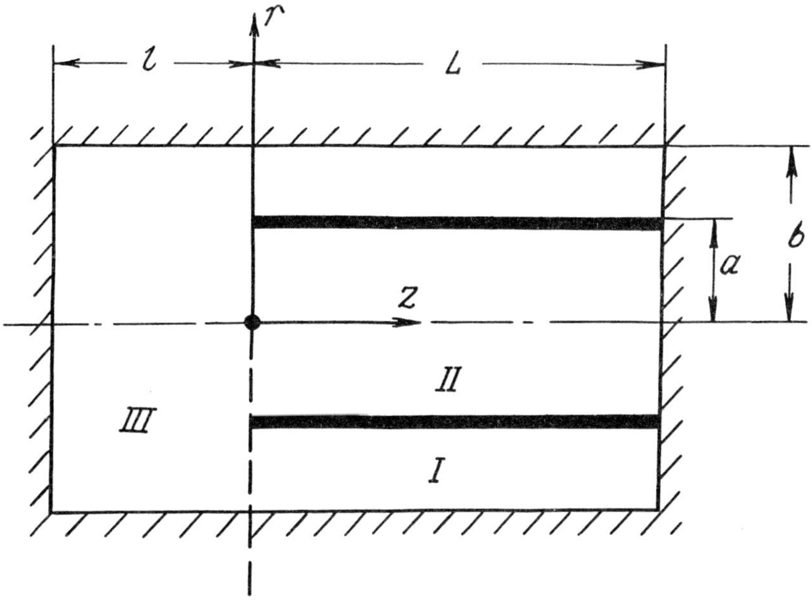

Figure 4.9 *Cylindrical resonator with coaxial insert*

lation of the field distributions of eigenmodes is rather complicated. This makes it difficult to use the energy-perturbation method to calculate Q (numerical integration is needed). Therefore, Rodionova and Slepyan (1986) resorted to considering the impedance problem and developing an effective algorithm for its solution. As a result they obtained a characteristic equation for complex eigenfrequencies, the solution of which by means of the impedance-perturbation method allows determination of Q. In this case an approximate analytical formula for Q has been obtained that can not be derived easily in any other way (for example, by the energy-perturbation method). The material of this section is based on the paper by Rodionova and Slepyan (1986).

Let us examine the TM_{0np} oscillations which are of most interest. The field components of such modes can be expressed in terms of the azimuthal component of the magnetic field $H_\phi(r, z)$ that satisfies the equation

$$\frac{\partial^2 H_\phi}{\partial r^2}+\frac{1}{r}\frac{\partial H_\phi}{\partial r}+\frac{\partial^2 H_\phi}{\partial z^2}+k^2 H_\phi-\frac{H_\phi}{r^2}=0 \tag{4.81}$$

and the boundary conditions

$$\left\{\frac{1}{r}\frac{\partial}{\partial r}(rH_\phi)-j\eta H_\phi\right\}\Bigg|_{\substack{r=b,\ -l\le z\le L\\ r=a-0,\ 0\le z\le L}}=0 \tag{4.82}$$

$$\left\{\frac{1}{r}\frac{\partial}{\partial r}(rH_\phi)+j\eta H_\phi\right\}\Bigg|_{r=a+0,\ 0\le z\le L}=0 \tag{4.83}$$

$$\left(\frac{\partial H_\phi}{\partial z} \mp j\eta H_\phi\right)\Bigg|_{\substack{z=L\\z=-l}} = 0 \qquad (4.84)$$

where $\eta = kZ_S/W_0$.

The solution of eqn. 4.81 should also meet the edge condition. We have to find the complex eigenvalues $k = k' + jk''$ of the eigenvalue problem given by eqns. 4.81–4.84.

At the first stage of solution, we use the partial-regions method. The partial regions are indicated by Roman numerals (Figure 4.9). In area I the field sought is expressed as a superposition of the TM_{0n} modes of the coaxial waveguide (with radii of pipes a and b) propagating in opposite directions. In regions II, III similar expansions are written in terms of the TM_{0n} modes of circular waveguides with radii a and b, correspondingly. The field matching is performed in the $z = 0$ plane. As a result we obtain the infinite homogeneous system of linear algebraic equations

$$\sum_{s=1}^{\infty} x_s\left(\frac{1-\mu_p\sigma_s}{\Gamma_p-\beta_s} + \frac{\sigma_s-\mu_p}{\Gamma_p+\beta_s}\right) = 0 \qquad (4.85)$$

where

$$\Gamma_{2n} = h_n \qquad \Gamma_{2n-1} = \gamma_n \qquad h_s = (\lambda_s^2 - k^2)^{1/2}$$

$$\gamma_s = (\chi_s^2 - k^2)^{1/2} \qquad \beta_s = (\vartheta_s^2 - k^2)^{1/2}$$

$$\mu_{2p-1} = \left(\frac{\gamma_p + j\eta}{\gamma_p - j\eta}\right)\exp(-2\gamma_p L)$$

$$\mu_{2p} = \left(\frac{h_p + j\eta}{h_p - j\eta}\right)\exp(-2h_p L)$$

$$\sigma_s = \left(\frac{\beta_s + j\eta}{\beta_s - j\eta}\right)\exp(-2\beta_s l)$$

$p = 1, 2, \ldots$, x_s being unknown coefficients.

The values λ_s, ϑ_s and χ_s are the roots of the transcendental equations

$$\lambda_s J_0(\lambda_s a) = j\eta J_1(\lambda_s a)$$

$$\vartheta_s J_0(\vartheta_s b) = j\eta J_1(\vartheta_s b)$$

$$\{\chi_s J_0(\chi_s b) - j\eta J_1(\chi_s b)\}\{\chi_s N_0(\chi_s a) + j\eta N_1(\chi_s a)\}$$
$$= \{\chi_s N_0(\chi_s b) - j\eta N_1(\chi_s b)\}\{\chi_s J_0(\chi_s a) + j\eta J_1(\chi_s a)\}$$

The homogeneous algebraic system given by eqn. 4.85 presents the eigenvalue problem nonlinear in the spectral parameter k. This system is characterised by a 'relative' and slow convergence, with low stability against roundoff errors. Because of this direct numerical solution by means of the truncation method is very difficult.

Regularisation of systems of this type is usually carried out using the method of semi-inversion (Shestopalov *et al.*, 1984) or the modified residue-calculus technique (Mittra and Lee, 1971). The latter was applied to simplified spectral problems of the form of eqn. 4.85, when $\sigma_s = 0$, $\mu_p = 0$, $p > 1$ (Al-Hakkak, 1978).

The comparison between these two methods was drawn by Shestopalov *et al.* (1984) for inhomogeneous boundary-value problems. It was established that the modified residue-calculus technique results in algebraic systems with the highest rate of truncation-method convergence. The approach presented in this Section is as effective as the modified-residue-calculus technique and, unlike the analysis given by Al-Hakkak (1978), is free from any of the limitations mentioned above.

The solution of eqn. 4.85 is sought in the form of a regularised substitution

$$x_s = d_s \sum_{m=1}^{\infty} \frac{F_m}{\Gamma_m - \beta_s} \tag{4.86}$$

where F_m are new unknown coefficients, and $\{d_s\}$ are nontrivial solutions of the system

$$\sum_{s=1}^{\infty} \frac{d_s}{\Gamma_m - \beta_s} = 0 \tag{4.87}$$

with algebraic asymptotic behaviour at $s \to \infty$.

It can be shown (see, for example, Shestopalov *et al.*, 1984), that

$$d_s = \operatorname{Res} f(w)|_{w=\beta_s} \tag{4.88}$$

where

$$f(w) = C \exp(w\chi/\pi) \prod_{m=1}^{\infty} \frac{(1 - w/\gamma_m)(1 - w/h_m)}{(1 - w/\beta_m)}$$

$\chi = (b-a)\ln\{b/(b-a)\} + a\ln(b/a)$ and C is an arbitrary constant.

Substituting eqn. 4.86 into eqn. 4.85 and using eqns. 4.87 and 4.88, we obtain the following homogeneous system of equations for F_m:

$$\sum_{m=1}^{\infty} F_m D_{pm} = 0 \tag{4.89}$$

where

$$D_{pm} = \delta_{pm} \left.\frac{df}{dw}\right|_{w=\Gamma_p} - \frac{\mu_p f(-\Gamma_p)}{\Gamma_p + \Gamma_m} + \sum_{s=1}^{\infty} \frac{\sigma_s d_s}{\Gamma_m - \beta_s} \left(\frac{1}{\Gamma_p - \beta_s} - \frac{\mu_p}{\Gamma_p + \beta_s} \right) \tag{4.90}$$

The characteristic equation for complex eigenfrequencies is

$$\phi(k) = \operatorname{Det} \| D_{pm} \| = 0 \tag{4.91}$$

Approximate values of eigenfrequencies are obtained from eqn. 4.91 after being reduced to the Nth order. As can be seen from eqns. 4.89 and 4.90 the matrix operator D is a Fredholm operator in some Hilbert space which guarantees the existence of the infinite-order determinant Det D. The proof of convergence of the approximate eigenvalues to accurate eigenvalues at $N \to \infty$ is based on the theory of finitely meromorphic operator-functions (Vainikko and Karma, 1974) and results obtained by Shestopalov (1980, 1983).

Let us apply the perturbation technique to eqn. 4.91 taking into consideration the fact that $|\eta/k| \ll 1$. The difficulty lies in the function $\phi(k)$ being expressed by the infinite determinant and this function, according to the perturbation technique, is supposed to be differentiated, expanded in Taylor series etc.

However, all necessary operations, as in Section 3.8, can easily be performed within the framework of the theory of determinants. Let us present D_{pn} at small $|\eta|$ as

$$D_{pn} \simeq D_{pn}^{(0)} + j\eta D_{pn}^{(1)}$$

where

$$D_{pn}^{(0)} = \lim_{\eta \to 0} D_{pn} \qquad D_{pn}^{(1)} = \left.\frac{dD_{pn}}{d\eta}\right|_{\eta=0}$$

The eigenvalue k to be found can be presented as $k = k_0 + \Delta k$, where k_0 satisfies eqn. 4.91 at $\eta = 0$ [Det $\| D_{pn}^{(0)} \| = \phi(k_0) = 0$] where Δk is a small complex correction due to finite conductivity of the walls. The Q-factor of the eigenoscillation under examination is expressed as

$$Q = \frac{k_0}{2\,\mathrm{Im}(\Delta k)}$$

where

$$\Delta k = -j\eta \frac{\sum_{s=1}^{\infty} \mathrm{Det}\, \| B_{ij}^{(s)} \|}{\sum_{s=1}^{\infty} \mathrm{Det}\, \| C_{ij}^{(s)} \|}$$

The elements of the matrices $B^{(s)}$ and $C^{(s)}$ can be determined according to the relationships

$$B_{ij}^{(s)} = C_{ij}^{(s)} = D_{ij}^{(0)}(k_0) \qquad (i \neq s)$$

$$B_{ij}^{(s)} = D_{ij}^{(1)}(k_0) \qquad (i = s)$$

$$C_{ij}^{(s)} = \left.\frac{dD_{ij}^{(0)}}{dk}\right|_{k=k_0} \qquad (i = s)$$

In practical evaluations, as mentioned above, the matrix D is truncated to the Nth order; in this case to find Δk we should calculate $2N$ determinants of the Nth order. When N is large, computation of these may take a long time and this technique can therefore be especially effective in combination with analytical methods of regularisation, resulting in rapidly converging systems of linear equations.

Let us consider the TM_{01p} oscillations which are of particular practical interest. In this case it is possible, with some assumptions, to obtain approximate analytical expressions for k_0 and Q. Let $l \gtrsim L$, $l \gtrsim b$ and $k_0 < \nu_{01}/b$. Then the oscillations under consideration are formed due to the reflection of the TEM mode from the coaxial-to-circular waveguide junction ($z = 0$) and the end wall ($z = L$) of the resonator (these modes are not actually affected by the other end wall $z = -l$). In the first approximation the characteristic equation is

$$D_{11}(k) = \left.\frac{df}{dw}\right|_{w=\Gamma_1} - \frac{\mu_1 f(-\Gamma_1)}{2\Gamma_1} = 0 \tag{4.92}$$

Applying the perturbation method (η small) to eqn. 4.92 we obtain the following formula for the dimensionless parameter q:

$$q \simeq \frac{2yx\{(2p+1)+2\gamma_0(k)\}\ln x^{-1}}{[(1+x)\{(2p+1)\pi+2\varepsilon(k_0)\}+yx\{1+y^2\beta(k_0)\}\ln x^{-1}]} \tag{4.93}$$

where $x = a/b$, $y = k_0 b$ and $p = 0, 1, 2, \ldots$. The values $\beta(k_0)$, $\gamma(k_0)$ and $\varepsilon(k_0)$ are expressed by

$$\beta(k_0) \simeq \sum_{s=1}^{\infty}\left[\frac{1}{v_{0s}^2}\left\{\frac{1}{(v_{0s}^2-y^2)^{1/2}} - \frac{x^2}{(v_{0s}^2-y^2x^2)^{1/2}}\right\} - \frac{1-x}{s^2\pi^2}\left\{\left(\frac{s\pi}{1-x}\right)^2 - y^2\right\}^{-1/2}\right] \tag{4.94}$$

$$\begin{aligned}\gamma(k_0) \simeq \sum_{s=1}^{\infty}\Bigg[&\left\{\frac{y}{(v_{0s}^2-y^2)^{1/2}} - \frac{yx}{(v_{0s}^2-y^2x^2)^{1/2}}\right\} \\ &- y\left\{\left(\frac{s\pi}{1-x}\right)^2 - y^2\right\}^{-1/2} + \tan^{-1}\left[y\left\{\left(\frac{s\pi}{1-x}\right)^2 - y^2\right\}^{-1/2}\right] \\ &+ \tan^{-1}\left\{\frac{yx}{(v_{0s}^2-y^2x^2)^{1/2}}\right\} - \tan^{-1}\left\{\frac{y}{(v_{0s}^2-y^2)^{1/2}}\right\}\Bigg]\end{aligned} \tag{4.95}$$

$$\begin{aligned}\varepsilon(k_0) \simeq \sum_{s=1}^{\infty}\Bigg(&\frac{y}{v_{0s}^2}(v_{0s}^2-y^2)^{1/2} - \frac{yx}{v_{0s}^2}(v_{0s}^2-y^2x^2)^{1/2} \\ &- \left(\frac{s\pi}{1-x}\right)^2 y\left\{\left(\frac{s\pi}{1-x}\right)^2 - y^2\right\}^{1/2} \\ &+ \tan^{-1}\left[y\left\{\left(\frac{s\pi}{1-x}\right)^2 - y^2\right\}^{-1/2}\right] + \tan^{-1}\left\{\frac{yx}{(v_{0s}^2-y^2x^2)^{1/2}}\right\} \\ &- \tan^{-1}\left\{\frac{y}{(v_{0s}^2-y^2)^{1/2}}\right\}\Bigg)\end{aligned} \tag{4.96}$$

The resonant wave number k_0 is determined from eqn. 4.91 at $\eta = 0$, $N = 1$, or, which is the same, from eqn. 4.92 at $\eta = 0$.

The results of calculation of q for the TM_{010} mode according to eqns. 4.93–4.96 are shown in Figure 4.10. Figure 4.10(i) shows the dependence of q on L/b for two values of the ratio a/b. When $L/b \to 0$ we have that $q \to y \to v_{01}$. At $L/b \lessapprox 1$ a sharp fall in this dependence is observed, which transfers at $L/b \gtrapprox 1$ into a smooth decrease. The dependence of q on $x = a/b$ at different L/b is shown in Figure 4.10(ii). The q coefficient is maximal at $x \simeq 0.3$ for any relationship between L and b.

Also of interest is the case when $l \ll L$, $l \ll b$. Then we can consider the resonator as quasistationary: area III in Figure 4.9 plays the role of capacitance and area I that of inductance (in this case the electric and magnetic fields are localised in areas III and I, respectively). Evaluation of the resonant frequency is simplest when the model of a coaxial insert with metallised end wall is used. Then the capacitance, neglecting edge effects, is expressed as $C \simeq \varepsilon_0 \pi a^2/2l$, the

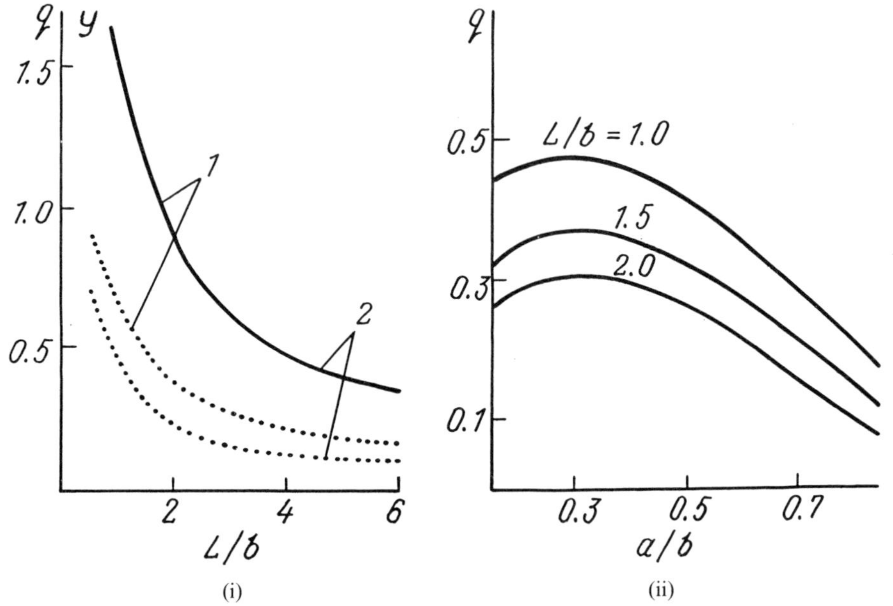

Figure 4.10 *Characteristics of fundamental mode in cylindrical resonator with coaxial insert*

(i) Normalised eigenfrequency and Q-factor against L/b

——— q

······· y

(ii) Normalised Q-factor against x

inductance being $\mathscr{L} = \mu_0 L \ln x^{-1}/2\pi$, and the resonant-wave number is

$$k_0 \simeq \left(\frac{\varepsilon_0 \mu_0}{\mathscr{L}C}\right)^{1/2} \simeq \left(\frac{4l}{La^2 \ln x^{-1}}\right)^{1/2}$$

The Q-factor of the resonator can be determined by means of the energy-perturbation method as the field structure of the eigenoscillation is relatively simple. For this case, eqn. 4.14 can be written as

$$Q = \frac{2}{\Delta_0} \frac{\int_V H_\phi^2 \, dV}{\int_S H_\phi^2 \, dS} \tag{4.97}$$

where by V we mean the volume of 'inductive' partial area I, and by S the part of its surface which is metallised. Supposing that $H_\phi = I/2\pi r$ (I being the total current) and integrating in eqn. 4.97, we obtain the following formula for Q:

$$Q = \frac{L \ln x^{-1}}{\Delta_0 \{\ln x^{-1} + (L/b)(1 + x^{-1})\}} \tag{4.98}$$

which is the final result of the analysis.

Experimental investigations of the resonator described have also been carried out. The structure of the experimental model and the block diagram of the measuring equipment are described in the paper by Rodionova and Nekhay (1981). The resonator frequency retuning was carried out by means of a metal ring moved in area III. The resonator surface was coated with silver and polished to the 12th surface-finish class to remove the microroughness.

Three experimental models were examined with different values of L/b and the same $a/b = 0.3$. At the upper operating frequency $f_0 = 2.14$ GHz the retuning ring was placed at a distance from the edge of the coaxial insert (the assumptions made when deriving eqn. 4.93 are fulfilled). At the lower operating frequency $f_0 = 1.07$ GHz the retuning ring was located near the edge of the coaxial insert and played the role of the second plate of a capacitor. The eigenoscillation here became quasistationary and its Q-factor is expressed by eqn. 4.98. Thus, the validity of eqns. 4.93 and 4.99 was verified using one experimental model.

For two values of f_0 the halfpower bandwidth $2\Delta f$ and the standing-wave ratio SWR_0 in resonance were measured. The experimental value of the Q-factor was determined from the results of measurements according to the formula

$$Q = (1 + SWR_0)\frac{f_0}{2\Delta f}$$

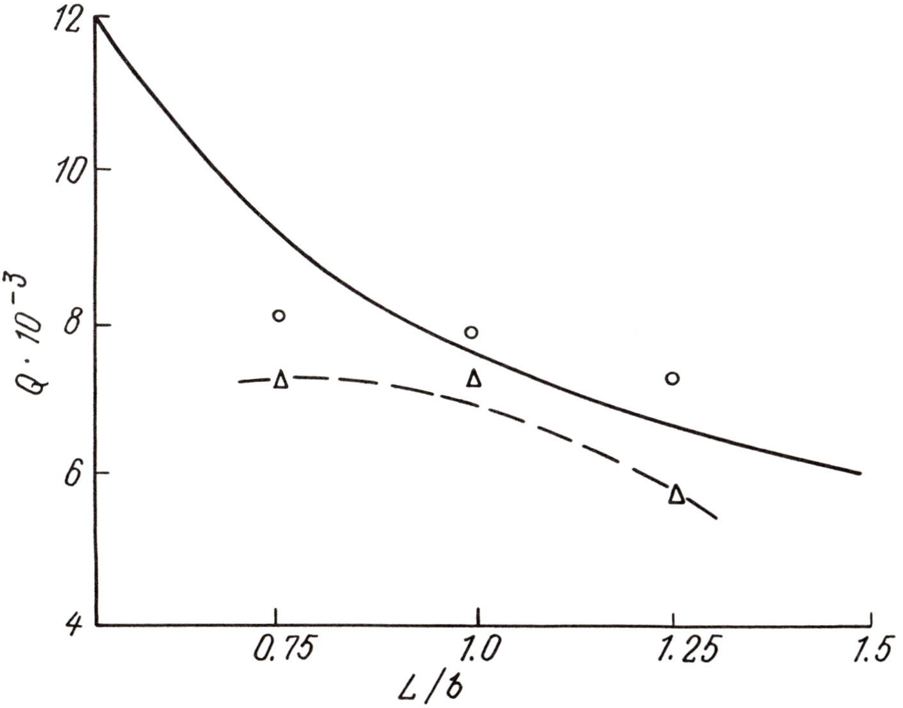

Figure 4.11 *Comparison of theoretical and experimental results for Q-factor of fundamental mode in cylindrical resonator with coaxial insert*

Figure 4.11 shows a comparison between the results of measurements and calculations (the solid line represents calculation according to eqns. 4.93–4.96, and the broken line estimation according to eqn. 4.98, circles and triangles: measurements at $f_0 = 2.14$ GHz and $f_0 = 1.07$ GHz, respectively). As can be seen from Figure 4.11, there is good agreement between the theoretical and experimental data.

Chapter 5

Electromagnetic-wave diffraction by finitely conducting comb-shaped structures

This Chapter is dedicated to the diffraction of plane waves by finitely conducting comb-shaped structures. Finite conductivity is taken into consideration by means of impedance boundary conditions. Periodic structures of comparatively simple configuration (array of halfplanes, symmetrical and nonsymmetrical lamellar gratings, echelettes) are examined as well as structures with complicated shape profiles. Highly effective numerical methods are described for the latter which are universal in relation to the structure configuration.

Approximate formulae and numerical results are given for a linear (per-period) power of heat losses. Its dependence on the angle of incidence, wavelength and configuration of the unit cell is analysed.

For the case of H-polarisation considerable attention is paid to the effect of abnormally small dissipation (the losses in a periodical structure may be smaller than those in a smooth surface of the same material). For E-polarised waves this does not occur. The physical mechanism of the absorption anomalies is analysed, and their practical use in reducing attenuation in microwave waveguides and resonators is discussed.

5.1 Diffraction of a plane wave by an array of impedance halfplanes: H-polarisation

In this section we shall consider the problem of a planewave diffraction by an array of finitely conducting halfplanes (Figure 5.1*a*). Let us assume halfplanes to be infinitely thin. It is permissible for the fields of electric polarisation ($\boldsymbol{E} = \{E_x, 0, E_z\}$, $\boldsymbol{H} = \{0, \psi, 0\}$) on condition that the thickness of halfplanes is small compared with the structure period and wavelength, but exceeds considerably the skin depth. Mathematical statement of the problem is as follows: we have to find the solution of a 2-dimensional Helmholtz equation

$$\nabla_{xz}^2 \psi + k^2 \psi = 0 \tag{5.1}$$

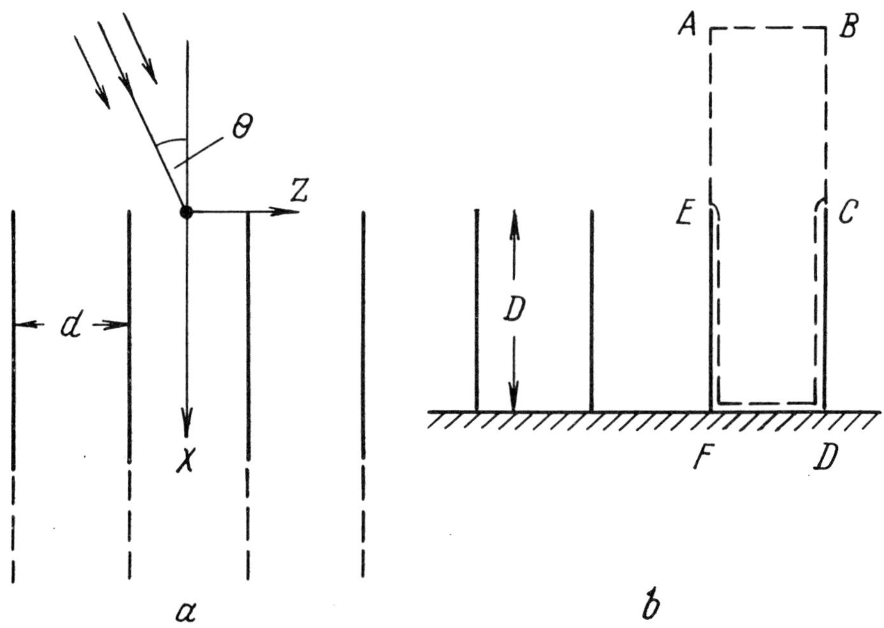

Figure 5.1 *Schematic representation of gratings*

a Parallel-plate grating
b Comb-shaped periodic structure

satisfying the impedance boundary condition

$$\left(\frac{\partial \psi}{\partial z} \pm j\eta\psi\right)\bigg|_{z=(m+1/2)d,\, 0\le x<\infty} = 0 \qquad (5.2)$$

($m = 0, \pm 1, \pm 2, \ldots$, and the signs '±' in eqn. 5.2 correspond to the right-hand and left-hand sides of the plates, $\eta = kZ_S/W_0$) and the edge condition. From the latter it follows that

$$\int_{\delta V} (k^2|\psi|^2 - |\text{grad}\, \psi|^2)\, dx\, dz < \infty$$

where δV is a bounded area in the plane $0xz$ containing the edge.

At $x \to \infty$ we impose the 'extinction' condition

$$\lim_{x\to\infty} \psi(x, z) = 0 \qquad (5.3)$$

and at $x \to -\infty$ ψ should satisfy the radiation condition [scattered field $\psi - \psi^0$ being a superposition of outgoing plane waves, $\psi^0(x, z) = \exp\{jk(x \cos \vartheta + z \sin \vartheta)\}$].

The problem of wave diffraction by the array of halfplanes may be considered as a key problem for a whole class of periodic structures. We shall describe the effective solution technique for this problem (Slepyan, 1980), which can easily be modified for comb-shaped structures, echelette gratings etc. (see below).

Let us present the field over the structure as an expansion in terms of Floquet's harmonics:

$$\psi(x, z) = \psi^0(x, z) + \sum_{p=-\infty}^{\infty} R_p \exp(j\alpha_p z + \Gamma_p x) \tag{5.4}$$

where $\alpha_p = 2(p\pi + u)/d$, $u = ka \sin \vartheta$, $a = d/2$ and $\Gamma_p = (\alpha_p^2 - k^2)^{1/2}$.

To satisfy the radiation condition a square-root branch in the expression for Γ_p has to be chosen according to the condition $\operatorname{Re} \Gamma_p > 0$ (if $\operatorname{Re} \Gamma_p = 0$, then $\operatorname{Im} \Gamma_p < 0$).

The field in the mth cell of the structure at $x > 0$ can be presented as a series in terms of eigenmodes of a plane-impedance waveguide of the width d:

$$\psi^{(m)}(x, z) = \sum_{n=0}^{\infty} \{\bar{T}_n^{(m)} \bar{\psi}_n(z - md) \exp(-\bar{\gamma}_n x) + \bar{\bar{T}}_n^{(m)} \bar{\bar{\psi}}_n(z - md) \exp(-\bar{\bar{\gamma}}_n x)\} \tag{5.5}$$

where $\bar{\psi}_n(z) = \cos(\bar{\beta}_n z)$, $\bar{\bar{\psi}}_n(z) = \sin(\bar{\bar{\beta}}_n z)$, $\bar{\gamma}_n = (\bar{\beta}_n^2 - k^2)^{1/2}$, $\bar{\bar{\gamma}}_n = (\bar{\bar{\beta}}_n^2 - k^2)^{1/2}$, and $\bar{\beta}_n$ and $\bar{\bar{\beta}}_n$ are the roots of the transcendental equations

$$-\bar{\beta}_n \sin(\bar{\beta}_n a) = j\eta \cos(\bar{\beta}_n a)$$

$$\bar{\bar{\beta}}_n \cos(\bar{\bar{\beta}}_n a) = j\eta \sin(\bar{\bar{\beta}}_n a)$$

Applicability of the field representation given by eqn. 5.5, unlike the case of perfectly conducting halfplanes, is not obvious and needs special substantiation. For a non-self-adjoint operator, corresponding to a homogeneous problem for an impedance-plane waveguide with a complex impedance of the walls, the eigenfunction set is generally speaking incomplete (see, for example, Keldysh, 1951). Only the eigenfunctions and associated functions form a complete set. Whether or not there will be associated functions depends on the values of the coefficient η in eqn. 5.2. In Section 3.6 some limitations on η are formulated: with these limitations there are no associated waves. Later we shall consider these limitations to be fulfilled, so the use of eqn. 5.5 will be quite relevant.

To fulfil the 'extinction' condition at $x \to \infty$ a square-root branch in expressions for $\bar{\gamma}_n, \bar{\bar{\gamma}}_n$ should be chosen in accordance with the conditions $\operatorname{Re} \bar{\gamma}_n > 0$, $\operatorname{Re} \bar{\bar{\gamma}}_n > 0$.

Owing to Floquet's conditions of pseudoperiodicity we have

$$\bar{T}_n^{(m)} = \bar{T}_n^{(0)} \exp(j2mu), \qquad \bar{\bar{T}}_n^{(m)} = \bar{\bar{T}}_n^{(0)} \exp(j2mu)$$

This makes it possible to examine only one unit cell of the structure.

Using the continuity of ψ and $\partial\psi/\partial x$ across the $x = 0$ plane and the orthogonality property of the functions $\{\bar{\psi}_n, \bar{\bar{\psi}}_n\}$ in the interval $|z| \leq a$, we come to the following dual system of linear algebraic equations of the first kind (SLAE-1) for unknown coefficients C_p:

$$\sum_{p=-\infty}^{\infty} \frac{C_p}{\Gamma_p - \bar{\gamma}_n} = \frac{1}{\Gamma_0 + \bar{\gamma}_n} \tag{5.6}$$

$$\sum_{p=-\infty}^{\infty} \frac{C_p(1 + \omega_p)}{\Gamma_p - \bar{\bar{\gamma}}_n} = \frac{1 + \omega_0}{\Gamma_0 + \bar{\bar{\gamma}}_n} \tag{5.7}$$

where

$$\omega_p = -\frac{2j\eta}{(\alpha_p + j\eta \cot u) \sin 2u}$$

$$C_p = \frac{(-1)^p R_p(\alpha_p + j\eta \cot u)}{k \sin \vartheta + j\eta \cot u}$$

According to the edge condition a physically correct solution of the SLAE-1 (eqns. 5.6 and 5.7) should belong to the class

$$\{C_n\} \in \tilde{l}_2 : \sum_{n=-\infty}^{\infty} |C_n|^2 (1 + |n|)^{-1} < \infty \tag{5.8}$$

As shown by Slepyan (1980), the SLAE-1 has a unique solution in $\tilde{l}_2$.

Direct use of the truncation method for the system given by eqns. 5.6 and 5.7 is not particularly effective because of the 'relative convergence' phenomenon (Mittra and Lee, 1971). Its essence is the following: if we take N_1 equations in eqn. 5.6 and N_2 equations in eqn. 5.7 with the total number of unknown quantities $N = N_1 + N_2$, then when N tends to infinity, the result will depend greatly on the ratio N_1/N_2. A correct limiting transition (when the condition given by eqn. 5.8 is fulfilled) can be obtained only at $N_1 = N_2$ (Ilyinsky and Slepyan, 1983). But even in this case the SLAE-1 is characterised by slow convergence of the truncation method and low stability against computational errors.

The SLAE-1 therefore needs an analytical regularisation (see Section 1.3). It is necessary to underline that the form of the SLAE-1 differs from the systems examined by Mittra and Lee (1971) and Shestopalov *et al.* (1984). This difference is caused by the impedance character of the boundary condition and is the reason why the procedures of analytical regularisation suggested by these authors cannot be applied directly to eqns. 5.6 and 5.7.

The regularisation procedure described in this Section is also based chiefly on the possibility of an exact solution for the matrix-convolution equations by means of the residue-calculus technique (Mittra and Lee, 1971, Shestopalov *et al.*, 1984). We employ a regularising substitution of a special type (see eqn. 5.9) but, unlike that in Section 1.3, the relationship between the coefficients sought and new unknowns in eqn. 5.9 is not linear but fraction-linear. Only this method of analytical regularisation allows us to obtain the system of linear algebraic equations of the second kind (SLAE-2) with the most simple expressions for matrix elements and free terms.

Let us present the solution of eqns. 5.6 and 5.7 in the form

$$C_p = \frac{P(\Gamma_p)}{P(-\Gamma_0)} d_p \tag{5.9}$$

where

$$P(w) = 1 + \sum_{m=1}^{\infty} \frac{F_m}{w - \bar{\bar{\gamma}}_m} \tag{5.10}$$

F_m being new unknown coefficients, and $\boldsymbol{d} = \{d_p\}$ is the solution of eqns. 5.6 and 5.7 at $\omega_p = 0$ ($\boldsymbol{d} \in \tilde{l}_2$).

The coefficients d_p are expressed as

$$d_p = \operatorname{Res} f(w)|_{w=\Gamma_p}$$

where the function $f(w)$ satisfies the following conditions:

(i) $f(w)$ is analytical everywhere in the complex w plane except at the points $w=\Gamma_p$, $w=-\Gamma_0$ ($p=0, \pm 1, \pm 2, \ldots$) where it has simple poles;
(ii) $f(\bar{\gamma}_n)=0$, $f(\bar{\bar{\gamma}}_n)=0$, $f'(w)\neq 0$ at $w=\bar{\gamma}_n, \bar{\bar{\gamma}}_n$;
(iii) $f(w)$ decreases algebraically at $w\to\infty$; and
(iv) $\operatorname{Res} f(w)|_{w=-\Gamma_0}=1$.

The function $f(w)$ can easily be expressed in terms of infinite products (Mittra and Lee, 1971, Shestopalov *et al.*, 1984).

It is worth noting that there is a relationship between the regularising-substitution method and the modified-residue-calculus technique. In fact, we can present

$$C_p = \operatorname{Res} \phi(w)|_{w=\Gamma_p}$$

$$\phi(w) = \frac{P(w)}{P(-\Gamma_0)} f(w)$$

The poles $w=\bar{\bar{\gamma}}_n$ of the function $P(w)$ describe a displacement of the zeros of $\phi(w)$ relative to the zeros of $f(w)$. However, explicit formulation of a condition, modifying condition (ii) for the function $f(w)$, and derivation of an expression for $\phi(w)$ seem in this case to be rather difficult.

The coefficients C_p determined by eqn. 5.9 satisfy eqn. 5.6 and the condition $C=\{C_p\}\in \tilde{l}_2$ as $|P(w)|<$ constant. The unknown coefficients F_m in eqn. 5.10 can be found from eqn. 5.7. Substituting eqn. 5.9 into eqn. 5.7 and changing (formally yet) the order of summation, we obtain the system of linear algebraic equations for F_m

$$\sum_{m=0}^{\infty} F_m \left\{ \sum_{p=-\infty}^{\infty} \frac{d_p(1+\omega_p)}{(\Gamma_p-\bar{\bar{\gamma}}_m)(\Gamma_p-\bar{\gamma}_n)} + \frac{1+\omega_0}{(\Gamma_0+\bar{\bar{\gamma}}_m)(\Gamma_0+\bar{\gamma}_n)} \right\} = \frac{1+\omega_0}{\Gamma_0+\bar{\gamma}_n} - \sum_{p=-\infty}^{\infty} \frac{d_p(1+\omega_p)}{(\Gamma_p-\bar{\gamma}_n)} \tag{5.11}$$

Let us transform eqn. 5.11 to a more convenient form. Consider the integral

$$I_{mn} = \frac{1}{2\pi j} \oint_{C_\infty} \frac{f(w)\,dw}{(w-\bar{\gamma}_n)(w-\bar{\bar{\gamma}}_m)}$$

The integration path C_∞ is a circle of infinitely large radius.[26] It can be easily seen that $I_{mn}=0$. On the other hand, I_{mn} can be evaluated according to the Cauchy residue theorem. Equating the sum of residues with zero, we come to the important identity

$$\sum_{p=-\infty}^{\infty} \frac{d_p}{(\Gamma_p-\bar{\bar{\gamma}}_m)(\Gamma_p-\bar{\gamma}_n)} + \frac{1}{(\Gamma_0+\bar{\bar{\gamma}}_m)(\Gamma_0+\bar{\gamma}_n)} + \delta_{mn}\Lambda_n = 0 \tag{5.12}$$

where $\Lambda_n = df/dw|_{w=\bar{\bar{\gamma}}_n}$. Using eqn. 5.12 and the properties of $f(w)$, it is easy to write eqn. 5.11 in a standard manner:

---

[26] From the mathematical point of view it would be more rigorous to consider C_∞ to be a so-called 'regular system of contours' (Evgrafov, 1965).

$$F_n + \sum_{m=0}^{\infty} F_m Q_{nm} = p_n \tag{5.13}$$

where

$$Q_{nm} = \frac{\tilde{Q}_{nm}}{\Lambda_n} \qquad p_n = \frac{\tilde{p}_n}{\Lambda_n}$$

$$\tilde{Q}_{nm} = \frac{\omega_0}{(\Gamma_0 + \bar{\bar{\gamma}}_m)(\Gamma_0 + \bar{\bar{\gamma}}_n)} + \sum_{p=-\infty}^{\infty} \frac{d_p \omega_p}{(\Gamma_p - \bar{\bar{\gamma}}_m)(\Gamma_p - \bar{\bar{\gamma}}_n)}$$

$$\tilde{p}_n = \frac{\omega_0}{\Gamma_0 + \bar{\bar{\gamma}}_n} - \sum_{p=-\infty}^{\infty} \frac{d_p \omega_p}{\Gamma_p - \bar{\bar{\gamma}}_n}$$

$$d_0 = -\exp\left(-\frac{2\Gamma_0 d}{\pi}\ln 2\right)\frac{\bar{\gamma}_0 - \Gamma_0}{\bar{\gamma}_0 + \Gamma_0}\prod_{m=1}^{\infty}\frac{\gamma_m - \Gamma_0}{\gamma_m + \Gamma_0}\exp\left(\frac{2\Gamma_0 d}{m\pi}\right)$$
$$\times \prod_{n=1}^{\infty}\frac{(\Gamma_n + \Gamma_0)(\Gamma_{-n} + \Gamma_0)}{(\Gamma_n - \Gamma_0)(\Gamma_{-n} - \Gamma_0)}\exp\left(-\frac{2\Gamma_0 d}{n\pi}\right)$$

$$d_p = -2\exp\left\{-(\Gamma_p + \Gamma_0)\frac{d}{\pi}\left(\ln 2 + \frac{1}{p}\right)\right\}$$
$$\times \frac{\Gamma_0(\bar{\gamma}_0 - \Gamma_p)(\Gamma_{-p} + \Gamma_0)}{(\Gamma_0 - \Gamma_p)(\bar{\gamma}_0 + \Gamma_p)(\Gamma_{-p} - \Gamma_p)}$$
$$\times \prod_{n=1}^{\infty}{}^{(p)}\frac{(\Gamma_n + \Gamma_0)(\Gamma_{-n} + \Gamma_0)}{(\Gamma_n - \Gamma_p)(\Gamma_{-n} - \Gamma_p)}\exp\left\{-\frac{(\Gamma_p + \Gamma_0)d}{n\pi}\right\}$$
$$\times \prod_{m=1}^{\infty}\frac{\gamma_m - \Gamma_p}{\gamma_m + \Gamma_0}\exp\left\{-\frac{(\Gamma_p + \Gamma_0)d}{m\pi}\right\} \qquad (|p| \geq 1)$$

$$\Lambda_n = -2\exp\left\{-(\bar{\bar{\gamma}}_n + \Gamma_0)\frac{d}{\pi}\left(\ln 2 - \frac{1}{2n+1}\right)\right\}$$
$$\times \frac{\Gamma_0 \bar{\gamma}_n \bar{\bar{\gamma}}_n (\gamma_0 - \bar{\bar{\gamma}}_n)}{(\bar{\gamma}_n + \Gamma_0)(\gamma_0 + \Gamma_0)(\Gamma_0^2 - \bar{\bar{\gamma}}_n^2)}$$
$$\times \prod_{m=1}^{\infty}{}^{(2n+1)}\frac{\gamma_m - \bar{\bar{\gamma}}_n}{\gamma_m + \Gamma_0}\exp\left(\frac{(\bar{\bar{\gamma}}_n + \Gamma_0)d}{m\pi}\right)$$
$$\times \prod_{p=1}^{\infty}\frac{(\Gamma_p + \Gamma_0)(\Gamma_{-p} + \Gamma_0)}{(\Gamma_p - \bar{\bar{\gamma}}_n)(\Gamma_{-p} - \bar{\bar{\gamma}}_n)}\exp\left\{-\frac{(\bar{\bar{\gamma}}_n + \Gamma_0)d}{p\pi}\right\}$$

Here $\gamma_{2m} = \bar{\gamma}_m$, $\gamma_{2m+1} = \bar{\bar{\gamma}}_m$, the symbol $\prod_{n=1}^{\infty}{}^{(p)}$ denotes that the pth term of the infinite product is omitted.

Following Neimark (1954), we have that at $n \to \infty$

$$\begin{aligned} \bar{\gamma}_n &\sim \bar{\beta}_n \sim n\frac{\pi}{a} + O(n^{-1}) \\ \bar{\bar{\gamma}}_n &\sim \bar{\bar{\beta}}_n \sim \frac{(2n+1)\pi}{2a} + O(n^{-1}) \end{aligned} \tag{5.14}$$

Taking into account eqn. 5.14, according to Mittra and Lee (1971), we obtain the asymptotic evaluation:

$$|\Lambda_n| = O(n^{-1/2}) \tag{5.15}$$

$$|d_p| = O(p^{-1/2}) \tag{5.16}$$

From eqns. 5.14–5.16 it follows that

$$|\tilde{Q}_{mn}| \sim O(m^{-1}n^{-1}) \tag{5.17}$$

$$|\tilde{p}_n| \sim O(n^{-1}) \tag{5.18}$$

Using eqns. 5.17 and 5.18, it can be proved that the system given by eqn. 5.13 is reduced to the Fredholm system by means of substitution of variables of the form $F_n = \tilde{F}_n \varepsilon_n$, where $\tilde{F}_n$ are new unknowns and $\varepsilon_n = O(n^\sigma)$, σ being an infinitely small positive number. This fact allows the following conclusions to be made:

(i) A solution of eqn. 5.13 does exist and is unique.
(ii) For any η, when there are no associated waves, the truncation method for eqn. 5.13 is converging.
(iii) The following asymptotic evaluation is valid:

$$|F_m| = o(m^{\sigma - 1/2})$$

If $\sin\vartheta = 0$ or $kd \sin\vartheta = m\pi$ then $\omega_p = 0$ (for any p). In this case it follows from eqn. 5.13 that $F_m = 0$, and we have an exact analytical solution of the problem considered:

$$C_p = d_p$$

If these conditions are not fulfilled then the SLAE-2 after truncation should be solved numerically. At a small η the iteration technique is also effective. The proof of its applicability will be given in Section 5.4.

5.2 Diffraction of a plane wave by an array of impedance halfplanes: E-polarisation

In Section 5.1 the diffraction by an impedance parallel-plate grating has been treated for the case of H-polarisation. We shall now consider the plane-wave diffraction by the same geometry for the E-polarised case. This case has essential differences: the Leontovich-impedance boundary condition is inapplicable for the model of infinitely thin plates (see Section 2.4). Direct analysis of structures with plates of finite thickness and taking into account the actual geometry of edges leads to considerable complication of the solution techniques. We therefore use the model of infinitely thin plates, employing the modified impedance condition described in Section 2.4 instead of the Leontovich condition. This allows us to apply highly effective analytical–numerical methods and take into account indirectly the edge-configuration influence on the value of the power absorbed.

The mathematical statement of the problem is identical to that in Section 5.1, but now $\psi = E_y(x, z)$. For convenience we rewrite the boundary condition in the form

$$\left(\psi \mp j\delta \frac{\partial \psi}{\partial z}\right)\Bigg|_{z=(m+1/2)d,\, 0\le x\le \infty} = 0 \tag{5.19}$$

where $\delta = Z_S/W_0 k$, Z_S is the modified impedance introduced in Section 2.4.

Let us solve this problem using the Wiener–Hopf technique. According to this method (Noble, 1958) we assume that $k = k' + jk''$, $k'' > 0$ (the solution for real k is obtained by letting $k'' \to 0$ at the end of analysis). Let us define the Fourier transform of the function $\psi(x, z)$ with respect to x as

$$\Phi(\alpha, z) = \int_{-\infty}^{\infty} \psi(x, z) \exp(j\alpha x)\, dx$$

Taking into account Floquet's condition of pseudoperiodicity, we obtain from eqn. 5.1

$$\Phi(\alpha, z) = \begin{cases} A(\alpha)\dfrac{\sin \gamma(d-\tilde{z})}{\sin \gamma d} + C(\alpha)\dfrac{\sin \gamma \tilde{z}}{\sin \gamma d}\exp(2ju) & (\tilde{z} \ge 0) \\[2ex] C(\alpha)\dfrac{\sin \gamma(d+\tilde{z})}{\sin \gamma d} - A(\alpha)\dfrac{\sin \gamma \tilde{z}}{\sin \gamma d}\exp(-2ju) & (\tilde{z} \le 0) \end{cases} \tag{5.20}$$

where $A(\alpha)$ and $C(\alpha)$ are unknown functions, $\gamma = (k^2 - \alpha^2)^{1/2}$, $u = ka \sin \vartheta$ and $\tilde{z} = z - a$.

Let us introduce the Fourier integrals as

$$\Phi_+(\alpha, z) = \int_0^{\infty} \psi(x, z) \exp(j\alpha x)\, dx$$

$$\Phi_-(\alpha, z) = \int_{-\infty}^{0} \psi(x, z) \exp(j\alpha x)\, dx$$

and $\Phi_+(\alpha, \pm a) = \lim_{z \to \pm a} \Phi_+(\alpha, z)$ (the symbol $\pm$ denotes the limit approached from the right and from the left, respectively).

It is obvious that the following relationships are true:

$$\Phi_+(\alpha, +a) + \Phi_-(\alpha, a) = A(\alpha) \tag{5.21}$$

$$\Phi_-(\alpha, -a) - \Phi_-(\alpha, a) = C(\alpha) \tag{5.22}$$

Differentiating eqn. 5.20 with respect to z we obtain

$$\Phi'_+(\alpha, +a) + \Phi'_-(\alpha, a) = \frac{\gamma \exp(2ju)}{\sin(\gamma d)} C(\alpha) - A(\alpha)\gamma \cot(\gamma d) \tag{5.23}$$

$$\Phi'_+(\alpha, -a) + \Phi'_-(\alpha, a) = \gamma \cot(\gamma d) C(\alpha) - A(\alpha)\frac{\gamma \exp(-2ju)}{\sin(\gamma d)} \tag{5.24}$$

where the prime indicates the derivative of a corresponding function with respect to z.

Finally, presenting the boundary condition given by eqn. 5.19 in terms of

Fourier transforms, we obtain another pair of formulae:

$$\Phi_+(\alpha, +a) - j\delta\Phi'_+(\alpha, +a) = \frac{1 - \delta k \sin\vartheta}{j(k\cos\vartheta + \alpha)} \tag{5.25}$$

$$\Phi_+(\alpha, -a) + j\delta\Phi'_+(\alpha, -a) = \frac{1 + \delta k \sin\vartheta}{j(k\cos\vartheta + \alpha)} \tag{5.26}$$

It is possible to show (see Noble, 1958) that functions with the subscript 'plus' are regular in the halfplane $\operatorname{Im}\alpha > -k''\cos\vartheta$, and functions with the subscript 'minus' in the halfplane $\operatorname{Im}\alpha < k''$. Then eqns. 5.21–5.26 can be considered as a system of functional equations for $\Phi'_\pm(\alpha, \pm a)$, $\Phi_\pm(\alpha, \pm a)$, $A(\alpha)$, $C(\alpha)$ that is correct in the strip $-k''\cos\vartheta < \operatorname{Im}\alpha < k''$. Excluding $A(\alpha)$ and $C(\alpha)$ from this system and introducing new unknown quantities

$$S_+(\alpha) = \Phi_+(\alpha, -a) - \Phi_+(\alpha, +a)$$

$$L_+(\alpha) = \Phi'_+(\alpha, +a) - \Phi'_+(\alpha, -a)$$

we come to the following system of Wiener–Hopf functional equations $(-k''\cos\vartheta < \operatorname{Im}\alpha < k'')$:

$$G(\alpha)S_+(\alpha) + j\delta P(\alpha)L_+(\alpha) = 2\Phi'_-(\alpha, a) + \frac{2k\delta\sin\vartheta}{k\cos\vartheta + \alpha} \tag{5.27}$$

$$T(\alpha)L_+(\alpha) + P(\alpha)S_+(\alpha) = 2\Phi_-(\alpha, a) + \frac{2}{k\cos\vartheta + \alpha} \tag{5.28}$$

where

$$G(\alpha) = \frac{\gamma\delta\sin(\gamma d)}{\cos(2u) - \cos(\gamma d)} - j$$

$$T(\alpha) = \delta + \frac{j\sin(\gamma d)}{\gamma\{\cos(2u) - \cos(\gamma d)\}}$$

$$P(\alpha) = \frac{\sin(2u)}{\cos(2u) - \cos(\gamma d)}$$

The diffraction field is expressed in terms of the solutions of eqns. 5.27 and 5.28 as

$$\psi(x, z) = \frac{1}{4\pi}\int_{-\infty + jb}^{\infty + jb}\left[\begin{Bmatrix}\sin\{\gamma(d - z)\} + \sin(\gamma z)\exp(2ju) \\ \sin\{\gamma(d + z)\} - \sin(\gamma z)\exp(-2ju)\end{Bmatrix}\right.$$
$$\left.\times \frac{L_+(\alpha)}{\gamma} - S_+(\alpha)\begin{Bmatrix}\exp(2ju)\cos(\gamma z) - \cos\{\gamma(d - z)\} \\ \cos\{\gamma(d + z)\} - \cos(\gamma z)\exp(-2ju)\end{Bmatrix}\right]$$
$$\times \frac{\exp(-j\alpha x)}{\cos(2u) - \cos(\gamma d)}\,d\alpha \tag{5.29}$$

The upper line in the large curly brackets corresponds to the case $z \geq 0$, the lower line to $z \leq 0$, where b is an arbitrary real number satisfying the inequality $-k''\cos\vartheta \leq b \leq k''$.

Thus, unlike the case of perfect conductivity investigated by Weinstein

(1969*b*), the problem under examination is reduced to the Wiener–Hopf matrix equation.

The exact solution of the system of functional equations given by eqns. 5.27 and 5.28 is possible only when $P(\alpha) = 0$, i.e. for $\vartheta = \sin^{-1}(n\pi/kd)$, $n = 0, \pm 1, \pm 2, \ldots$. In this case the system is divided into two independent equations. Now we proceed to examine the normal incidence of a plane wave ($\vartheta = 0$).[27]

So, let us assume that $\sin\vartheta = 0$ in eqns. 5.27 and 5.28. From eqn. 5.27 it follows that

$$S_+(\alpha) = \Phi'_-(\alpha, a) = 0$$

Then eqn. 5.28 can be written as

$$g(\alpha)L_+(\alpha) - 2\Phi_-(\alpha, a) = \frac{2}{j(\alpha + k)} \tag{5.30}$$

where $g(\alpha) = \gamma^{-1}\cot(\gamma a) - j\delta$. The solution of eqn. 5.30 can be obtained in a closed form by the factorisation method. The final result is (Noble, 1958)

$$L_+(\alpha) = -\frac{2j}{(\alpha + k)g_-(-k)g_+(\alpha)} \tag{5.31}$$

where $g_\pm(\alpha)$ are functions obtained as a result of factorisation of the function $g(\alpha)$.

The function $g(\alpha)$ is a ratio of two meromorphic functions and, in principle, can be factorised by the method of infinite products. However, the slow convergence of infinite products obtained in this case and the need to calculate numerically a large number of zeros of the function $g(\alpha)$ in the complex α plane make this method practically inapplicable. Much more effective is the integral representation of $g_\pm(\alpha)$ in the form suggested by Noble (1958).

From here on, let us confine our interest to the case $kd < \pi$. Substituting eqn. 5.31 into eqn. 5.29 and deforming the integration path over the poles of the integrand in the upper halfplane, according to the Cauchy residue theorem we obtain for $x < 0$:

$$\psi(x, z) = \exp(jkx) + R(k)\exp(-jkx) + \cdots$$

where dots represent an infinite sum of Floquet's higher evanescent harmonics.

The reflection coefficient is expressed as

$$R(k) = R_0 = -\exp\{-\mu(k)\} \tag{5.32}$$

where

$$\mu(k) = \frac{2k}{\pi j}\int_0^\infty \ln\left((k^2 - \beta^2)a\left[\frac{\cot\{(k^2-\beta^2)^{1/2}a\}}{(k^2-\beta^2)^{1/2}} - j\delta\right]\right)\frac{d\beta}{\beta^2 - k^2}$$

It is easy to see that if $\mathrm{Re}\,Z_S = 0$ then $|R| = 1$ (there is no absorption in the halfplanes). If $\mathrm{Re}\,Z_S > 0$ then the magnitude of the reflection coefficient characterises losses in the structure. Taking into consideration the fact that

[27] When the Wiener–Hopf matrix equation cannot be solved exactly it can be reduced to a SLAE-2 or to a system of Fredholm integral equations of the second kind.

$2(\mathrm{Im}\,\mathcal{Z}_S)^2 \gg (\mathrm{Re}\,\mathcal{Z}_S)^2$ and neglecting the terms $O(|\mathcal{Z}_S|^2)$, we have

$$|R| \simeq 1 - v(k)$$

where

$$v(k) = \frac{\mathrm{Re}\,\mathcal{Z}_S}{\pi W_0}(I_1 + I_2) \tag{5.33}$$

$$I_1 = 2\int_0^\infty \left(\coth u - \frac{u\,\mathrm{Im}\,\mathcal{Z}_S}{kaW_0}\right)^{-1} \mathrm{d}u$$

$$I_2 = 2\int_0^\infty \left[\frac{1}{\{u^2 + (ka)^2\}^{1/2}} - \frac{1}{u}\right]\frac{\mathrm{d}u}{\coth u} - 2\int_0^{ka} \frac{\tan u\,\mathrm{d}u}{\{(ka)^2 - u^2\}^{1/2}}$$

Here it is relevant to recall the remark made at the end of Section 2.4 on the problem of choosing the sign with the imaginary part of the impedance $\mathcal{Z}_S$ determined by eqns. 2.27 and 2.28. When the choice of the sign is wrong, fictitious surface waves occur (as mentioned in Section 2.4). This may also lead to a physical nonsense for the problem considered here: even at $kd < \pi$ and $\mathrm{Re}\,\mathcal{Z}_S = 0$, $|R| \neq 1$, i.e. the magnitude of the reflection coefficient does not characterise the heat losses completely.

Let us proceed to the evaluation of the integrals in eqn. 5.33. The integral I_1, as can be easily seen, is reduced to that evaluated in a closed form (see Ilyinsky and Slepyan, 1983). To determine I_2 let us examine the complex u plane with the cut between the points $\pm jka$ along the imaginary axis. We can

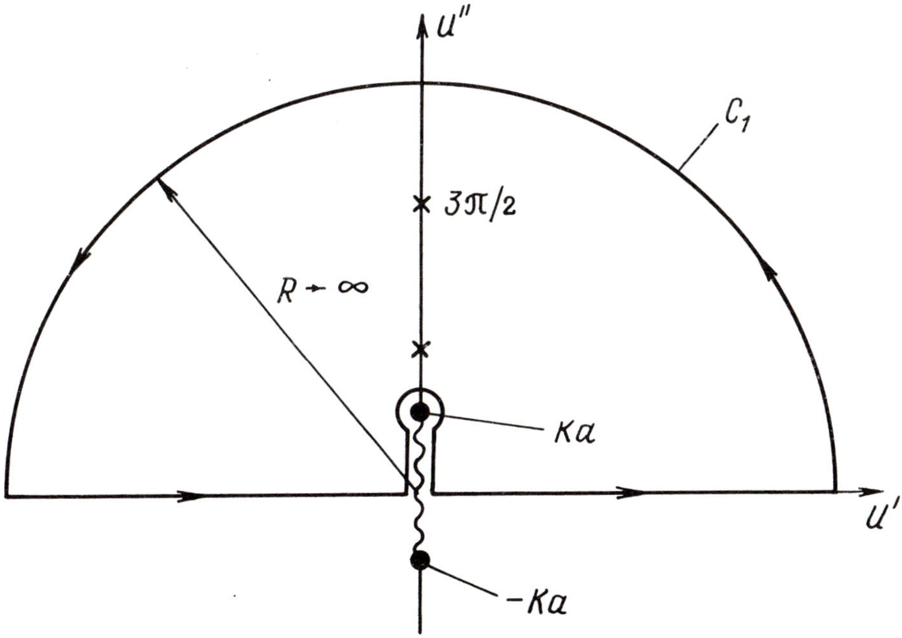

Figure 5.2 *Integration path for calculating I_2*

establish that

$$I_2 = \int_{C_1} \left[\frac{u}{\{u^2 + (ka)^2\}^{1/2}} - 1 \right] \frac{\tanh u \, du}{u} \tag{5.34}$$

C_1 being the contour of Figure 5.2. As analytical continuation of the integrand in eqn. 5.34 tends to zero at $|u| \to \infty$, the integral in eqn. 5.34 can be expressed through the sum of residues at the poles $u = j(n + \frac{1}{2})\pi$, $n = 0, 1, \ldots, \infty$. Finally, in this case we obtain for $v(k)$

$$v(k) = \frac{2 \operatorname{Re} Z_S}{\pi W_0} \left(C + \ln\left(\frac{4 W_0 ka}{\pi |Z_S|}\right) + \sum_{n=0}^{\infty} \left[\frac{1}{\{(n+\frac{1}{2})^2 - (ka/\pi)^2\}^{1/2}} - \frac{1}{n+\frac{1}{2}} \right] \right) \tag{5.35}$$

In accordance with eqns. 2.27 and 2.28 for the rounded edge we shall consider $|Z_S| \simeq k\rho W_0$ (ρ being the curvature radius near the edge), and $|Z_S| \simeq 0.025 kdW_0$ for the edge geometry near rectangular, where d is the thickness of plates, and $C = 0.5772\ldots$ is Euler's constant.

It is necessary to underline the fact that the argument of the logarithmic function in eqn. 5.35 involves geometrical parameters characterising the edge configuration. This fact points to the inapplicability of the Leontovich boundary condition and the usual impedance-perturbation method. Eqn. 5.35 coincides with that obtained by Weinstein *et al.* (1986) for an array of impedance halfplanes of finite thickness. Note that the analysis made by Weinstein *et al.* (1986) and Zhurav (1987) is based on a rather complicated variant of the Wiener–Hopf approximate technique. The use of modified impedance-boundary conditions imposed on infinitely thin surfaces appears to be much simpler.

In Figure 5.3 are given the results of computation according to eqn. 5.35. Note that the absorption in such an array considerably exceeds that in a smooth metal surface of the same material.

In the case of oblique incidence, as it was mentioned above, no exact solution of the problem has so far been obtained. A simple approximate solution for an arbitrary angle of incidence ϑ can be found when the structure period is small compared with the wavelength. In this case the equivalent surface impedance of the structure should be independent of ϑ.

The reflection coefficient $R(\vartheta)$, according to Nefedov and Sivov (1977), is expressed through the equivalent impedance Z^h by the relationship

$$R(\vartheta) = -\frac{W_0 - Z^h \cos \vartheta}{W_0 + Z^h \cos \vartheta}$$

On the other hand, Z^h is expressed in terms of $R(0)$ by the formula

$$Z^h = -\frac{R(0) + 1}{R(0) - 1} W_0$$

where $R(0)$ is the reflection coefficient for the case of normal incidence determined by eqn. 5.32.

These two formulae allow us to express $R(\vartheta)$ through $R(0)$; the final result

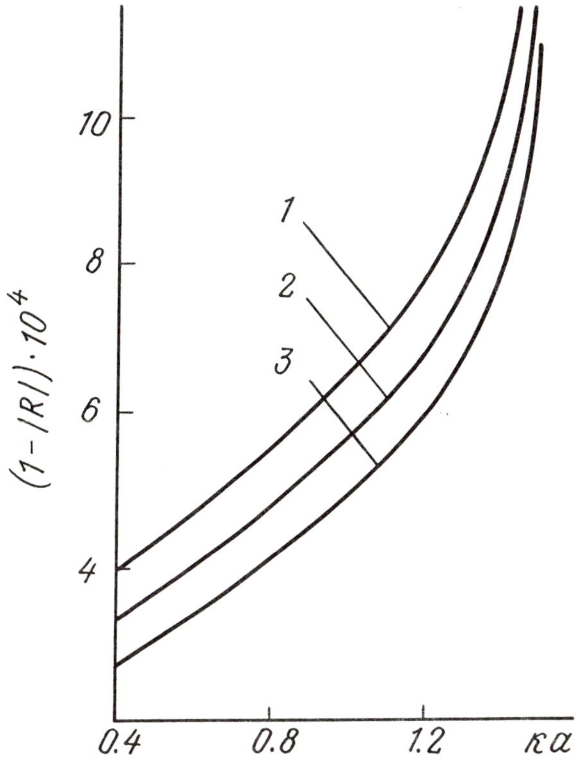

Figure 5.3 *Absorption in parallel-plate grating with rounded edges*

E polarisation, $Z_S(k_{min})/W_0 = 10^{-4}$

1 $a/\rho = 0.0075$
2 $a/\rho = 0.015$
3 $a/\rho = 0.03$

can be written as

$$R(\vartheta) = \frac{p - R(0)}{pR(0) - 1} \tag{5.36}$$

where $p = \tan^2(\vartheta/2)$.

5.3 Diffraction by a finitely conducting comb-shaped structure

Figure 5.1*b* shows the configuration of the structure under study and defines the notation. The formulation of the problem is completely similar to that in Section 5.1: it is necessary to find a solution of eqn. 5.1 satisfying the impedance boundary conditions

$$\left(\frac{\partial\psi}{\partial z}\pm j\eta\psi\right)\bigg|_{z=(m+1/2)d,\, D\geq x\geq 0}=0 \tag{5.37}$$

$$\left(\frac{\partial\psi}{\partial x}-j\eta_1\psi\right)\bigg|_{x=D,\,|z|\leq\infty}=0 \tag{5.38}$$

the edge condition and the radiation condition at $x\to-\infty$.

Such a mathematical statement of the problem describes also a similar acoustic problem if by ψ, η, η_1 we mean the corresponding physical values. This is why, with the reservation mentioned, all further considerations will be applicable also for acoustic waves.

Let us consider briefly the basic steps of the solution procedure. First we pass from the original boundary-value problem to the infinite system of linear algebraic equations writing down the field expansions in partial areas and performing field matching at the interface $x=0$. This system is of the first kind (SLAE-1) and is ineffective for a direct solution, i.e. it is an intermediate step. Then we transform the SLAE-1 into the infinite system of linear algebraic equations of the second kind (SLAE-2) using the regularising substitution technique described in Section 5.1. Finally, we derive approximate formulae for a physical analysis.

Let us represent the field above the structure, as in Section 5.1, in the form of an infinite series in terms of Floquet's harmonics. The field in the mth cell can be expanded into a series in terms of eigenmodes of a plane impedance waveguide of the width d, propagating in opposite directions:

$$\psi^{(m)}(x,z)=\sum_{n=0}^{\infty}\phi_n(z-md)\,T_n^{(m)}\{\exp(-\gamma_n x)+\rho_n\exp(\gamma_n x)\} \tag{5.39}$$

where

$$\rho_n=\frac{\gamma_n+j\eta_1}{\gamma_n-j\eta_1}\exp(-2\gamma_n D)\qquad(\gamma_{2n}=\bar{\gamma}_n,\ \gamma_{2n+1}=\bar{\bar{\gamma}}_n)$$

the functions $\phi_n(z)$ being determined in Section 5.1.

Let us assume that the η coefficient satisfies the conditions defined in Section 3.6. Then there will be no associated waves and the use of the expansion given by eqn. 5.39 is relevant.

The expressions for the field given by eqns. 5.4 and 5.39 satisfy the Helmholtz equation, radiation condition and impedance boundary conditions on the structure surface. Because of the condition of pseudoperiodicity, $T_n^{(m)}=\mathrm{T}_n^{(0)}\exp(2jmu)$, and we can consider just one unit cell of the structure (for example, with $m=0$). The unknown coefficients R_p, $T_n^{(0)}$ can be found from the condition of continuity of ψ and $\partial\psi/\partial x$ across the $x=0$ plane at the interval $|z|\leq a$. That leads to the following system of functional equations for R_p and $T_n^{(0)}$:

$$\exp(jkz\sin\vartheta)+\sum_{p=-\infty}^{\infty}R_p\exp(j\alpha_p z)=\sum_{n=0}^{\infty}T_n^{(0)}\phi_n(z)(1+\rho_n) \tag{5.40}$$

$$-jk\cos\vartheta\exp(jkz\sin\vartheta)-\sum_{p=-\infty}^{\infty}\Gamma_p R_p\exp(j\alpha_p z)$$
$$=\sum_{n=0}^{\infty}\gamma_n\mathrm{T}_n^{(0)}\phi_n(z)(1-\rho_n) \tag{5.41}$$

If there are no associated waves, the function $\phi_n(z)$ form a complete orthogonal system in $L_2(a, -a)$ (Keldysh, 1951), so that on multiplying the right-hand and left-hand sides of eqns. 5.40 and 5.41 by $\phi_p(z)$ $(p = 0, 1, 2, \ldots)$ and integrating over the structure period, we obtain a system of linear algebraic equations equivalent to eqns. 5.40 and 5.41:

$$\left.\begin{aligned} &M_{0s} + \sum_{p=-\infty}^{\infty} R_p M_{ps} = T_s^{(0)}(1+\rho_s)n_s \\ &-jk\cos\vartheta M_{0s} - \sum_{p=-\infty}^{\infty} \Gamma_p R_p M_{ps} = T_s^{(0)}(1-\rho_s)n_s\gamma_s \end{aligned}\right\} \qquad (5.42)$$

Here

$$M_{p,2s} = \frac{2(-1)^p \cos(\bar{\beta}_s a)\sin u}{\alpha_p^2 - \bar{\beta}_s^2}(\alpha_p + j\eta\cot u)$$

$$M_{p,2s+1} = -\frac{2j(-1)^p \sin(\bar{\bar{\beta}}_s a)\cos u}{\alpha_p^2 - \bar{\bar{\beta}}_s^2}(\alpha_p - j\eta\tan u)$$

$$n_s = \int_{-a}^{a} \phi_s^2(z)\, dz$$

Excluding $T_n^{(0)}$ in eqn. 5.42 and introducing new variables according to the formula

$$C_p = (-1)^p R_p \frac{\alpha_p + j\eta\cot u}{k\sin\vartheta + j\eta\cot u}$$

we obtain the system of linear algebraic equations for C_p

$$\sum_{p=-\infty}^{\infty} C_p \frac{1+\omega_{pn}}{1+\omega_{0n}}\left(\frac{1}{\Gamma_p - \gamma_n} + \frac{\rho_n}{\Gamma_p + \gamma_n}\right) = \frac{1}{\Gamma_0 + \gamma_n} + \frac{\rho_n}{\Gamma_0 - \gamma_n} \qquad (5.43)$$

where

$$\omega_{p,2n} = 0, \qquad \omega_{p,2n+1} = \frac{Q_1}{\alpha_p + Q_2}$$

$$Q_1 = \frac{2j\eta}{\sin(2u)}, \qquad Q_2 = j\eta\cot u$$

To fulfil the edge condition, the vector $\boldsymbol{C} = \{C_p\}$ has to belong to the class of numerical sequences meeting the condition given by eqn. 5.8.

The linear system obtained is a SLAE-1 of a more general form than that examined in Section 5.1. We can also apply for its regularisation the procedure developed in Section 5.1. Through this procedure we reduce eqn. 5.43 to the SLAE-2 allowing an effective numerical and analytical investigation of the problem.

The amplitudes of waveguide modes in the grooves $T_n^{(0)}$ can be expressed through the coefficients C_p according to the recalculation formula resulting from eqn 5.40.

For the diffraction problem formulated in this Section a theorem of uniqueness has been proved by Slepyan and Slepyan (1980*a*). From this theorem there follows a theorem of uniqueness for eqn. 5.43 in the class of numerical sequences defined by eqn. 5.8.

Then we easily obtain the relationship

$$-\lim_{x\to\infty}\operatorname{Im}\int_{-a}^{a}\psi^*\frac{\partial\psi}{\partial x}\,dz=\operatorname{Re}\int_{EFDC}\eta(s)\,|\psi(s)|^2\,ds \tag{5.44}$$

where $\eta(s)=\eta$ on FE and CD and $\eta(s)=\eta_1$ on FD ($s\in EFDC$).

Substituting eqn. 5.4 into eqn. 5.44, we obtain the result

$$-\operatorname{Im}\left(\frac{W_0}{2k}\sum_{p=-N_-}^{N_+}(\delta_{p0}-|R_p|^2)\Gamma_p d\right)=P \tag{5.45}$$

where $N_\pm$ is the number of propagating Floquet's harmonics (positive and negative, respectively) and

$$P=\frac{W_0}{2k}\operatorname{Re}\int_{EFDC}\eta(s)\,|\psi(s)|^2\,ds \tag{5.46}$$

P being the linear per-period dissipated power.

Eqn. 5.45 reflects the law of energy conservation. It shows that part of the energy of the incident wave is distributed between the propagating Floquet's harmonics and the remainder is dissipated in the comb-shaped structure, turning into the Joule heat.

We continue now with the description of the regularisation procedure. We shall present the solution of the SLAE-1 as a fraction–linear substitution given by eqn. 5.9 with

$$P(w)=1+\sum_{m=1}^{\infty}\frac{F_m}{w-\gamma_m} \tag{5.47}$$

The difference between eqns. 5.47 and 5.10 is in the fact that the summation in eqn. 5.47 is carried out in all m and not only in indices corresponding to antisymmetrical modes.

Substituting eqns. 5.8 and 5.47 into eqn. 5.43 and changing the order of summation,[28] we obtain

$$\begin{aligned}
&\sum_{p=-\infty}^{\infty}d_p(1+\omega_{pn})\left(\frac{1}{\Gamma_p-\gamma_n}+\frac{\rho_n}{\Gamma_p+\gamma_n}\right)+\sum_{m=0}^{\infty}F_m\sum_{p=-\infty}^{\infty}d_p(1+\omega_{pn})\\
&\quad\times\left\{\frac{1}{(\Gamma_p-\gamma_n)(\Gamma_p-\gamma_m)}+\frac{\rho_n}{(\Gamma_p+\gamma_n)(\Gamma_p-\gamma_m)}\right\}\\
&=(1+\omega_{p0})\left(\frac{1}{\Gamma_0+\gamma_n}+\frac{\rho_n}{\Gamma_0-\gamma_n}\right)-\sum_{m=0}^{\infty}F_m(1+\omega_{p0})\\
&\quad\times\left\{\frac{1}{(\Gamma_0+\gamma_n)(\Gamma_0+\gamma_m)}+\frac{\rho_n}{(\Gamma_0-\gamma_n)(\Gamma_0+\gamma_m)}\right\}
\end{aligned} \tag{5.48}$$

We shall need a few mathematical identities. In a way similar to that in

[28] The correctness of this operation can be substantiated with the aid of the evaluations given by eqns. 5.14–5.18.

Section 5.1, let us consider the integral

$$I_{mn} = \frac{1}{2\pi j} \oint_{C_\infty} \frac{f(w)\, dw}{(w + \gamma_n)(w - \gamma_m)}$$

From the asymptotic behaviour of $f(w)$ at $|w| \to \infty$ it follows that $I_{mn} = 0$. Calculating I_{mn} by the Cauchy residue theorem, we obtain the identity

$$\sum_{p=-\infty}^{\infty} \frac{d_p}{(\Gamma_p + \gamma_n)(\Gamma_p - \gamma_m)} + \frac{1}{(\Gamma_0 - \gamma_n)(\Gamma_0 + \gamma_m)} = \frac{f(-\gamma_n)}{\gamma_n + \gamma_m} \tag{5.49}$$

In addition, we shall use eqn. 5.12 and the well known formulae, proved by Mittra and Lee (1971):

$$\frac{1}{\Gamma_0 - \gamma_n} - \sum_{p=-\infty}^{\infty} \frac{d_p}{\Gamma_p + \gamma_n} = f(-\gamma_n) \tag{5.50}$$

$$\frac{1}{\Gamma_0 + \gamma_n} = \sum_{p=-\infty}^{\infty} \frac{d_p}{\Gamma_p - \gamma_n} \tag{5.51}$$

Using eqns. 5.12, 5.47, 5.48 and 5.49, we proceed from eqn. 5.41 to the following system of linear algebraic equations for $\boldsymbol{F} = \{F_m\}$ (Slepyan and Slepyan, 1980*a*):

$$(I + B)\boldsymbol{F} = \boldsymbol{t} \tag{5.52}$$

where

$$B_{mn} = -\frac{1}{\Lambda_n}\left\{\frac{\omega_{0n}}{\Gamma_0 + \gamma_m}\left(\frac{1}{\Gamma_0 + \gamma_n} + \frac{\rho_n}{\Gamma_0 - \gamma_n}\right) + \sum_{p=-\infty}^{\infty} \frac{d_p \omega_{pn}}{\Gamma_p - \gamma_m}\left(\frac{1}{\Gamma_p - \gamma_n} + \frac{\rho_n}{\Gamma_p + \gamma_n}\right) + \frac{\rho_n f(-\gamma_n)}{\gamma_n + \gamma_m}\right\}$$

$$\Lambda_n = \left.\frac{df}{dw}\right|_{w=\gamma_n} \tag{5.53}$$

The free terms t_n are expressed as

$$t_n = -\frac{1}{\Lambda_n}\left\{\frac{\omega_{0n}}{\Gamma_0 + \gamma_n} + \frac{\rho_n \omega_{0n}}{\Gamma_0 - \gamma_n} - \sum_{p=-\infty}^{\infty} d_p \omega_{pn}\left(\frac{1}{\Gamma_p - \gamma_n} + \frac{\rho_n}{\Gamma_p + \gamma_n}\right) + \rho_n f(-\gamma_n)\right\} \tag{5.54}$$

The infinite system of linear algebraic equations given by eqn. 5.52 is the result required.[29]

Let us consider the properties of the matrix B as a linear operator in the Hilbert space of numeric sequences. Using eqns. 5.15 and 5.16, we can show,

[29] If the side walls of the cells are perfectly conducting ($\eta = 0$), then the system of linear equations given by eqn. 5.52 coincides with that obtained by the modified residue-calculus technique (Mittra and Lee, 1971).

as in Section 5.1, that the following asymptotic evaluations for B_{nm}, t_n are correct:

$$|B_{mn}| \sim O(m^{-1}n^{-1/2}) \tag{5.55}$$

$$|t_n| \sim O(n^{-1/2}) \tag{5.56}$$

Then, it is easy to show that the system given by eqn. 5.52 is a Fredholm system and its solution can be obtained by the truncation method. In this case

$$|F_m| \sim o(m^{-1/2})$$

and this allows us to draw a conclusion that $\boldsymbol{C}$ expressed through $\boldsymbol{F}$ by eqns. 5.9, 5.47 satisfy the condition given by eqn. 5.8. Thus, a full mathematical substantiation of the formal solution of the problem has been implemented.

5.4 Perturbation technique

First, we shall present some auxiliary information. Vector $\boldsymbol{t}(\eta)$ will be called analytical in η in the area $\mathscr{D}$ if all its components $t_i(\eta)$ are analytical in the area $\mathscr{D}$. Similarly, the matrix operator $B(\eta)$ will be called analytical if $B_{ik}(\eta)$ are analytical for any i, k. The following theorem (see Evgrafov, 1965) is correct:

Theorem 5.1: a sum of uniformly convergent series of analytical functions is an analytical function at every interior point of that set where the series converges uniformly.

Now, let us examine eqn. 5.52 in the light of perturbation theory. We extract the operator B_0, which corresponds to the case of perfect conductivity of the ridges, as a main part of B, i.e. $B_0 = \lim_{|\eta|\to 0} B(\eta)$. Assuming that $k \neq n\pi/d$, we present eqn. 5.52 as

$$[I + (B_0 + I)^{-1}\{B(\eta) - B_0\}]\boldsymbol{F} = (B_0 + I)^{-1}\boldsymbol{t}(\eta) \tag{5.57}$$

Let us construct Neumann's series for the vector $\boldsymbol{F}$

$$\boldsymbol{F}(\eta) = \sum_{\nu=0}^{\infty} (I + B_0)^{-\nu}\{B(\eta) - B_0\}^{\nu}(B_0 + I)^{-1}(-1)^{\nu}\boldsymbol{t}(\eta) \tag{5.58}$$

The operator $(I + B_0)^{-1}$ is bounded, $(I + B_0)^{-1}(B - B_0)$ being completely continuous in l_2^{σ} $(0 < \sigma < 1)$[30] and its norm can be as small as required at sufficiently small η.

Taking into account theorem 5.1, Weierstrass' criterion of uniform convergence (Korn and Korn, 1968) and evaluations given by eqns. 5.55 and 5.56, we find that the vector function $\boldsymbol{F}(\eta)$ expressed by eqn. 5.58 is analytical in the vicinity of the point $\eta = 0$, if $B(\eta)$ and $\boldsymbol{t}(\eta)$ are analytical in this vicinity. Let us prove now that $B(\eta)$ and $\boldsymbol{t}(\eta)$ determined by eqns. 5.53 and 5.54 are really analytical in the neighbourhood of zero.

The functions $\gamma_n(\eta)$ at $k \neq n\pi/d$ are analytical.[31] Considering this fact we can

---

[30] The symbol l_2^{σ} means a Hilbert space of numerical sequences; the scalar product of two vectors $\boldsymbol{x}$ and $\boldsymbol{y}$ in l_2^{σ} is defined as $(\boldsymbol{x}, \boldsymbol{y}) = \sum_{m=0}^{\infty} x_m y_m^* (1+m)^{-\sigma}$.

[31] An exception can be made for the point η_0 corresponding to a two-fold root of the waveguide dispersion equation. Earlier, such points have been excluded from consideration. As is shown in Section 3.6, there is a circle of a finite radius where $\eta \neq \eta_0$.

easily [see Evgrafov (1965)] establish the analyticity of $\Lambda(\eta)$, $s(\eta) = f(-\gamma_n)$ and $d_p(\eta)$ on the Riemann surface of the logarithmic function. The analyticity of $B(\eta)$, $\boldsymbol{t}(\eta)$ follows from theorem 5.1 and Weierstrass' criterion mentioned above.

Thus, $\boldsymbol{F}(\eta)$ can be presented as a Taylor series converging in some neighbourhood of the point $\eta = 0$. Expanding $\boldsymbol{t}(\eta)$ and $B(\eta)$ into a Taylor series with the point $\eta = 0$ as a centre of the circle of convergence, and multiplying the series in eqn. 5.58, we obtain an expansion for $\boldsymbol{F}(\eta)$ in the form of a series in integer powers of η. According to the theorem of uniqueness for analytical functions (Evgrafov, 1965) this series will be a Taylor series for $\boldsymbol{F}(\eta)$.

It follows then that the coefficients $R_n(\eta)$, $T_n^{(0)}(\eta)$ and the function $\psi(x, z, \eta)$ for any values of x and z are analytical in the vicinity of the point $\eta = 0$. This will not be true, however, for the derivatives of ψ with respect to x and z, as after a term-by-term differentiation the series in eqn. 5.4 converges nonuniformly about the points $x = 0$, $z = (m + \frac{1}{2})d$ $(m = 0, \pm 1, \pm 2, \ldots)$. In fact, the analysis shows that at the points $x = 0$, $z = (m + \frac{1}{2})d$ the analyticity of grad $\psi(\eta)$ is broken.

For electrically polarised fields for metals in microwave and millimetre-wave bands the condition $|\eta| \ll k$ is fulfilled. Therefore, it is reasonable to consider the losses as a small perturbation. The analysis presented in this Section rigorously substantiates the validity of such an approach when there are no associated waves and the frequencies are not too near to the cutoff frequencies for the eigenmodes in the grooves. Otherwise such an approach is not correct. In a single-mode range $(kd < \pi)$ the energy absorption can be characterised by both the magnitude of the reflection coefficient $|R_0|$ and the linear per-period dissipated power in the comb-shaped structure. These values are connected by the formula

$$P = \tfrac{1}{2} d \cos \vartheta (1 - |R_0|^2) W_0 \tag{5.59}$$

resulting from eqn. 5.44 if $kd < \pi$. In practice, it is enough to determine $|R_0|$ up to the terms of order η. It is possible, however, to calculate P directly and to find $|R_0|$ from eqn. 5.59. Here, when calculating P, we can use the field distribution corresponding to the case of perfect conductivity (zero-order approximation of the perturbation method). Otherwise, to calculate $|R_0|$ directly from eqn. 5.52, it is necessary to carry out complicated computations taking into account small quantities of the first order in η.

Unfortunately, for fields of magnetic polarisation a similar simplification is impossible even for small losses. This can be explained by the breaking of analyticity of $R_n(\eta)$, $T_n^{(0)}(\eta)$ in the neighbourhood of an infinite point. The physical sense of such a situation was discussed in Section 5.2. Here, we note once again that this greatly complicates loss calculations for cases of magnetic and mixed polarisation.

5.5 Effect of abnormally small absorption in periodic structures

Let us consider the incidence of a plane wave of electric polarisation on a finitely conducting comb-shaped structure. We assume that the depth of the structure is sufficiently large and its period is such that all E_{0n} modes in the grooves are evanescent (the E_{01} mode is not very close to the cutoff). Mathematically these assumptions are formulated as

$$\mu \gtrsim 1 \tag{5.60}$$

$$\Upsilon < \pi \qquad \exp\{-\mu(\pi^2 - \Upsilon^2)^{1/2}\} \ll 1 \tag{5.61}$$

$$\pi^2 - \Upsilon^2 \gg \mathrm{Re}\, Z_S / W_0 \Upsilon \tag{5.62}$$

where $\mu = D/d$, $\Upsilon = kd$ and Z_S is the surface impedance of the material of the ridges. Taking into account the inequality $|Z_S|/W_0 \ll 1$, in accordance with the results of Section 5.4, we shall use the perturbation theory. In other words, when calculating the linear (per-period) lost power P, we substitute the distribution of the magnetic field corresponding to the case of perfect conductivity of the grating material into eqn. 5.46. Then for P we obtain the expression (Slepyan and Slepyan, 1979)

$$P = \frac{\mathrm{Re}\, Z_S}{2} d \left[2\chi \int_{-1}^{1} \hat{f}(\zeta)\, \mathrm{d}\zeta + \mu \int_0^1 \{F_+(\xi) + F_-(\xi)\}\, \mathrm{d}\xi \right] \tag{5.63}$$

where

$$\hat{f}(\zeta) = \left| \sum_{n=0}^{\infty} T_n \exp(-h_n \mu) \cos\left\{ \frac{n\pi}{2} (\zeta - 1) \right\} \right|^2$$

$$F_\pm(\zeta) = \left| \sum_{n=0}^{\infty} T_n \varepsilon_n^\pm [\exp(-h_n \mu \xi) + \exp\{h_n \mu (\xi - 2)\}] \right|^2$$

$$h_n = \{(n\pi)^2 - \Upsilon^2\}^{1/2} \qquad (n \geq 1) \qquad h_0 = -j\Upsilon$$

$$\varepsilon_n^+ = 1 \qquad \varepsilon_n^- = (-1)^n \qquad \chi = |\eta_1|/|\eta|$$

T_n is the complex amplitude of the nth eigenmode in the groove, calculated on the assumption of perfect conductivity of the structure material.

In agreement with the conditions presented by eqns. 5.60–5.62 we can neglect the terms with $n \geq 1$ in the series for $F_\pm(\xi)$ and $\hat{f}(\zeta)$. This physically corresponds to the fact that only the contribution of the TEM mode to the power absorbed is taken into account. The reason for this simplification is a considerable decrease in T_n when n increases. In addition, the fields of the evanescent E_{0n} modes are exponentially damped with respect to x. This causes the terms ignored in eqn. 5.63 after integration to acquire an additional order of smallness.

Integrating in eqn. 5.63 and performing some simplifications, we obtain the following approximate formula:

$$P = 2d\, \mathrm{Re}\, Z_S \left(|T_0|^2 \left[\chi + \mu \left\{ 1 + \frac{\sin(2\Upsilon\mu)}{2\Upsilon\mu} \right\} \right] \right) \tag{5.64}$$

Thus, what is left is to determine the magnitude of the coefficient T_0.

In the zeroth order of the perturbation theory we have for T_n

$$T_n = -\frac{(2-\delta_{n0})v_n(-1)^n \hat{F}_n \exp(2h_n\mu)}{\left(1 - \sum_{m=0}^{\infty} \frac{\hat{F}_m d}{\Gamma_0 d + h_m}\right)} \left.\frac{df}{dw}\right|_{w=h_n/d,\, \eta=0} \tag{5.65}$$

where

$$v_n = \begin{Bmatrix} \sin(\Upsilon \sin\vartheta/2), & n=0,2,\ldots \\ j\cos(\Upsilon\sin\vartheta/2), & n=1,3,\ldots \end{Bmatrix} \qquad \hat{F}_n = F_n|_{\eta=0}$$

The coefficients $\hat{F}_n$ are determined from eqn. 5.52 in which the terms of the higher order of smallness in $\eta/k = Z_S/W_0$ have been ignored. It is possible to find an approximate analytical solution of eqn. 5.52 when the conditions given by eqns. 5.60–5.62 are fulfilled (in this case the 'reflection' of the higher E_{0n} modes from the $x = D$ plane will be negligibly small).[32] Then ignoring the exponentially small terms in eqn. 5.52 and performing some transformations, we can easily obtain the following formula for T_0:

$$T_0 = \frac{\Upsilon(\cos\vartheta - 1)\exp\left\{-\dfrac{j\Upsilon(1-\cos\vartheta)\ln 2}{\pi}\right\} H(-j\Upsilon)}{v_0 \tan\vartheta \left[1 - p\exp\left\{2j\Upsilon\left(\mu - \dfrac{\ln 2}{\pi}\right)\right\}\dfrac{H(-j\Upsilon)}{H(j\Upsilon)}\right] H(-j\Upsilon\cos\vartheta)} \tag{5.66}$$

where $p = \tan^2(\vartheta/2)$. The function $H(w)$ is expressed as

$$H(w) = \prod_{q=1}^{\infty}\left(1 - \frac{w}{h_q}\right)\left(1 - \frac{w}{\Gamma_q d}\right)^{-1}\left(1 - \frac{w}{\Gamma_{-q} d}\right)^{-1} \tag{5.67}$$

Now we can find the magnitude of T_0 from eqns. 5.66 and 5.67. Considering identities (they can easily be checked)

$$|H(j\Upsilon)|^2 = \frac{\tan^2\vartheta\{\cos(\Upsilon\sin\vartheta) - \cos\Upsilon\}\Upsilon}{2\sin^2\{(\Upsilon\sin\vartheta)/2\}\sin\Upsilon}$$

$$|H(j\Upsilon\cos\vartheta)|^2 = \frac{|H(j\Upsilon)|^2}{\Upsilon^2\sin\vartheta} 4\sin^2\left(\frac{\Upsilon\sin\vartheta}{2}\right)$$

we obtain for $|T_0|^2$ the formula

$$|T_0|^2 = \frac{(1-p)^2}{1 + p^2 - 2p\cos\Omega(\Upsilon)} \tag{5.68}$$

where

$$\Omega(\Upsilon) = 2\Upsilon\left(\mu - \frac{\ln 2}{\pi}\right) + 2\sum_{q=1}^{\infty}\left\{\sin^{-1}\left(\frac{\Upsilon}{2q\pi + \Upsilon\sin\vartheta}\right) + \sin^{-1}\left(\frac{\Upsilon}{2q\pi - \Upsilon\sin\vartheta}\right) - \sin^{-1}\left(\frac{\Upsilon}{q\pi}\right)\right\} \tag{5.69}$$

[32] The term 'reflection' in reference to evanescent modes is rather conditional; thus we use here the quotation marks.

Let us analyse the formulae obtained, taking for simplicity $\chi = 1$. The ohmic losses in the comb-shaped structure can be characterised by the coefficient

$$\alpha = \frac{P(\Upsilon, \vartheta, \mu)}{\tilde{P}} = \frac{1 - |R_0|^2}{1 - |\tilde{R}_0|^2} \tag{5.70}$$

where $\tilde{R}_0 = \lim_{D \to 0} R_0$ is the reflection coefficient for a smooth surface of the same material, $\tilde{P} = \lim_{D \to 0} P(\Upsilon, \vartheta, \mu)$ being the linear (per-interval equal to the structure period) power of losses in such a surface.

Let us consider first the case of normal incidence. Then $|T_0| = 1$, and, taking into consideration eqns. 5.64 and 5.70, we obtain

$$\alpha \simeq \left[1 + \mu\left\{1 + \frac{\sin(2\Upsilon\mu)}{2\Upsilon\mu}\right\}\right] \tag{5.71}$$

It is clear from eqn. 5.71 that at any Υ, μ $\alpha > 1$, i.e. the absorption in the comb-shaped structure at normal incidence is greater than that in a halfspace of the same material with a smooth surface. This result is rather obvious from the physical point of view: at normal incidence the reflection from the ridges is extremely small.[33] This is why the absorption in the grooves' bottoms is practically the same as that in a halfspace without ridges. In addition, there are additional conducting surfaces: ridges which significantly increase the total losses.

In Figures 5.4–5.6 are shown the dependencies $\alpha(\vartheta)$ ($\Upsilon, \mu = $ constant), $\alpha(\mu)$ ($\Upsilon, \vartheta = $ constant) and $\alpha(\Upsilon)$ ($\mu, \vartheta = $ constant). They are calculated by means of eqns. 5.64, 5.68 and 5.69 at $\chi = 1$. It can be seen from Figure 5.4 that $\alpha(\vartheta)$ is monotonically decreasing with increasing ϑ.[34] Physically, such a trend in the dependence $\alpha(\vartheta)$ can be explained by the fact that the field is 'forced out' of the comb-shaped structure in the case of oblique incidence. This process gets more intensive with increasing angle of incidence. As a result the currents induced on the metal surfaces decrease and this leads to diminishing of the losses. At $\vartheta \to \pi/2$ the amplitudes of all surface waves $|R_n|$ ($n \geq 1$) are strictly equal to zero, and the reflected wave accurately compensates the incident one. This is the Rayleigh diffraction anomaly characterised by the fact that the longitudinal wave number of the fundamental space harmonic Γ_0 is transformed into null. The Rayleigh anomaly can be considered as an extreme case of the field 'forcing out' from the periodic structure: 'forcing out to infinity'. A similar mechanism is typical for a plane wave scattered by a rough surface: if the wave incident is at a shallow angle, it in practice does not 'feel' any roughness while any other waves are scattered intensively.

Nevertheless, the use of the Rayleigh anomaly alone is not an effective method of loss reduction in electrodynamic systems as it provides small ohmic losses only at grazing incidence. However, there is one more mechanism for 'forcing out' the field; to understand the physical nature of this we shall refer to eqn. 5.66. Let us present eqn. 5.66 as

---

[33] In the case of perfect conductivity there is none.

[34] Such reduction takes place up to the angle $\vartheta \simeq \cos^{-1}(|\eta|/k)$.

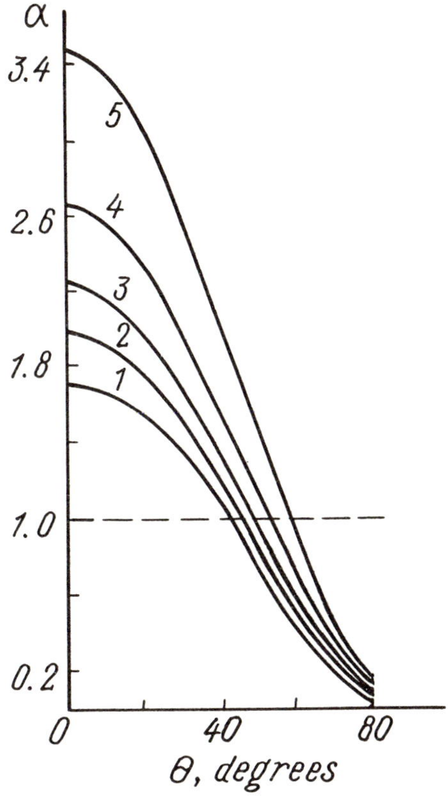

Figure 5.4 *Absorption in comb-shaped structure against angle of incidence*

H polarisation

1 $\Upsilon = 2.2,\ \mu = 0.71$
2 $\Upsilon = 1.57,\ \mu = 1.0$
3 $\Upsilon = 1.26,\ \mu = 1.25$
4 $\Upsilon = 0.94,\ \mu = 1.66$
5 $\Upsilon = 0.63,\ \mu = 2.5$

$$T_0 = \frac{s_0}{1 - r_0 \exp(2jkD)} = s_0\{1 + r_0 \exp(2jkD) + r_0^2 \exp(4jkD) + \cdots\} \tag{5.72}$$

where

$$s_0 = \frac{\Upsilon(\cos\vartheta - 1)\exp\left\{-\dfrac{j\Upsilon(1-\cos\vartheta)\ln 2}{\pi}\right\} H(-j\Upsilon)}{v_0 \tan\vartheta H(-j\Upsilon\cos\vartheta)}$$

$$r_0 = p\exp\left(\frac{2jkd\ln 2}{\pi}\right)\frac{H(-j\Upsilon)}{H(j\Upsilon)}$$

more rigorous analysis considering the higher modes (see below) leads to an even faster increase in dependencies $\alpha(\Upsilon)$ at $\Upsilon > \pi/2$ (broken lines in Figure 5.6).

Numerical evaluations show that when the conditions given by eqns. 5.60–5.62 are fulfilled, we can neglect the infinite series in eqn. 5.69. Then we obtain the following simple formula for the dissipated power:

$$P \simeq \left[1 + \mu\left\{1 + \frac{\sin(2\Upsilon\mu)}{2\Upsilon\mu}\right\}\right] \frac{(\sqrt{2})d(1-p)^2|Z_S|}{1 + p^2 - 2\cos\left\{2\Upsilon\left(\mu - \frac{\ln 2}{\pi}\right)\right\}} \tag{5.74}$$

To check on the accuracy of the approximate analytical formulae obtained and also lift the restrictions given by eqns. 5.60–5.62, the problem under consideration has been solved numerically. The coefficients T_n have been determined in accordance with eqn. 5.65. The SLAE-2 given by eqn. 5.52 was solved in a zeroth-order approximation of the perturbation method by means of truncation up to the Nth order. The dissipated power was calculated with the aid of eqn. 5.60 but in the series for $\hat{f}(\zeta)$ and $F_{\pm}(\xi)$ the first N terms were retained. Specific calculations have been carried out for $N = 10$ which, as shown by numerical experiments consisting in the variation of N, ensures, in this range of parameters, an accuracy within 0.1%. The results of the numerical solution are shown in Figure 5.5 (insert) and in Figure 5.6 (broken lines). Comparison of these data with the results obtained by means of eqns. 5.64, 5.68, 5.69 or eqn. 5.74 shows that at $\mu \geq 1$, $\Upsilon \leq \pi/2$ they coincide within the accuracy of the graphical presentation. At $\Upsilon > \pi/2$ or $\mu < 1$ a slight difference is observed which can be explained by the increasing effect of the higher modes in the grooves. Note that the qualitative character of the dependencies $\alpha(\Upsilon)$, calculated by approximate analytical formulae, also remains correct at $\Upsilon > \pi/2$.

A similar effect also takes place in acoustics: to reduce the energy losses at the reflection of a sound wave from a rigid absorbing surface, the latter should be corrugated. The theory and formulae for calculation in the acoustic case are the same.

We return now to the case of E-polarised wave diffraction by the parallel-plate grating analysed in Section 5.2. At $kd < \pi$ and, even more so, at $kd \ll 1$, the electrodynamic characteristics of such structures will not differ much from those of lamellar gratings with sufficiently deep grooves (all modes in the grooves are evanescent). Using eqn. 5.59 (which is correct also for the case of E-polarisation) and eqn. 5.35 for $|R|$, we obtain

$$P(\vartheta) = \frac{(1-p^2)\cos\vartheta}{\Gamma^2 p^2 - 2\Gamma p \cos\gamma + 1} P(0) \tag{5.75}$$

where Γ and γ are the magnitude and phase of the reflection coefficient.

To evaluate $P(\vartheta)$, the quantities Γ and γ can be taken assuming that the structure material is perfectly conducting. Then, according to Weinstein (1969*b*), for $kd \ll 1$

$$\Gamma \simeq 1 \qquad \gamma \simeq \gamma_0 \simeq \pi + \frac{2kd \ln 2}{\pi} \tag{5.76}$$

The analysis of eqns. 5.75 and 5.76 shows that at $\vartheta \to \pi/2$ (grazing incidence) $P(\vartheta)$ sharply falls down to zero. Here, the Rayleigh diffraction anomaly demonstrates itself as being independent of the wave polarisation. However, in this case there is no interference mechanism of the field 'forcing out', as is typical for electrically polarised waves. This is why the losses in the structure under consideration are higher than those in the smooth surface of the same material. In fact, introducing the coefficient $\alpha(\vartheta) = P(\vartheta)/P_0(\vartheta)$ [$P(\vartheta)$ being the linear (per-interval d) dissipated power in a smooth surface of the same material illuminated by magnetically polarised plane wave with unit amplitude], we obtain from eqn. 5.76

$$\alpha(\vartheta) \simeq \frac{1+p^2}{1+p^2-2p\cos\gamma_0}\,\alpha(0) \tag{5.77}$$

Taking into consideration the fact that $\alpha(0) > 1$ (see Section 5.2), from eqn. 5.77 it follows that $\alpha(\vartheta) > 1$ for all ϑ. In other words, under the conditions discussed above, there is no abnormally small dissipation effect for E-polarised waves.

5.6 Absorption in inclined comb-shaped structures and echelettes

The configuration of the structures under consideration and main designations are shown in Figure 5.7. The echelette grating is considered as a limiting case of an inclined comb-shaped structure ($D \to 0$). Let us limit the scope of our interests to the analysis of the H-polarisation. We shall use the energy-perturbation method according to which the linear (per-period of the structure) dissipated power P is evaluated by the formula

$$P = \frac{1}{2}\int_L \operatorname{Re} Z_S |\psi|^2\, \mathrm{d}l \tag{5.78}$$

where ψ describes the current density distribution on the surface of a perfectly conducting structure of the same geometry, and L is a contour AEFO corresponding to one unit cell of the structure (see Figure 5.7).

In this case for $x > 0$ we have

$$\psi(x, z) = \exp\{jk(z\sin\vartheta - x\cos\vartheta)\} + \sum_{p=-\infty}^{\infty} R_p \exp(j\alpha_p z - \Gamma_p x) \tag{5.79}$$

where $\alpha_p = (2p\pi + u)/d \sec\beta$, $u = kd\sin\vartheta\sec\beta$ and $\Gamma_p = (\alpha_p^2 - k^2)^{1/2}$.

For $-D - md\tan\beta < \zeta < -md\tan\beta$ and $md < \xi < (m+1)d$ (m being the number of the cell), we have

$$\psi^{(m)}(\zeta, \xi) = \sum_{n=0}^{\infty} T_n^{(m)} \cos\left\{n\pi\left(\frac{\xi}{d} - m\right)\right\} (\exp\{\gamma_n(\zeta + md\tan\beta)\} + \exp[-\gamma_n\{\zeta + 2D + (m+2)d\tan\beta\}]) \tag{5.80}$$

where $\gamma_n = \{(n\pi/d)^2 - k^2\}^{1/2}$ and $T_n^{(m)} = T_n^{(0)}\exp(jmu)$.

Matching of the expansions given by eqns. 5.79 and 5.80 is performed with the aid of a 2-dimensional version of the second Green's theorem for the triangle area OAB (Mittra and Lee, 1971, Shestopalov *et al.*, 1984). As a result we obtain an infinite system of linear algebraic equations for the coefficients R_p and the

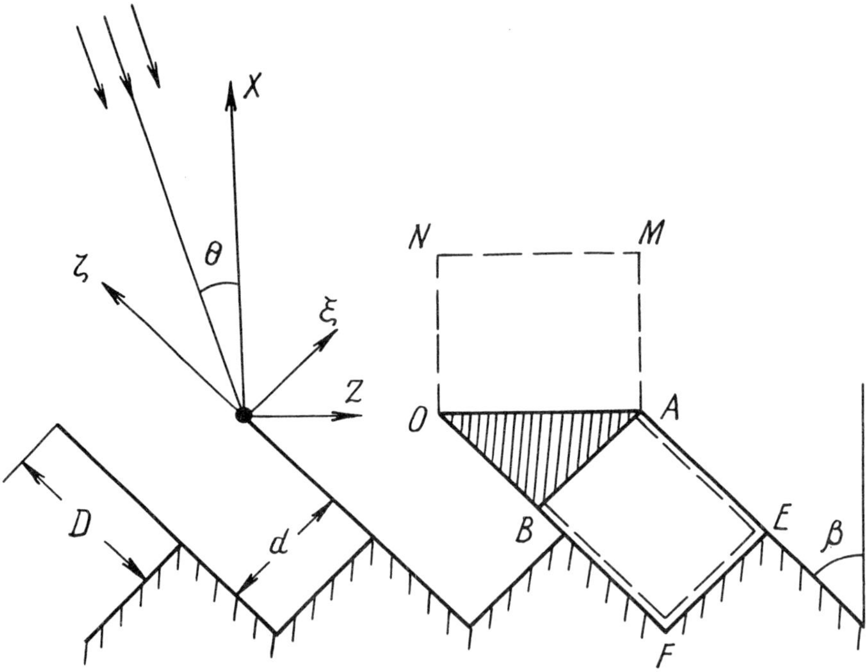

Figure 5.7 *Inclined comb-shaped structure*

recalculating formula for $T_n^{(m)}$. Solution of this system and evaluation of the coefficients $T_n^{(m)}$ are carried out through the modified residue–calculus technique. For a conventional comb-shaped structure (corresponding to $\beta \to 0$ in Figure 5.7) the representation of ψ by eqn. 5.80 is correct over all the contour L and calculation of P is reduced to substitution of eqn. 5.80 into eqn. 5.78 and numerical integration, as in Section 5.5. In the case considered $\beta \neq 0$, and the problem therefore arises how to evaluate ψ on the segment OB [representations given by eqns. 5.79 and 5.80 in the triangle area OAB are not correct (Shestopalov *et al.*, 1984)]. This difficulty can be avoided using the second Green identity for some area including the OAB triangle. This area can be the polygon ONMAEF (see Figure 5.7) which is convenient from the computational point of view. In this case we obtain, according to papers by Slepyan (1984) and Slepyan and Slepyan (1986)

$$\psi(\zeta)|_{OB} = \int_{\mathscr{L}} \left(\psi \frac{\partial G}{\partial \boldsymbol{n}} - G \frac{\partial \psi}{\partial \boldsymbol{n}} \right) \mathrm{d}l \tag{5.81}$$

where $\mathscr{L}$ represents the boundary of the area ONMAEF excluding the segment OF,

$$G = \{\mathcal{N}_0(kR_1) + \mathcal{N}_0(kR_2)\} \quad \text{and} \quad R_{1,2} = \{(\zeta - \zeta')^2 + (\xi \pm \xi')^2\}^{1/2}$$

The integral in eqn. 5.81 is evaluated numerically and ψ and $\partial\psi/\partial\boldsymbol{n}$ are calculated from eqn. 5.79 on the segment AEF and from eqn. 5.80 on all other parts of the contour $\mathscr{L}$.

For an echelette grating a more effective method is possible (Slepyan and Slepyan, 1986). Apart from the structure under consideration an auxiliary structure of this type with the same period and $\beta' = \pi/2 - \beta$ is considered (in fact it is the same structure but in performing the matching of modal expansions we introduce a virtual-plane waveguide on its other facet). In this case $\psi(\zeta)|_{OB}$, for the structure under consideration, equals $\psi(\xi)|_{BA}$ in the auxiliary problem and hence is represented by an expansion in terms of eigenmodes of a virtual $(D \to 0)$ plane waveguide with the width $d' = d \tan \beta$.

Let us present some results of calculations obtained with the aid of the technique described. The parameters of the gratings are such that all space harmonics except those which are mirror-reflected are nonpropagating. In all figures the α coefficient is defined by eqn. 5.70.

In Figure 5.8 the dependencies of α on the angle of incidence ϑ are shown for echelette gratings with different $X = kd \sec \beta$. As for lamellar gratings considered in Section 5.5, the effect of abnormally small dissipation $(\alpha < 1)$ is observed. However, in echelette gratings it occurs at larger angles of incidence. A similar regularity is typical for inclined comb-shaped structures.

Figures 5.9 and 5.10 show the coefficient α as a function of X at different ϑ for inclined comb-shaped gratings. Unlike conventional comb-shaped structures, for some values of ϑ the dependence $\alpha(X)$ at small ϑ is increasing (curves 1–4 in Figure 5.9). The area $X > \pi \sec \beta$ in Figure 5.10 is of considerable interest (in this area two waveguide modes are propagating). For this regime, a large increase in losses is observed and sharp oscillations of α take place. Physically such behaviour of α can be explained by the interference phenomena for different modes propagating in the parallel-plate waveguides.

Figure 5.11 shows the dependence of α on the echelette blaze angle β at two values of the angle of incidence. At $\vartheta = 80°$, $\beta_{opt} = 45°$ corresponds to the regime of minimal absorption. For $\vartheta = 0°$ the situation is different: $\alpha_{min} = 1$ at $\beta \to 90°$, and at $\beta = 45°$ the coefficient α is maximal.

To conclude, we would like to dwell upon one methodological question. For all the structures examined above, in the case of H-polarisation the energy-perturbation method has been used. It is based on the integration of current-density distributions evaluated in approximation of perfect conductivity (eqns. 5.46 and 5.78). This approach led to rather convenient computational algorithms (most of the difficulties arose with inclined comb-shaped structures—see eqn. 5.81). Let us note here that constructive methods for solving the problems directly in impedance formulation do exist: the semi-inversion technique for echelette gratings (Shestopalov *et al.*, 1984), and the method of regularising substitution for inclined comb-shaped structures (Slepyan, 1984).

For E-polarisation the energy-perturbation method does not lead to a satisfactory solution. First, its direct application is impossible for all structures except echelette gratings—a modification described in Section 2.4 has to be carried out. In addition, the algorithms obtained in such a manner need a rather large number of computations. Thus, for example, in Section 5.2 the induced current density is presented as a Fourier integral. Its evaluation and the subsequent numerical integration are very consuming of machine time. Of course, instead of a Fourier integral we can turn to an expansion in terms of eigenmodes for

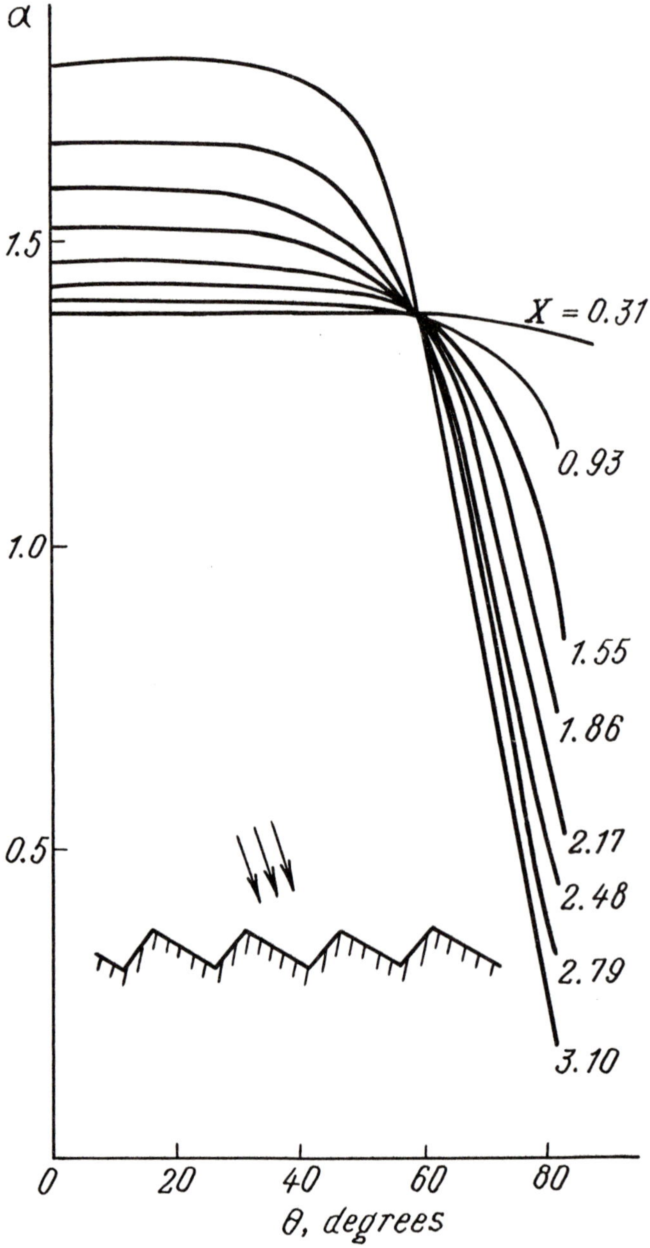

Figure 5.8 *Absorption in echelette grating against angle of incidence*
H polarisation, $\beta = 60°$

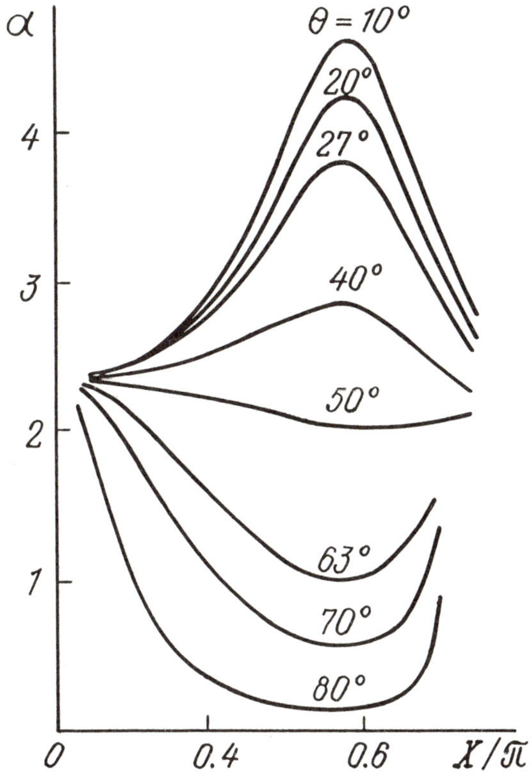

Figure 5.9 *Absorption in inclined comb-shaped structure against X*
H polarisation, $\beta = 60°$, $D/d = 1$

the current density but the convergence of this expansion would be rather slow. The reason is the singular behaviour of the current density at the edge vicinity (for H-polarisation there are no such singularities). Therefore for the E-polarised fields it is preferable to solve the diffraction problem directly with the use of the impedance formulation (see, for example, Section 5.2).

5.7 Diffraction by a complex-shaped periodic structure: integral equation method

So far we have examined absorption processes in periodic structures with special configurations which allowed us to use special highly effective numerical–analytical methods. The analysis of wave diffraction by periodic structures of an arbitrary configuration is of considerable interest. Development of analysis methods suitable for any grating profile would enable us, specifically, to solve problems of optimisation of the shape of periodic structures to minimise the power absorbed. Certainly, such methods can only be numerical, based on the modern achievements in computational mathematics.

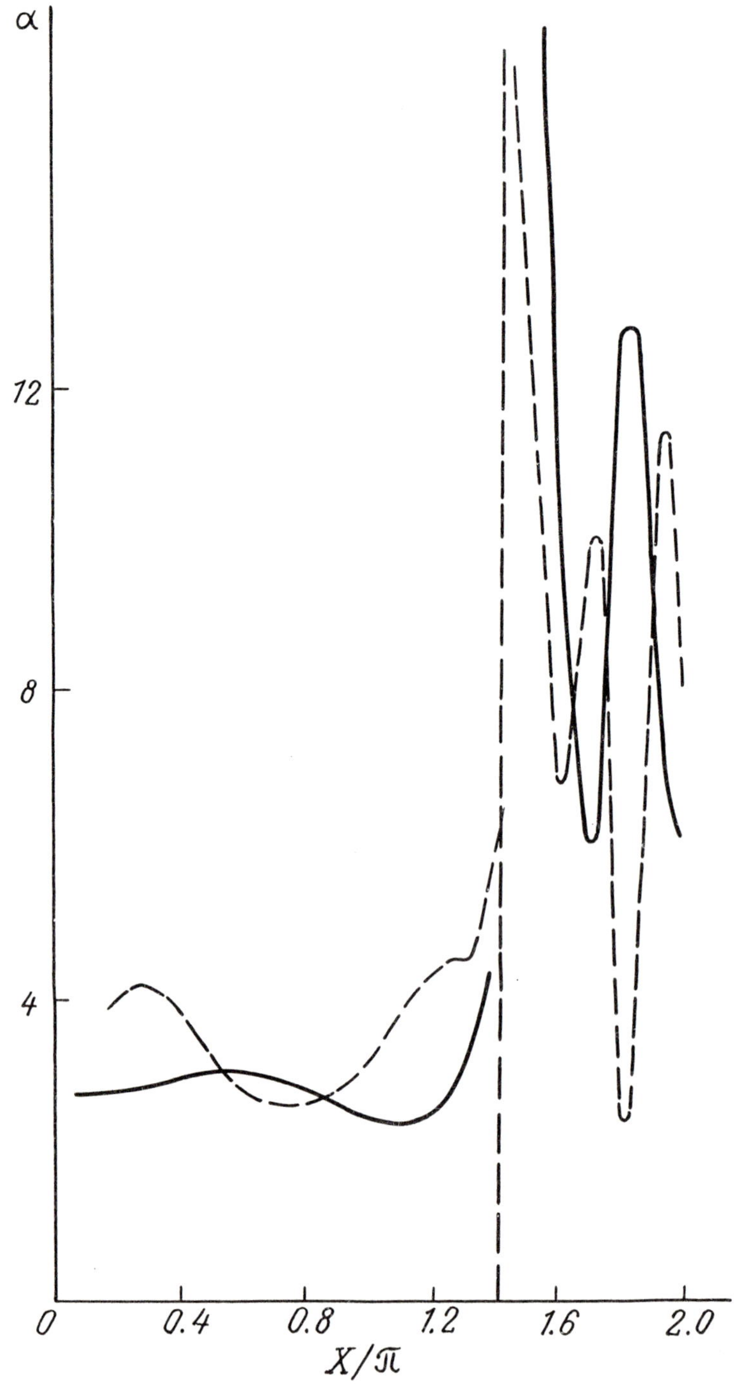
α
12
8
4
0
0.4
0.8
1.2
1.6
2.0
X/π

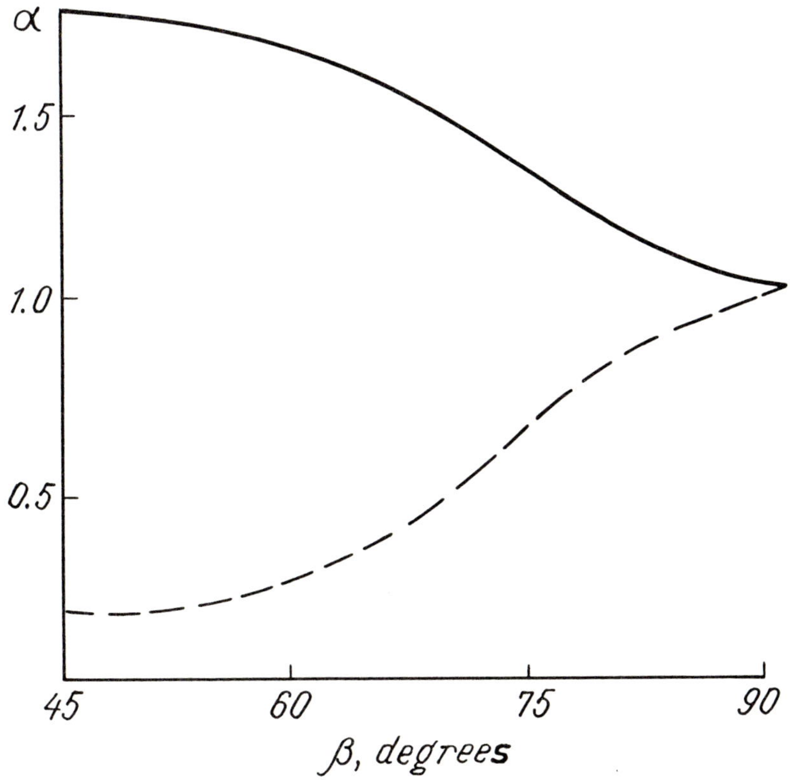

Figure 5.11 *Absorption in echelette grating as a function of blaze angle*

H polarisation, $X = 1.79$
——— $\vartheta = 0$
- - - - $\vartheta = 80°$

The configuration of the structures under consideration and main designations are shown in Figure 5.12. It is necessary to distinguish between transmitting [Figure 5.12(i)] and reflecting [Figures 5.12(ii) and (iii)] gratings. In this Section we dwell mainly upon the transmission structures. The method of integral equations based on potential theory will be used. This method has already been used extensively in mathematical physics and mathematical theory of diffraction, in particular Colton and Kress (1983), Parton and Perlin (1977), Poggio and Miller (1973) and Galishnikova and Ilyinsky (1987). For impedance-boundary

---

Figure 5.10 *Absorption in inclined comb-shaped structure against X/π*

H polarisation, $\vartheta = 0$, $\beta = 45°$
——— $D/d = 1.666$
- - - - $D/d = 1.0$

conditions, use of this technique is characterised by some specific features, which will be discussed in this Section.

Let us consider a problem of diffraction of electromagnetic field by an infinitely wide grating with period d obtained by translation of a cylinder with cross-section S, bounded by the contour C, along the z axis. Now let the incident excitation be simultaneously created by a line source at the point $M(\tilde{x}, \tilde{z})$ and the plane-wave set of the type

$$\psi^0(x, z) = \sum_n C_n^0 \exp(\Gamma_n x + j\alpha_n z)$$

where $\Gamma_n = (\alpha_n^2 - k^2)^{1/2}$ ($\mathrm{Im}\,\Gamma_n < 0$; $\mathrm{Re}\,\Gamma_n > 0$ if $\mathrm{Im}\,\Gamma_n = 0$), and α_n and C_n^0 are given constants. This form of the incident field is typical of diffraction electronics, in particular problems of radiation of charged particles moving across a periodic structure [Smith-Purcell effect, Shestopalov (1976)].

The problem of wave scattering by such a periodic structure consists in solving the Helmholtz equation for the total field

$$\nabla^2\psi + k^2\psi = -2\pi\delta(M, \tilde{M}) \tag{5.82}$$

which satisfies the following boundary conditions on the structure surface:

$$\frac{\partial\psi}{\partial \boldsymbol{n}}(P) - j\eta(P)\psi(P) = 0 \qquad (P \in C) \tag{5.83}$$

In addition, the radiation condition should be added; the analytical form for the radiation condition will be formulated later.

One more peculiar feature of this statement of the problem is a nonperiodical excitation of the periodic structure owing to which Floquet's conditions of pseudoperiodicity cannot be used. This difficulty can be overcome with the aid of the discrete Fourier transform.

Let us reduce the original diffraction problem to that in a strip:

$$D_1 = \left\{ -\frac{d}{2} \leq z \leq \frac{d}{2} \qquad -\infty < x < \infty \right\}$$

Let us introduce the transform (Ilyinsky, 1973)

$$U(\tau, x, z) = \sum_{n=-\infty}^{\infty} \psi(x, z + nd)\exp(-jn\tau) \tag{5.84}$$

The function $U(\tau, x, z)$ is defined for any quadratically integrable function $\psi(x, z)$ in D_1 and satisfies Floquet's condition of pseudoperiodicity with the parameter τ:

$$U(\tau, x, z + d) = \exp(j\tau) \sum_{n=-\infty}^{\infty} \psi(x, z + nd)\exp(-jn\tau)$$
$$= \exp(j\tau)U(\tau, x, z) \tag{5.85}$$

Because eqn. 5.85 is true, to determine $\psi(x, z)$ at any point of the space it is enough to determine $U(\tau, x, z)$ in the strip D_1. The functions $\psi(x, z)$ and $U(\tau, x, z)$ are related by the formula

$$\psi(x, z) = \frac{1}{2\pi}\int_0^{2\pi} U(\tau, x, z)\, d\tau \tag{5.86}$$

The function $U(\tau, x, z)$ satisfies the equation

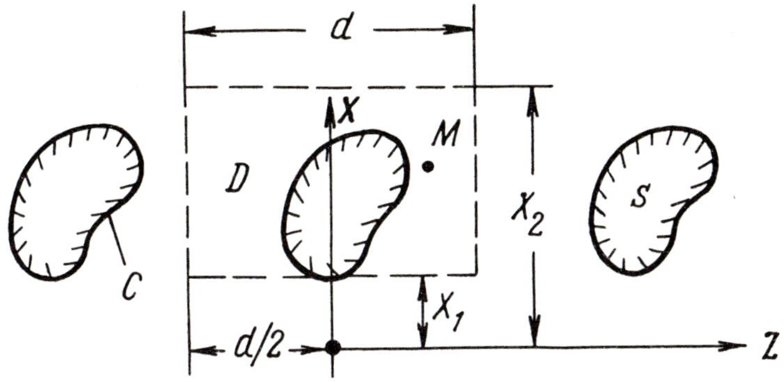

(i)

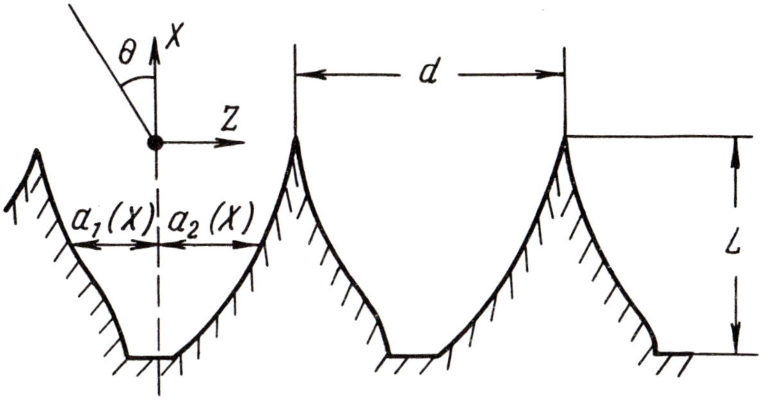

(ii)

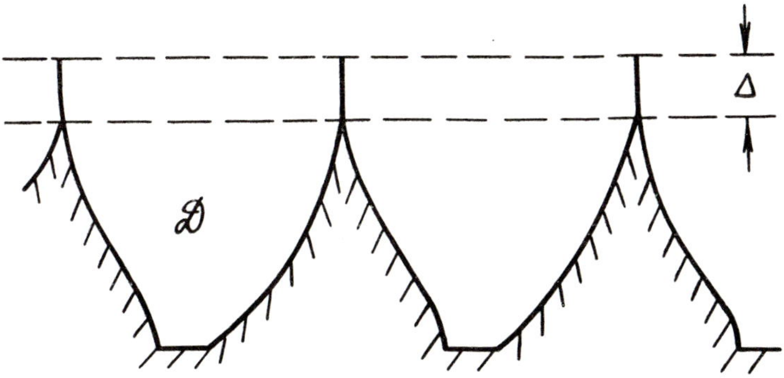

(iii)

Figure 5.12 *Geometry of complex-shaped periodic structures*

(i) Transmission grating
(ii) and (iii) Reflecting gratings

$$\nabla^2_{xy} U(\tau, x, z) + k^2 U(\tau, x, z) = -2\pi \sum_{n=-\infty}^{\infty} \delta(x - \tilde{x})\delta(z - \tilde{z} + nd) \exp(-jn\tau) \quad (5.87)$$

the boundary condition given by eqn. 5.83, and the radiation condition. The latter can be formulated as follows: let us consider the strip $x_1 \leq x \leq x_2$ containing the line source and the body S. The function $U(\tau, x, z)$ at $x \leq x_1$ and $x \geq x_2$ can be presented as

$$U(\tau, x, z) = \sum_{n=-\infty}^{\infty} T_n(\tau)\psi_n(\tau, z) \exp(\Gamma_n x) \qquad x \leq x_1 \quad (5.88)$$

$$U(\tau, x, z) = U_0(\tau, x, z) + \sum_{n=-\infty}^{\infty} R_n(\tau)\psi_n(\tau, z) \exp(-\Gamma_n x) \qquad x \geq x_2 \quad (5.89)$$

where

$$\psi_n(\tau, z) = \exp(j\alpha_n z)/(\sqrt{d}) \qquad \alpha_n = (\tau + 2\pi n)/d$$

$U_0(\tau, x, z)$ being the transform defined by eqn. 5.84 for the incident field $\psi^0(x, z)$. The coefficients $R_n(\tau)$ and $T_n(\tau)$ are unknown and should be determined. Eqns. 5.88 and 5.89 can be regarded as an analytical form of the radiation condition.

To obtain the integral equation for the function $U(\tau, x, z)$ let us introduce Green's function $g(M, P)$ satisfying Floquet's condition and the equation

$$\nabla^2 g(M, P) + k^2 g(M, P) = -2\pi\delta(M, P) \quad (5.90)$$

at any point of the space. The solution of eqn. 5.90 can be expressed as

$$g(M, P) = \frac{j\pi}{2} \sum_{n=-\infty}^{\infty} H_0^{(1)}[k\{(x_M - x_P)^2 + (z_M + nd - z_P)^2\}^{1/2}] \exp(-jn\tau) \quad (5.91)$$

This function satisfies the following conditions of pseudoperiodicity:

$$\left.\begin{aligned} g(x_M, z_M, x_P, z_P) &= \exp\{j(n-m)\tau\} g(x_M, z_M + md, x_P, z_P + nd) \\ \frac{\partial}{\partial z_M} g(x_M, z_M, x_P, z_P) &= \exp(j\tau) \frac{\partial}{\partial z_M} g(x_M, z_M - d, x_P, z_P) \\ \frac{\partial}{\partial z_P} g(x_M, z_M, x_P, z_P) &= \exp(-j\tau) \frac{\partial}{\partial z_P} g(x_M, z_M, x_P, z_P - d) \end{aligned}\right\} \quad (5.92)$$

Applying the second Green identity to the functions $U(\tau, x, z)$ and $g(M, P)$ in the area D_1 [Figure 5.12(i)] and taking into account the conditions of pseudoperiodicity (eqns. 5.85 and 5.92) and also the radiation condition (eqns. 5.88 and 5.89), we obtain an integral representation for the total field

$$U(\tau, M) = \frac{1}{2\pi} \int_C \left\{ g(M, P) \frac{\partial U(\tau, P)}{\partial \boldsymbol{n}_P} - U(\tau, P) \frac{\partial g(M, P)}{\partial \boldsymbol{n}_P} \right\} dl_P + g(M, \tilde{M}) + U_0(\tau, M) \qquad M \in D \quad (5.93)$$

$\boldsymbol{n}$ being the unit normal vector directed out of S. Eqn. 5.93 enables us to find $U(\tau, M)$ everywhere in the area D in terms of $U(\tau, P)$ and $\partial U(\tau, P)/\partial \boldsymbol{n}_P$ on C. Assuming that $\eta(P) \not\equiv 0$, $M \in C$ and using eqn. 5.83 and the properties of simple-layer and double-layer potentials (Tikhonov and Samarsky, 1977), we obtain the integral equation for $\partial U(\tau, P)/\partial \boldsymbol{n}_P$

$$\frac{1}{2}\frac{j}{\eta(P)}\frac{\partial U(\tau, P)}{\partial \boldsymbol{n}_P} + \frac{1}{2\pi}\int_C \left\{\frac{j}{\eta(P')}\frac{\partial g(P, P')}{\partial \boldsymbol{n}_{P'}} + g(P, P')\right\}\frac{\partial U(\tau, P')}{\partial \boldsymbol{n}_{P'}}\, dl_{P'}$$
$$= -g(P, \tilde{M}) - U_0(\tau, P) \qquad P \in C \qquad \eta(P) \not\equiv 0 \qquad (5.94)$$

If $\eta(P) \not\equiv \infty$, we similarly obtain the integral equation for $U(\tau, P)$

$$\frac{U(\tau, P)}{2} - \frac{1}{2\pi}\int_C \left\{j\eta(P')g(P, P') - \frac{\partial g(P, P')}{\partial \boldsymbol{n}_{P'}}\right\} U(\tau, P')\, dl_{P'}$$
$$= g(P, \tilde{M}) + U_0(\tau, P) \qquad P \in C \qquad \eta(P) \not\equiv \infty \qquad (5.95)$$

Eqn. 5.94 at $\eta(P) \equiv \infty$ and eqn. 5.95 at $\eta(P) \equiv 0$ turn into integral equations for a perfectly conducting grating in cases of magnetic and electric polarisation, respectively. Eqns. 5.94 and 5.95 can be written in a general form

$$\mu(P) + \frac{1}{2\pi}\int_C K(P, P')\mu(P')\, dl_{P'} = F(P) \qquad (5.96)$$

where $\mu(P)$ is an unknown function.

To construct a numerical algorithm for the solution of eqn. 5.96 accurate enough for various shapes of the contour C, it is necessary to approximate the integral operator in such a manner that changing of the contour will not influence the accuracy of approximation. The difficulty of natural integration along the contour C is in the fact that for an arbitrary contour the co-ordinates of the points P and P', on which the kernel in eqn. 5.96 depends, cannot be expressed in terms of the length of the arc without any quadrature computation. For this reason let us introduce inside S an auxiliary contour C_1 consisting of straight lines and segments of circles. Let $r = r_0(\phi)$ be an equation of the contour C in the cylindrical polar co-ordinates r, ϕ with the origin at the point O situated inside the contour. Let us connect the system of co-ordinates $O\xi\zeta$ with the contour C_1. The ξ axis makes the angle ρ with the x axis of initial Cartesian-co-ordinate system. Let us assume, for simplicity, that the origins of the systems Oxz and $O\xi\zeta$ coincide. Taking into account the character of the contour C, we can choose the angle ρ and the dimensions for the contour C_1. If δ is the angle in the polar system of co-ordinates with the ξ axis, then $\phi = \rho + \delta$, and for the contour C_1 it is easy to express the angle δ in terms of the length of the arc s, i.e. $\delta = \delta(s)$. Establishing a one-to-one correspondence of the points $P \in C$ and $P \in C_1$ having the same polar azimuth ϕ, we obtain a parametric description of the contour C as a function of the arc length of the contour C_1:

$$\left.\begin{aligned} r &= r_0\{\rho + \delta(s)\} = R(s) \\ x &= R(s)\cos\{\rho + \delta(s)\} \\ z &= R(s)\sin\{\rho + \delta(s)\} \end{aligned}\right\} \qquad (5.97)$$

The arc elements for the contours C and C_1 are related as follows:

$$dl_C = \{R^2(s)(\delta_s')^2 + (R_s')^2\}^{1/2}\, ds_{C_1} = I(s)\, ds_{C_1} \tag{5.98}$$

Replacing the integration contour C in eqn. 5.96 by C_1 and denoting the kernel by $K(s_P, s_{P'})$ (s_P and $s_{P'}$ are the values of s for the points P and P', respectively), we obtain

$$\mu(s_P) + \frac{1}{2\pi}\int_{C_1} K(s_P, s_{P'}) I(s_{P'}) \mu(s_{P'})\, ds_{P'} = F(s_P) \tag{5.99}$$

To solve the integral equation given by eqn. 5.99 numerically, the Krylov–Bogolubov technique (Kantorovich and Krylov, 1958) is used. In the framework of this method the integral equation is reduced to an algebraic system of linear equations. For this purpose let us divide the contour C_1 by points $s_0, s_1, \ldots, s_N$ ($s_0 = s_N$) into N subintervals approximating $\mu(s_P)$ as a piecewise-constant function. Computing the integrals by a quadrature formula, we can write eqn. 5.99 in the form

$$\frac{1}{2\pi}\sum_{j=0}^{N-1} \mu(s_{P_{j+1/2}}) \int_{s_{P_j}}^{s_{P_{j+1}}} K(s_P, s_{P'}) I(s_{P'})\, ds_{P'} + \mu(s_P) = F(s_P) \tag{5.100}$$

where

$$s_{P_{j+1/2}} = s_{P_j} + \tfrac{1}{2}(s_{P_{j+1}} - s_{P_j})$$

Assuming that $s_P = s_{P_{j+1/2}}$ in eqn. 5.100, we obtain a system of linear algebraic equations for $\mu(s_{P_{j+1/2}})$:

$$\sum_{j=0}^{N-1} A_{ij}\mu(s_{P_{j+1/2}}) = F(s_{P_{i+1/2}}) \tag{5.101}$$

where $i = 0, 1, \ldots, N-1$, and

$$A_{ij} = \frac{1}{2\pi}\int_{s_{P_j}}^{s_{P_{j+1}}} K(s_{P_{i+1/2}}, s_P) I(s_P)\, ds_P + \delta_{ij}$$

The accuracy of solving the linear system obtained depends on the accuracy of computation of the coefficients A_{ij} defined by the behaviour of $K(s_P, s_{P'})$ on the contour C_1.

Green's function expressed by eqn. 5.91 has a logarithmic singularity when the arguments coincide; when calculating the matrix elements in eqn. 5.101 this singularity should be extracted, i.e. we need to present Green's function as

$$g(P, P') = g_1(P, P') + \ln \Phi(P, P') \tag{5.102}$$

where $g_1(P, P')$ is a regular function, and $\lim_{P\to P'} \Phi(P, P') = 0$.

With the aid of Poisson summation formula we present $g(P, P')$ in the form

$$g(P, P') = \frac{\pi}{d}\sum_{n=-\infty}^{\infty} \frac{\exp\{j\alpha_n(z_P - z_{P'}) - \Gamma_n|x_P - x_{P'}|\}}{\Gamma_n}$$

Let us consider the function $\hat{g}(P, P')$, having the singularity $\ln(r_{PP'})^{-1}$ at

$P = P'$ similar to $g(P, P')$:

$$\hat{g}(P, P') = \exp\left(j\tau \frac{z_P - z_{P'}}{d}\right) \sum_{n=1}^{\infty} \cos\left(2n\pi \frac{z_P - z_{P'}}{d}\right) \times \exp\left(-2n\pi \frac{|x_{P'} - x_P|}{d}\right) \frac{1}{n}$$

Summing the infinite series in this expansion, we have

$$\hat{g}(P, P') = \exp\left(j\tau \frac{z_P - z_{P'}}{d}\right) \times \left\{-\ln 2 + \frac{\pi}{d}|x_{P'} - x_P| + \frac{1}{2}\ln \Phi(P, P')\right\}$$

where $\Phi(P, P')$ are written in the form

$$\Phi(P, P') = \left[\sinh^2\left\{\frac{\pi}{d}(x_P - x_{P'})\right\} + \sin^2\left\{\frac{\pi}{d}(z_P - z_{P'})\right\}\right]^{-1/2}$$

Using the formulae written previously, we can present Green's function in the form of eqn. 5.102.

Let us examine $\partial g(P, P')/\partial \boldsymbol{n}_P$, the normal derivative of Green's function expressed by eqn. 5.102. It can be shown [see Tikhonov and Samarsky (1977)] that, if the curvature $\chi(P)$ of the contour C is continuous, then the normal derivative of Green's function has no singularity and

$$\lim_{P \to P'} \frac{\partial g(P, P')}{\partial \boldsymbol{n}_{P'}} = \frac{1}{2}\chi(P)$$

The series involved in the expression for $\partial g_1(P, P')/\partial \boldsymbol{n}_{P'}$ have an order of convergence of $O(n^{-1})$. Therefore, in spite of the fact that $\partial g_1(P, P')/\partial \boldsymbol{n}_{P'}$ is a regular function, it is necessary to improve their convergence. One possible way to do it is described by Galishnikova and Ilyinsky (1987). Let us note that the kernel singularity is completely determined by the logarithmic singularity of Green's function.

Thus while the elements of the matrix A_{ij} $(i \neq j)$ in eqn. 5.101 are calculated, the integrals are evaluated by means of ordinary quadrature formulae with the accuracy required. The integrals of the singular functions which appear when $i = j$ are calculated in the following manner:

$$\int_{s_{P_j}}^{s_{P_{j+1}}} \exp\left\{\frac{j\tau}{d}(z_{P_{i+1/2}} - z_{P'})\right\} I(s_{P'}) \ln \Phi(P_{i+1/2}, P')\, ds_{P'}$$
$$= \int_{s_{P_j}}^{s_{P_{j+1}}} \exp\left\{\frac{j\tau}{d}(z_{P_{i+1/2}} - z_{P'})\right\} I(s_{P'}) \times \ln\{\Phi(P_{i+1/2}, P')|s_{P'} - s_{P_{i+1/2}}|\}\, ds_{P'}$$
$$- \int_{s_{P_j}}^{s_{P_{j+1}}} \exp\left\{\frac{j\tau}{d}(z_{P_{i+1/2}} - z_{P'})\right\} I(s_{P'}) \ln|s_{P'} - s_{P_{i+1/2}}|\, ds_{P'} \quad (5.103)$$

The first integral in eqn. 5.103 can be computed to the desired accuracy and the second is expressed in a closed form:

$$\int_{s_{P_j}}^{s_{P_{j+1}}} \exp\left\{\frac{j\tau}{d}\left(z_{P_{i+1/2}} - z_{P'}\right)\right\} I(s_{P'}) \ln|s_{P'} - s_{P_{i+1/2}}| \, \mathrm{d}s_{P'}$$

$$= \exp\left\{\frac{j\tau}{d}\left(z_{P_{i+1/2}} - z_{P_{j+1/2}}\right)\right\} I\left(s_{P_{j+1/2}}\right)$$

$$\times \{(s_{P_{j+1/2}} - s_{P_{i+1/2}})(\ln|s_{P_{j+1/2}} - s_{P_{i+1/2}}| - 1)$$

$$- (s_{P_j} - s_{P_{i+1/2}})(\ln|s_{P_j} - s_{P_{i+1/2}}| - 1)\}$$

Solving the system given by eqn. 5.101, we find $\mu(s)$ in the points of the contour C_1 and, as a consequence, the unknown functions in eqn. 5.94 or eqn. 5.95. By means of eqn. 5.93 we can determine $U(\tau, M)$ at any interior point of the strip D_1. Then, using eqns. 5.85 and 5.86, we can find the total field $\psi(x, z)$ at any point of the space.

It is possible to show that the diffraction field in the far zone is determined by the coefficients $T_n(\tau)$ and $R_n(\tau)$, corresponding to propagating space harmonics. According to eqn. 5.88 the total field $\psi(x, z)$ in the area $x < 0$ can be presented in the form

$$\psi(x, z) = \frac{1}{2\pi} \sum_{m=-\infty}^{\infty} \int_0^{2\pi} T_m(\tau)\psi_m(\tau, z) \exp\{\Gamma_m(\tau)x\} \, \mathrm{d}\tau$$

$$= \frac{1}{2\pi} \sum_{m=-\infty}^{\infty} \int_{2\pi m}^{2\pi(m+1)} T_m(\tau' - 2\pi m)\psi_m(\tau' - 2\pi m, z)$$

$$\times \exp\{\Gamma_m(\tau' - 2\pi m)x\} \, \mathrm{d}\tau'$$

where $\tau' = \tau + 2\pi m$.

Let us introduce a variable $-\infty < \lambda < \infty$ by means of the formulae

$$\lambda d = \begin{cases} \tau' + 2\pi n & m = n, n > 0 \\ \cdots & \\ \tau' & m = 0 \\ \cdots & \\ \tau' - 2\pi n & m = -n, n > 0 \end{cases}$$

The expression for $\psi(x, z)$ can be written as

$$\psi(x, z) = \frac{\sqrt{d}}{2\pi} \int_{-\infty}^{\infty} T(\lambda) \exp\{-j\gamma(\lambda)x + j\lambda z\} \, \mathrm{d}\lambda \tag{5.104}$$

where $\gamma(\lambda) = (k^2 - \lambda^2)^{1/2}$,

$$T(\lambda) = \begin{cases} T_m(\lambda d - 2\pi m) & 2\pi m < \lambda d \le 2\pi(m+1), m > 0 \\ \cdots & \\ T_0(\lambda d) & 0 \le \lambda d \le 2\pi \\ \cdots & \\ T_{-m}(\lambda d + 2\pi m) & -2\pi m < \lambda d \le -2\pi(m+1), m > 0 \end{cases} \tag{5.105}$$

The representation given by eqns. 5.104 and 5.105 is the Fourier transform of the solution sought.

To determine the radiation characteristics in the far zone let us introduce the cylindrical polar co-ordinates

$$x = r\cos\phi, \quad z = r\sin\phi, \quad \frac{\pi}{2} < \phi < \frac{3\pi}{2}, \quad (x, z) \in D_1$$

Using the saddle-point method, we obtain the following asymptotic representation of $\psi(x, z)$ when r tends to infinity:

$$\lim_{r\to\infty} \psi(r, \phi) = \left(\frac{1}{2\pi kr}\right)^{1/2} F_T(\phi)\exp\left\{j\left(kr + \frac{\pi}{4}\right)\right\}$$

Thus at $r \to \infty$ the field, transmitted through the grating, is a cylindrical inhomogeneous wave diverging from the periodic structure. Using eqn. 5.105, we can express the radiation pattern $F_T(\phi)$ in terms of amplitudes of the propagating space harmonics:

$$F_T(\phi) = \begin{cases} d^{1/2}(-k\cos\phi)\,T_n(kd\sin\phi - 2\pi n) & 2\pi n \le kd\sin\phi < kd \\ \cdots & \\ d^{1/2}(-k\cos\phi)\,T_0(kd\sin\phi) & 0 \le kd\sin\phi \le 2\pi \\ \cdots & \\ d^{1/2}(-k\cos\phi)\,T_{-n}(kd\sin\phi + 2\pi n) & -kd < kd\sin\phi \le -2\pi n \end{cases} \tag{5.106}$$

where

$$\frac{\pi}{2} < \phi < \frac{3\pi}{2} \qquad n = \left[\frac{kd}{2\pi}\right]$$

the symbol $[a]$ standing for the greatest integer in a.

A similar expression can be obtained for the reflected field. Designating the radiation pattern for the reflected field as $F_R(\phi)$, we obtain

$$F_R(\phi) = \begin{cases} d^{1/2}k\cos\phi R_n(kd\sin\phi - 2\pi n) & 2\pi n \le kd\sin\phi < kd \\ \cdots & \\ d^{1/2}k\cos\phi R_0(kd\sin\phi) & 0 \le kd\sin\phi \le 2\pi \\ \cdots & \\ d^{1/2}k\cos\phi R_{-n}(kd\sin\phi + 2\pi n) & -kd < kd\sin\phi \le -2\pi n \end{cases} \tag{5.107}$$

where $-\pi/2 < \phi < \pi/2$.

The total flux of the reflected and transmitted energy in the areas $x \le x_1$ and $x \ge x_2$ is expressed in terms of the amplitudes of propagating space harmonics. The following energy identity is valid:

$$\frac{1}{2\pi}\int_0^{2\pi}\left\{-\sum_{n\in N'(\tau)} \operatorname{Im}\Gamma_n(|T_n(\tau)|^2 + |R_n(\tau)|^2)\right.$$

$$+\int_C \operatorname{Re}\eta^{-1}\left|\frac{\partial U(\tau,x,z)}{\partial \boldsymbol{n}}\right|^2 \mathrm{d}s\Bigg\}\,\mathrm{d}\tau$$

$$=\int_0^{2\pi}\operatorname{Im}U(\tau,\tilde{M})\,\mathrm{d}\tau-\frac{1}{2\pi}\int_0^{2\pi}\sum_{n\in N'(\tau)}\operatorname{Im}\Gamma_n|B_n^0(\tau,x_2)|^2\,\mathrm{d}\tau \qquad (5.108)$$

where

$$B_n^0(\tau,x)=\int_{-d/2}^{d/2}U_0(\tau,x,z)\psi_n^*(\tau,z)\,\mathrm{d}z$$

Summation in eqn. 5.108 is taken over the set $\mathcal{N}'(\tau)$: $n\in\mathcal{N}'(\tau)$ if $k^2-\alpha_n^2>0$, i.e. $\mathcal{N}'(\tau)$ is a set of integer indices corresponding to the numbers of propagating space harmonics. Using eqns. 5.93 and 5.94 we can show that under the condition of the surface impedance being small for a sufficiently smooth contour C, the losses can be taken into account in the framework of the energy-perturbation method. In other words, we can solve the problem on the assumption of perfect conductivity and then, having the currents distribution on the surface of the structure, determine the dissipated power in accordance with eqn. 5.46.

A modification of this method for the case of reflecting gratings [Figures 5.12(ii) and (iii)] is obvious. The approach described has been tested for a series of problems. For example, periodic structures made of rectangular and circular bars have been examined for cases of excitation by plane waves and line sources (Ilyinsky and Slepyan, 1983, Galishnikova and Ilyinsky, 1987). Here we shall present the absorption characteristics of periodic structures excited by H-polarised plane waves. In this case the algorithm described above can be simplified essentially: there is no need to use the τ-transform defined by eqn. 5.84. To calculate the induced current density and then the dissipated power we simply solve eqn. 5.94 assuming that $g(P,\tilde{M})=0$ and $\tau=kd\sin\vartheta$.

As in Section 5.5, it is possible to characterise the absorption in a periodic structure by the dimensionless coefficient α defined by eqn. 5.70. In Figures 5.13 and 5.14 the dependencies of α on the angle of incidence are shown for a comb-shaped structure with ridges of finite thickness (solid lines). Similar dependencies for structures with infinitely thin ridges, determined according to the method of Section 5.5, are presented in Figure 5.13 by broken lines. When the ridges are rather thin ($h/D\lesssim 0.04$), the results are very close to the corresponding broken-line curves. However, for $h/D=0.1$ the influence of the ridges thickness on the absorbed power is already noticeable, especially for small angles of incidence (curve 1 in Figure 5.13).

Note that, using the integral-equation method, we cannot diminish infinitely the thickness of the ridges h. Otherwise we may observe the computational instabilities caused by the fact that the matrix of the system given by eqn. 5.100 is close to singular. Thus, for parameters corresponding to Figure 5.13 the numerical algorithm described is unstable when $h/D\lesssim 0.02$. In this case the trend of the dependence $\alpha(\vartheta)$ becomes nonmonotonic and the small-scale oscillations of current-density distributions occur. These oscillations do not have any physical sense and are caused entirely by the accumulation of computational errors. Thus, when using this method for structures with thin ridges, we can recommend analysis using the current-density distributions (Galishnikova and Ilyinsky, 1987).

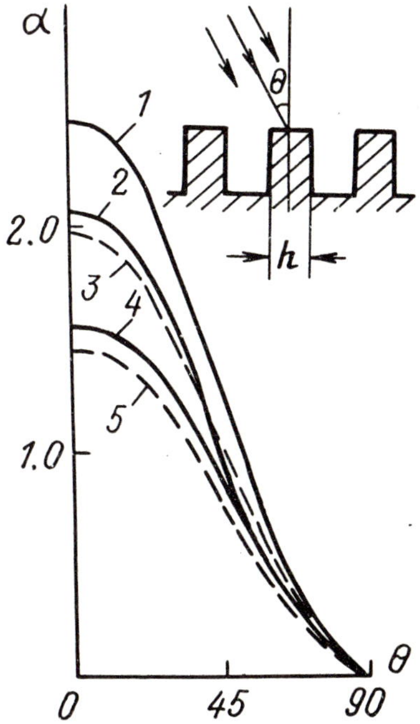

Figure 5.13 *Absorption in a comb-shaped structure with ridges of finite thickness against angle of incidence*

H polarisation, $\mu = 1$

1 $h/D = 0.1$, $\Upsilon = 1.57$
2 $h/D = 0.04$, $\Upsilon = 1.57$
3 $h = 0$, $\Upsilon = 1.57$
4 $h/D = 0.04$, $\Upsilon = 2.2$
5 $h = 0$, $\Upsilon = 2.2$

Investigation of losses in the structure shown in inserts to Figures 5.13 and 5.14 has been carried out by Gavrielides and Peterson (1979). They presented computational formulae and some numerical results for both polarisations though they did not describe the effect of abnormally small dissipation.

In Figure 5.14, specifically, are shown the dependencies $\alpha(\vartheta)$ in the neighbourhood of the points of origin of higher propagating harmonics (curves 4 and 5). A break in curve 5 can be clearly seen, similar to the well known Wood anomalies for diffraction characteristics.

Figure 5.15 shows the absorption coefficient α as a function of the incident angle ϑ for a structure with semicircular corrugations. For curves 1 and 2 a fairly smooth trend is observed in the area of moderate ϑ and a sharp fall within the narrow range of large values of ϑ. When $\Gamma_{-1} = 0$ the characteristic break also occurs (curve 3) but the peak height is considerably smaller than that in Figure 5.14.

In this Chapter we have analysed dissipation characteristics for different reflecting periodic structures: gratings with infinitely thin ridges and those of a

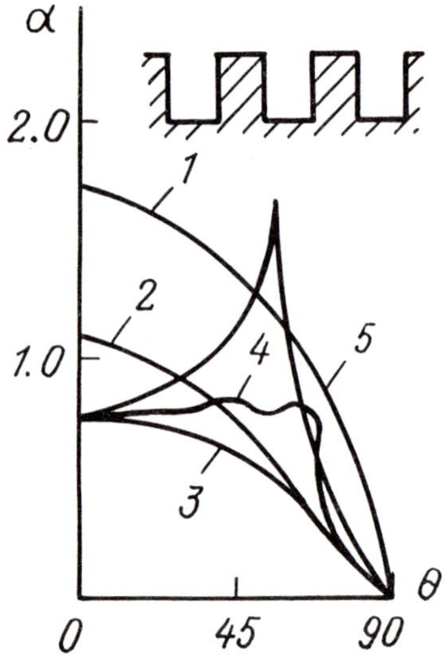

Figure 5.14 *As for Figure 5.13*

$\mu = 0.65$

1 $h/D = 1,\ \Upsilon = \pi$
2 $h/D = 0.5,\ \Upsilon = \pi$
3 $h/D = 0.25,\ \Upsilon = \pi$
4 $h/D = 0.25,\ \Upsilon = 1.05\pi$
5 $h/D = 0.25,\ \Upsilon = 1.1\pi$

finite thickness, inclined comb-shaped structures, echelette gratings and gratings with semicircular corrugations. The results obtained allow us to make general physical conclusions, to which we would like to draw the reader's attention:

(i) The phenomenon of abnormally small dissipation for H-polarised waves is a universal physical effect: it takes place for structures with any form of the period at sufficiently large angles of incidence. However, the strength of this effect (specifically, the range of angles where it is observed) depends greatly on the form of the grating profile.

(ii) The peaks of absorbed power, as with Wood anomalies for diffraction characteristics, are in the neighbourhood of the points of origin of higher propagating space harmonics. Their intensity is also determined by the shape of the grating.

5.8 Diffraction by a complex-shaped periodic structure: Galerkin's incomplete method with semi-inversion

In this Section we continue our discussion of problems of wave diffraction by periodic structures with complicated configurations. We shall present another mathematically rigorous method which is in many cases more effective than the approach based on integral equations. This method has been developed in a

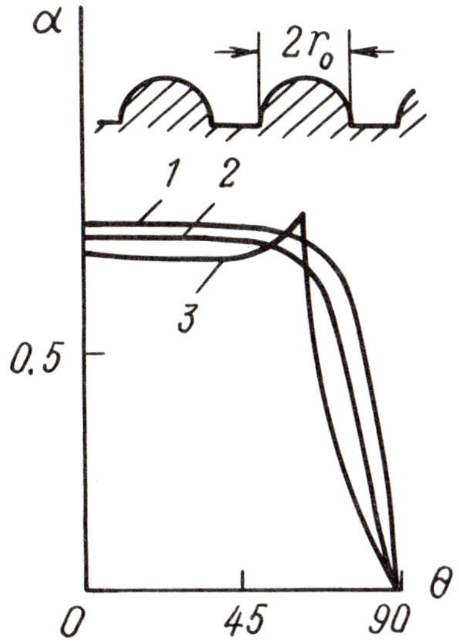

Figure 5.15 *Absorption in comb-shaped structure with semicircular ridges against angle of incidence*

H polarisation, $2r_0/d = 0.5$

1 $\Upsilon = 1.57$
2 $\Upsilon = 2.2$
3 $\Upsilon = 3.2987$

number of works by the authors of this book [Ilyinsky *et al.* (1987), Kopenkin *et al.* (1987, 1988, 1989), Ilyinsky *et al.* (1989), Slepyan (1990), Kuraev *et al.* (1991)].

The geometry of the problem and main designations are shown in Figure 5.12(ii). Let the periodic structure be infinitely wide and homogeneous along the y axis. Let us consider the case of excitation by an electrically polarised plane wave. The magnetic field of the incident wave is

$$\boldsymbol{H}^0 = \hat{i}_y \psi^0 = \hat{i}_y \exp\{jk(x \cos \vartheta + z \sin \vartheta)\}$$

The magnetic field scattered by the structure has a single component $\psi(x, z) = H_y(x, z)$ satisfying the Helmholtz equation

$$\nabla^2_{xz}\psi + k^2\psi = 0 \tag{5.109}$$

To calculate the linear (per-period) dissipated power we shall use the energy-perturbation method:[35]

$$P = \frac{1}{2} \operatorname{Re} Z_S \int_C |\psi|^2 \, dC \tag{5.110}$$

[35] Note that the use of the energy-perturbation method is not the only possible way to calculate the absorption characteristics of the structures considered: the approach in question can be easily transferred to the case of impedance-boundary conditions.

where C is a contour of one period and ψ corresponds to a model of perfectly conducting structure. To determine ψ we have to find the solution of eqn. 5.109 satisfying Neumann's boundary condition

$$\left.\frac{\partial\psi}{\partial \boldsymbol{n}}\right|_C = 0 \tag{5.111}$$

In addition, $\psi(x, z)$ should satisfy the radiation condition at $x \to +\infty$ and the edge condition.

The main idea of this approach is a synthesis of Galerkin's incomplete method and the method of semi-inversion. At the first stage the original diffraction problem is reduced to a boundary-value problem for an infinite-dimensional system of ordinary differential equations. This problem is ill-posed: it cannot be solved satisfactorily by means of truncation to a finite-dimensional problem and use of the shooting technique. The reason is that the matrix operators in the boundary conditions are noninvertible, as for matrix operators in the SLAE-1 to which diffraction problems in areas of simpler shapes are reduced [Mittra and Lee (1971), Shestopalov *et al.* (1984)]. As a result, as for such systems, computational instabilities arise and the relative-convergence phenomenon is observed. Note that the convergence is rather slow even when the method of truncation is correct. Compared with the approaches based on the direct solving of the SLAE-1 the task is redoubled by the fact that the discretisation errors are added and they may be essential.

To regularise the problem it is necessary to extract for the inverse in a closed form the matrix operators of convolution type in the boundary conditions as was done earlier in relation to infinite algebraic systems of the first kind (Shestopalov *et al.*, 1984). The boundary-value problem obtained can then be solved numerically by the truncation method with the use of the shooting technique.

Let us consider auxiliary diffraction structure shown in Figure 5.12(iii). It is composed of sections of parallel-plate waveguides of width $d = a_1(0) + a_2(0)$ and length Δ linked to the original structure at the $x = 0$ plane. We shall pass to the limit $\Delta \to 0$ in the final algorithm.

As the first step let us use the method of partial regions; the division into subareas is shown in Figure 5.12(iii) by broken lines. The field over the structure can be presented as Floquet's expansion

$$\psi(x, z) = \psi^0(x, z) + \sum_{p=-\infty}^{\infty} R_p \exp\{j\alpha_p z - \Gamma_p(x - \Delta)\} \tag{5.112}$$

where R_p are the coefficients sought, $\alpha_p = 2\pi p/d + k \sin \vartheta$ and $\Gamma_p = -j(k^2 - \alpha_p^2)^{1/2}$. The representation given by eqn. 5.112 satisfies the radiation conditions at $x \to \infty$. The field at $0 \le x \le \Delta$ can be expanded in a series in terms of eigenmodes of a parallel-plate waveguide of the width d:

$$\psi(x, z) = \sum_{n=0}^{\infty} \cos\left\{\frac{n\pi}{d}\left(z - \frac{d}{2}\right)\right\} [A_n^{(1)} \exp\{\gamma_n(x - \Delta)\}$$
$$+ A_n^{(2)} \exp\{-\gamma_n(x - \Delta)\}] \tag{5.113}$$

where $A_n^{(1,2)}$ are unknown coefficients and $\gamma_n = \{(n\pi/d)^2 - k^2\}^{1/2}$. In the area $-L \le x \le 0$, due to its nonco-ordinate form, let us use Galerkin's incomplete method. In accordance with this method, instead of eqns. 5.109 and 5.111, we can write a system of projection relations in the area $\mathscr{D}$

$$\int_{-a_1(x)}^{a_2(x)} (\nabla_{xz}^2\psi + k^2\psi)\phi_m(x, z)\, dz = \Delta_m(x) \tag{5.114}$$

where

$$m = 0, 1, \ldots, \phi_m(x, z) = \cos\left[\frac{m\pi\{z - a_2(x)\}}{a_1(x) + a_2(x)}\right]$$

$$\Delta_m(x) = \left(\phi_m \frac{\partial\psi}{\partial \boldsymbol{n}}\right)\Bigg|_{z=-a_1(x)} \left\{1 + \left(\frac{da_1}{dx}\right)^2\right\}^{1/2} + \left(\phi_m \frac{\partial\psi}{\partial \boldsymbol{n}}\right)\Bigg|_{z=a_2(x)} \left\{1 + \left(\frac{da_2}{dx}\right)^2\right\}^{1/2}$$

Eqn. 5.114 results from the 2-dimensional version of the second Green identity applied to the functions ψ and ϕ_m in the cross-section of the area $\mathscr{D}$ by the plane $x =$ constant.

Let us expand ψ into a series in terms of ϕ_m:

$$\psi(x, z) = \sum_{n=0}^{\infty} P_n(x)\phi_n(x, z) \tag{5.115}$$

Substituting eqn. 5.115 into eqn. 5.114 we arrive at a system of ordinary differential equations for $P_n(x)$ which in the vector–matrix designations can be presented as

$$\boldsymbol{P}''(x) + H(x)\boldsymbol{P}'(x) + B(x)\boldsymbol{P}(x) = 0 \tag{5.116}$$

the prime denoting the derivative with respect to x, and $\boldsymbol{P}(x) = \{P_n(x)\}$. The elements of the matrices $H(x)$, $B(x)$ are expressed as

$$B_{mn}(x) = \frac{1}{\mathcal{N}_m(x)}\left\{2\int_{-a_1(x)}^{a_2(x)} \frac{\partial\phi_n}{\partial x}\phi_m\, dz + (-1)^{n-m}a_1'(x) + a_2'(x)\right\}$$

$$\mathcal{N}_m(x) = \int_{-a_1(x)}^{a_2(x)} \phi_m^2\, dz$$

$$H_{mn}(x) = h_m^2(x)\delta_{mn} + \frac{1}{\mathcal{N}_m(x)}\int_{-a_1(x)}^{a_2(x)} \frac{\partial^2\phi_n}{\partial x^2}\phi_m\, dz$$

$$h_m^2(x) = k^2 - \left(\frac{m\pi}{a_1(x) + a_2(x)}\right)^2$$

Note that Galerkin's incomplete method is used not as a means for solving the boundary-value problem as a whole but as a means of presentation of the function sought in one of the partial regions. Usually, Galerkin's incomplete method is introduced as a finite-dimensional approximation of the boundary-value problem and in this connection a finite number of equities is retained in eqn. 5.114 and the same number of terms in eqn. 5.115 is taken into account. It is essential that our approach implies that eqn. 5.116 is of infinite order, and it is thereby actually equivalent to the original boundary-value problem given by eqns. 5.109 and 5.111. The truncation of eqn. 5.116 will be carried out at the next stage of the solution.

Boundary conditions for eqn. 5.116 at the $x=0$ plane result from the continuity of ψ and $\partial\psi/\partial x$ across $x=0$ and $x=\Delta$. For $x=0$, in accordance with Surattean *et al.* (1985), we obtain the relations

$$\boldsymbol{P}'(0) \pm \Gamma \boldsymbol{P}(0) + \xi^{+} \boldsymbol{P}(0) = \pm 2\Gamma E^{\pm} \boldsymbol{A}_{1,2} \tag{5.117}$$

where $\boldsymbol{A}_{1,2}$ are vectors of unknown coefficients $A_n^{(1,2)}$, Γ and $E^{\pm}$ are infinite-dimensional diagonal matrices with the elements $\delta_{mn}\gamma_n$ and $\delta_{mn}\exp(\mp\gamma_n\Delta)$, respectively, and ξ^{+} is a matrix with elements expressed as

$$\xi^{+}_{mn} = \lim_{x\to -0} \left\{ \frac{1}{N_m(x)} \int_{-a_1(x)}^{a_2(x)} \frac{\partial\phi_m}{\partial x} \phi_n \, dz \right\} \tag{5.118}$$

where the symbol -0 stands for the limit approached from the left.

Matching of expansions for ψ and $\partial\psi/\partial x$ defined by eqns. 5.112 and 5.113 in the $x=\Delta$ plane can be carried out by a method similar to that used by Mittra and Lee (1971). In this case, excluding R_p, we arrive at an infinite set of linear equations for $A_n^{(1,2)}$:

$$\sum_{n=0}^{\infty} \frac{v_n A_n^{(1)}}{\Gamma_s - \gamma_n} + \sum_{n=0}^{\infty} \frac{v_n A_n^{(2)}}{\Gamma_s + \gamma_n} = -j\delta_{0s} d \tan\vartheta \tag{5.119}$$

where $v_{2n} = \sin(u/2)$, $v_{2n+1} = j\cos(u/2)$, $u = kd\sin\vartheta$.

Eqn. 5.119 can be written in a more compact matrix form

$$G^{-}\boldsymbol{A}_1 + G^{+}\boldsymbol{A}_2 = \boldsymbol{b} \tag{5.120}$$

where $G^{\pm}$ are matrix operators with the elements $G^{\pm}_{ns} = v_n(\Gamma_s \pm \gamma_n)^{-1}$, and $\boldsymbol{b}$ is a vector with the components $b_n = -j\delta_{n0} d\tan\vartheta$.

Taking into account the fact that $\boldsymbol{A}_1$ and $\boldsymbol{A}_2$ are expressed by eqn. 5.117, then eqn. 5.119 (or eqn. 5.120) is the boundary condition to be found.

The boundary condition at $x=-L$ follows from eqn. 5.111. In fact, from eqn. 5.111 we can proceed to equivalent projection relations

$$\int_{-a_1(-L)}^{a_2(-L)} \left.\frac{\partial\psi}{\partial x}\right|_{x=-L} \phi_m(-L)\, dz = 0 \qquad (m = 0, 1, 2, \ldots, \infty)$$

Substituting into these equities the expansion given by eqn. 5.115, we obtain the boundary condition at $x=-L$ in the form

$$(\boldsymbol{P}' + \xi^{-}\boldsymbol{P})|_{x=-L} = 0 \tag{5.121}$$

where the matrix ξ^{-} is expressed in a manner similar to eqn. 5.118 but for the case when $x=-L+0$ (the symbol $+0$ denoting the right-hand limit).

Note that eqns. 5.116, 5.119 and 5.121, generally speaking, do not form a problem having a unique solution. They should therefore be supplemented by the edge condition. If, according to this condition, we demand a finiteness of electromagnetic energy in any bounded area containing an edge, then, as in Section 5.1, we arrive at the condition

$$\sum_{n=0}^{\infty} (1+n)(|A_n^{(1)}|^2 + |A_n^{(2)}|^2) < \infty \tag{5.122}$$

where $A_n^{(1,2)}$ are defined by eqn. 5.117. Together with this condition, eqns. 5.116,

5.119 and 5.121 form a 2-point boundary-value problem for an infinite-dimensional system of ordinary differential equations. However, such a problem cannot be solved numerically, e.g. by means of truncation to the finitely-dimensional problem and use of the shooting method. This method does not lead to successive results as the boundary-value problem is ill-posed because the boundary condition at $x=0$ contains the matrix convolution operator G^- and the solution procedure involves its numerical inversion. The operator G^- is responsible for the unfavourable computational properties of the algorithm obtained in such a manner (slow and 'relative' convergence of the truncation method, computational instabilities etc.). In fact the situation here is very similar to that occurring in Sections 5.1 and 5.3 for systems of linear algebraic equations of the first kind (see also Section 1.4). Thus, regularisation of the 2-point boundary-value problem given by eqns. 5.116, 5.119, 5.121 and 5.122 is needed. It can be provided by means of analytical inversion of the operator G^- in eqn. 5.119.

Let us dwell upon the procedure of semi-inversion in more detail. Let us transpose the term with $A_n^{(2)}$ to the right-hand side of eqn. 5.119 and consider the relationship obtained as an equation for $A_n^{(1)}$. Taking into account the edge condition expressed by eqn. 5.122, this equation can be solved exactly with the aid of the residue-calculus technique (Mittra and Lee, 1971, Shestopalov *et al.*, 1984). In this case we obtain

$$A_n^{(1)} = c_n v_n^{-1} - \sum_{m=0}^{\infty} d_n^{(m)} v_m v_n^{-1} A_m^{(2)} \tag{5.123}$$

where $\{c_n\}$, $\{d_n^{(m)}\}$ are the solutions of the systems

$$\sum_{n=0}^{\infty} \frac{c_n}{\Gamma_s - \gamma_n} = -j\delta_{0s} d \tan \vartheta$$

$$\sum_{n=0}^{\infty} \frac{d_n^{(m)}}{\Gamma_s - \gamma_n} = \frac{1}{\Gamma_0 + \gamma_m}$$

($s = 0, \pm 1, \ldots, \pm\infty$, $m = 0, 1, \ldots, \infty$) having an algebraic asymptote at $n \to \infty$. The coefficients c_n, $d_n^{(m)}$ are expressed in a closed form through the infinite products; the corresponding formulae are well known and we shall not present them here (see Mittra and Lee, 1971, and Shestopalov *et al.*, 1984).

The procedure described above can be written in a matrix form. Using eqn. 5.120 we can write eqn. 5.123 as

$$\boldsymbol{A}_1 + (G^-)^{-1} G^+ \boldsymbol{A}_2 = (G^-)^{-1} \boldsymbol{b} \tag{5.124}$$

Combining eqns. 5.117 and 5.124, we obtain a regularised boundary condition. Note that eqn. 5.124 is similar to systems of the second kind obtained within the framework of the semi-inversion method (see Section 1.3).

The 2-point boundary-value problem given by eqns. 5.116, 5.121 and 5.124 can be solved by the standard numerical method (truncation to the finite order $2N$ and use of shooting method). At this stage of the solution we can put $\Delta = 0$, and thereby return to the initial geometry of the structure. We can also explain the purpose of introducing the auxiliary geometry with $\Delta \neq 0$. Because of this the convolution matrix operator in the boundary condition given by eqn. 5.119 has been extracted in the explicit form. Such a formulation of the boundary condition makes possible the application of the regularising technique

6.1 Diffraction by a finitely conducting comb-shaped structure with a layered dielectric filling; resonance absorption

Let us consider the incidence of the electrically polarised plane wave

$$\psi^0(x, z) = \exp\{jk(\sqrt{\tilde{\varepsilon}})(x \cos \vartheta + z \sin \vartheta)\}$$

from the halfspace $x < 0$, filled by a homogeneous lossless medium ($\tilde{\varepsilon} = \varepsilon_0 \tilde{\varepsilon}_1$, $\mu = \mu_0$) on a metallic finitely conducting comb-shaped structure in an N-layer dielectric medium (Figure 6.1).[37] The ith layer is characterised by the thickness $\Delta_i - \Delta_{i-1}$ and permittivity $\varepsilon_0 \tilde{\varepsilon}_i$ (all layers, for simplicity, are considered to be isotropic, nonabsorbing and nonmagnetic), $i = 1, 2, \ldots, N$. It is assumed that the ridge thickness is significantly smaller than the wavelength, structure period or thickness of each layer, but significantly exceeds the skin depth. The energy losses in the comb-shaped structure can then be taken into account within the framework of skin effect theory, in a method similar to that in Section 5.3. The linear (per-period) dissipated power in the periodic structure is expressed by eqn. 5.46.

To calculate P, as in Section 5.5, we use a perturbation technique (Z_S small),

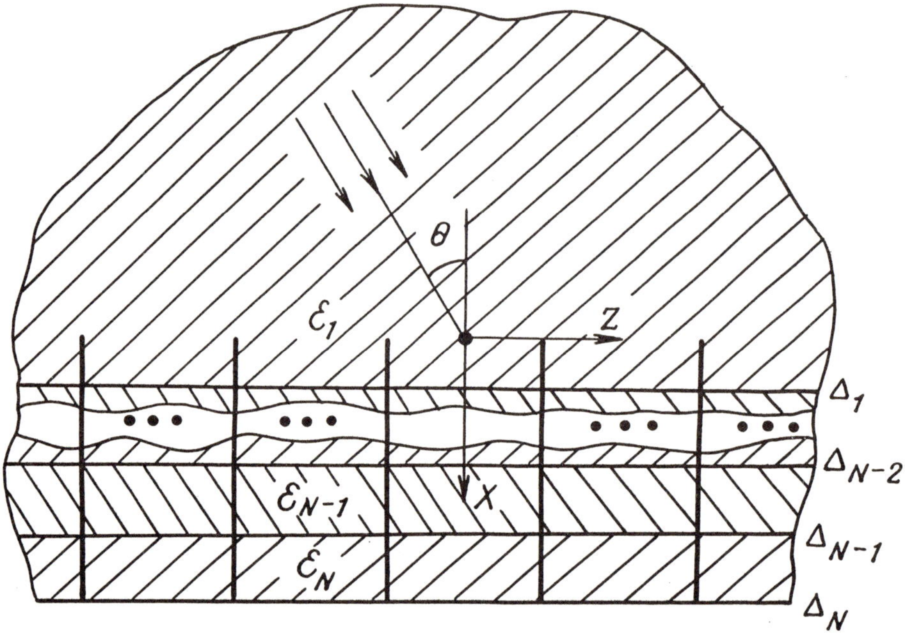

Figure 6.1 *Comb-shaped structure in layered dielectric medium*

[37] Similar problems for cases of perfect conductivity of the periodical structures were examined by Amitay *et al.* (1972), Shestopalov *et al.* (1986) and Andrenko and Shestopalov (1979), where resonance effects of 'reflection–transmission' were studied. This class of problems is of interest because there are both traditional (polarisers, antenna radomes, dielectric parameters measuring equipment) and new applications (for example, logic elements based on such periodic structures).

i.e. as a first approximation we replace the actual $|\psi|^2$ by the value corresponding to infinite conductivity. This can be done for frequencies sufficiently remote from the cutoff ones ($\omega_{in} = n\pi c/d\sqrt{\tilde{\varepsilon}_i}$), which will be assumed throughout. Presenting the field in each layer as an expansion in terms of forward and backward eigenmodes of the plane waveguide with a corresponding dielectric filling, we obtain the formula for dissipated power

$$P = \frac{\mathrm{Re}\,Z_S}{2}\left[4\int_{-d/2}^{d/2} \hat{f}(z)\,\mathrm{d}z + \sum_{i=1}^{N}\int_{\Delta_{i-1}}^{\Delta_i} \{F_i^+(x) + F_i^-(x)\}\,\mathrm{d}x\right] \tag{6.1}$$

where

$$F_i^{\pm}(x) = \left|\sum_{n=0}^{\infty} T_{n,i}\begin{Bmatrix}1\\(-1)^n\end{Bmatrix}\{\exp(h_{n,i}x) + \rho_{n,i}\exp(-h_{n,i}x)\}\right|^2$$

$$\hat{f}(z) = \left|\sum_{n=0}^{\infty} T_{n,N}\cos\left\{n\pi\left(\frac{z}{d}-\frac{1}{2}\right)\right\}\exp(-h_{n,N}\Delta_N)\right|^2$$

$h_{n,i} = \{(n\pi/d)^2 - k^2\tilde{\varepsilon}_i\}^{1/2}$ and $\Delta_0 = 0$. The coefficient $T_{n,i}$ is the complex amplitude of the nth forward eigenmode in the ith dielectric layer, and $\rho_{n,i}$ is the reflection coefficient of the nth mode from the interface between the ith and the $i+1$th layers. The coefficients $T_{n,i}$ and $\rho_{n,i}$ are expressed by recurrent relationships in terms of $T_{n,1}$ and $\rho_{n,N}$ ($\rho_{n,N}$ being the reflection coefficient of the nth mode from the metal plane $x = \Delta_N$). These relationships are obtained using the continuity of field tangential components across the layer interfaces and have the form:

$$\rho_{n,i} = \frac{\gamma_{n,i}^{-} + \rho_{n,i+1}\gamma_{n,i}^{+}\exp(-2h_{n,i+1}\Delta_i)}{\gamma_{n,i}^{+} + \rho_{n,i+1}\gamma_{n,i}^{-}\exp(-2h_{n,i+1}\Delta_i)}\exp(-2h_{n,i}\Delta_i) \tag{6.2}$$

$$\gamma_{n,i}^{\pm} = h_{n,i}\tilde{\varepsilon}_{i+1} \pm h_{n,i+1}\tilde{\varepsilon}_i$$

$$\rho_{n,N} = \exp(-2h_{n,N}\Delta_N) \tag{6.3}$$

$$T_{n,i+1} = T_{n,i}\left\{\frac{\exp(-h_{n,i}\Delta_i) + \rho_{n,i}\exp(h_{n,i}\Delta_i)}{\exp(-h_{n,i+1}\Delta_i) + \rho_{n,i}\exp(h_{n,i+1}\Delta_i)}\right\} \tag{6.4}$$

To calculate the coefficients $T_{n,1}$ we use the modal-expansion technique. The field in halfspace $x < 0$ can be written as a series in terms of space (Floquet) harmonics (see eqn. 5.4). The field in the region $0 < x < \Delta_1$ is presented as a series in terms of eigenmodes of the plane waveguide with dielectric filling of permittivity $\varepsilon = \varepsilon_0\tilde{\varepsilon}_1$. The amplitudes of forward modes are equal to $T_{n,1}$, those of the backward modes are $\rho_{n,1}T_{n,1}$ ($\rho_{n,1}$ being the reflection coefficient of the nth waveguide mode from the set of dielectric slabs lying on the metal plane $x = \Delta_N$). The use of the continuity of the tangential field components across $x = 0$ (as in Section 5.3) leads to a system of functional equations connecting the amplitudes of Floquet's harmonics R_p with the coefficients $T_{n,1}$. Then, performing certain transformations, we obtain the system of linear algebraic equations

$$\sum_{p=-\infty}^{\infty} \tilde{\alpha}_p(-1)^p R_p\left(\frac{1}{\tilde{\Gamma}_p - h_{n,1}} + \frac{\rho_{n,1}}{\tilde{\Gamma}_p + h_{n,1}}\right)$$
$$= k(\sqrt{\tilde{\varepsilon}_1}) \sin \vartheta \left(\frac{1}{\tilde{\Gamma}_0 + h_{n,1}} + \frac{\rho_{n,1}}{\tilde{\Gamma}_0 - h_{n,1}}\right) \qquad (6.5)$$

where

$$\tilde{\Gamma}_p = (\alpha_p^2 - k^2)^{1/2}$$
$$\alpha_p = 2(p\pi + \tilde{u})/d$$
$$\tilde{u} = kd(\sqrt{\tilde{\varepsilon}_1}) \sin \vartheta$$

The solution sought is to belong to the numerical-sequence space

$$\sum_{p=-\infty}^{\infty} (1+|p|)|R_p|^2 < \infty \qquad (6.6)$$

which provides a correct field behaviour near the edge. The coefficients $T_{n,1}$ are expressed through R_p using the formula[38]

$$\frac{1}{2v_n}(-1)^n T_{n,1} h_{n,1} d(1-\rho_{n,1}^2)(1+\delta_{n0})$$
$$= \sum_{p=-\infty}^{\infty} \tilde{\alpha}_p(-1)^p R_p\left(\frac{1}{\tilde{\Gamma}_p + h_{n,1}} + \frac{\rho_{n,1}}{\tilde{\Gamma}_p - h_{n,1}}\right)$$
$$- k(\sqrt{\tilde{\varepsilon}_1}) \sin \vartheta \left(\frac{1}{\tilde{\Gamma}_0 - h_{n,1}} + \frac{\rho_{n,1}}{\tilde{\Gamma}_0 + h_{n,1}}\right)$$

where $v_{2n} = \sin \tilde{u}$, $v_{2n+1} = j \cos \tilde{u}$.

Now we solve eqn. 6.5 in the required space of the numerical sequences defined by eqn. 6.6 and determine $T_{n,1}$ using the modified residue-calculus technique (Mittra and Lee, 1971). The latter allows us to reduce eqn. 6.5 to an infinite system of the second kind with simple matrix elements. As the methods of obtaining such systems are standard, we shall present only the final equations without going much into detail. The unknown coefficients $T_{n,1}$ are expressed by

$$T_{n,1} = -(2-\delta_{n0}) v_n (-1)^n F_n f(-h_{n,1})(\sqrt{\tilde{\varepsilon}_1})$$
$$\times \left\{ h_{n,1} s_n d \left(1 - \sum_{m=0}^{\infty} \frac{F_m}{h_{m,1} + \tilde{\Gamma}_0}\right)\right\}^{-1} \qquad (6.7)$$

where $s_n = \rho_{n,1} f(-h_{n,1})/\Lambda_n$, $\Lambda_n = \mathrm{d}f/\mathrm{d}w|_{w=h_{n,1}}$, and $f(w)$ is expressed as in Section 5.1 after the replacements $\gamma_n \rightarrow h_{n,1}$ and $\Gamma_n \rightarrow \tilde{\Gamma}_n$. The coefficients F_n satisfy the infinite system of linear algebraic equations

$$F_m - s_m \sum_{n=0}^{\infty} \frac{F_n}{h_{m,1} + h_{n,1}} = -s_n \qquad (6.8)$$

Our computational algorithm thus has four basic stages:

[38] A system of linear algebraic equations for the unknown coefficients $T_{n,1}$ can be obtained, which allows us to avoid calculation of R_n. However, this system seems less appropriate for the modified-residue-calculus-technique application.

(i) solution of eqn. 6.8 using the truncation method;
(ii) calculation of $T_{n,1}$ ($n = 1, 2, \ldots, N$) according to eqn. 6.7;
(iii) calculation of the coefficients to be used in eqn. 6.1 through the recurrent eqns. 6.2–6.4;
(iv) calculation of the linear (per-period) dissipated power P using eqn. 6.1.

The system given by eqn. 6.8 at $\Delta_1 \neq 0$ is of the second kind, which follows from the evaluation $s_m = O\{\exp(-m\Delta_1)\}$.[39]

Let us consider in more detail the case of a 2-layer medium, when eqns. 6.2–6.4 can be reduced to the simple explicit expressions

$$\rho_{n,1} = -\frac{h_{n,2}\tanh\{h_{n,2}(\Delta_2 - \Delta_1)\} - n_{12}^2 h_{n,1}}{h_{n,2}\tanh\{h_{n,2}(\Delta_2 - \Delta_1)\} + n_{12}^2 h_{n,1}} \exp(-2h_{n,1}\Delta_1) \tag{6.9}$$

$$T_{n,2} = T_{n,1}\left[\frac{\exp(-h_{n,1}\Delta_1) + \rho_{n,1}\exp(h_{n,1}\Delta_1)}{2\cosh\{h_{n,2}(\Delta_2 - \Delta_1)\}}\right]\exp(h_{n,2}\Delta_2) \tag{6.10}$$

$n_{12} = (\tilde{\varepsilon}_2/\tilde{\varepsilon}_1)^{1/2}$ being the relative refractive index.

The energy losses, as usual, are characterised by the factor

$$\alpha(\vartheta, d, \Delta_1, \Delta_2, n_{12}) = P(\vartheta, d, \Delta_1, \Delta_2, n_{12})/\tilde{P}$$

where $\tilde{P} = \lim_{\Delta_2 \to 0} P$.

(a) $n_{12} > 1$: In Figure 6.2 are shown the dependencies of α on Δ_2/d for the angle of incidence $\vartheta = 60°$, $n_{12} = \sqrt{10}$ and various values of kd. The interface between the dielectric layers is in the $x = \Delta_1$ plane ($\Delta_1 = 0.05d$). These curves, like those for homogeneous media, are wavy in shape; however, contrary to the case of homogeneous media (see Section 5.5), $\alpha \gg 1$ for all values of Δ_2/d. Thus when a substance of larger refractive index is used to fill the grooves, we have a considerable increase in energy losses. This result can be physically explained by the fact that the dielectric layer with $\varepsilon_2 > \varepsilon_1$ serves as a matching transformer between the medium with ε_1 and the corrugated metal surface.

The idea of filling the grooves with substances of larger refractive index looked promising for the development of compact low-loss corrugated waveguides and resonators. However, as the calculations showed, this idea proved irrelevant: reduced sizes result in the loss of the desirable effect (decrease in dissipation due to the surface corrugation).

In Figure 6.3 the dependencies $\alpha(\vartheta)$ for incomplete ($\Delta_1 = \Delta_2/2$) grooves filling with the dielectric ($n_{12} = \sqrt{10}$, $\Delta_1/d = 1$) are shown for two values of kd. These dependencies differ from the corresponding curves for homogeneous media, having a more smooth trend at small and moderate angles ($\vartheta \lesssim 40° - 60°$) and a shift of the zone of sharp decrease in α towards the region of large angles of incidence.

(b) $n_{12} < 1$: In Figure 6.4 are shown the dependencies of α on Δ_2/d at $\vartheta = 60°$, $n_{12} = \frac{1}{3}$, $\Delta_1 = 0.05d$ and different values of kd. Figure 6.5 shows the frequency dependencies of α at $\vartheta = 60°$, $\Delta_2 = 4d$ for two values of n_{12}. We note that these

[39] At $\Delta_1 = 0$, the truncation method is also applicable. This follows from the fact that the matrix operator on the left-hand side of eqn. 6.8 belongs to the operator class with a spectrum inside the unit circle (Sirenko, 1983).

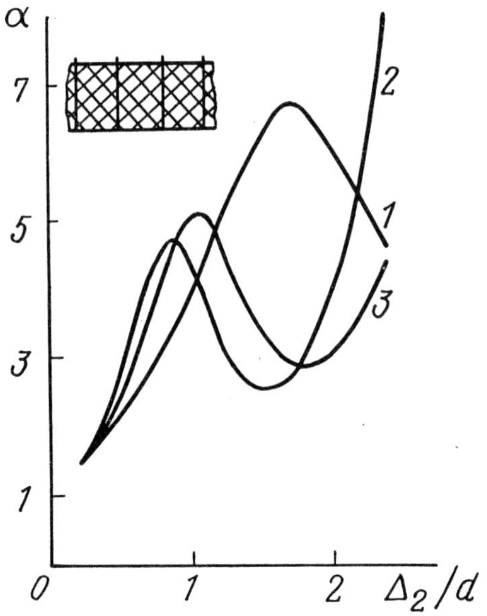

Figure 6.2 *Absorption in a comb-shaped structure as a function of* Δ_2/d

H polarisation

1 $\Upsilon = 0.298$
2 $\Upsilon = 0.5961$
3 $\Upsilon = 0.4967$

sets of parameters correspond to the condition of total internal reflection. The comb-shaped structure, as it can be seen from the above data, distorts this phenomenon at certain groove depths. This is indicated by the absorption resonances: sharp peaks of the curves $\alpha(\Delta_2/d)$. Similar data for $n_{12} = 1/\sqrt{3}$ are shown in Figure 6.6.

In Figure 6.7 are shown the dependencies $\alpha(\vartheta)$ at $n_{12} = \frac{1}{3}$, $\Delta_2/d = 3$, $\Delta_1 = 0$ and different values of kd. As for homogeneous media, $\alpha(\vartheta)$ is diminishing monotonically with ϑ increasing up to $\vartheta \simeq \cos^{-1}(|\eta|/k)$ at any values of $\tilde{\varepsilon}_2$ and kd. It is necessary, however, to note that the specific trends of the curves $\alpha(\vartheta)$ at different kd vary to a larger extent than for homogeneous media.

The most important result of the above consideration is evidently the effect of the resonance-wave absorption in comb-shaped structures (Figures 6.4 and 6.5). It is necessary to stress that this phenomenon results from the joint effect of the periodic structure and media interface. It occurs only at $n_{12} < 1$; though at $n_{12} \geq 1$ the absorption variations (dependent on the structure depth) can be observed, the absorption peaks are washed out and the range of variation of α is considerably smaller.

The effect of resonance absorption in the comb-shaped structures is applicable in many cases. It can be used, for example, for the suppression of parasitic modes in waveguides and resonators.

The case of negative resistance Re $Z_S < 0$ is of special interest, as it corresponds

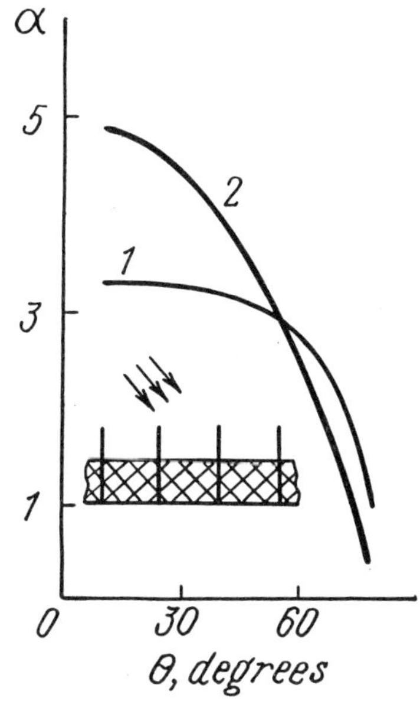

Figure 6.3 *Absorption in comb-shaped structure with dielectric layer against angle of incidence*

H polarisation

1 $\Upsilon = 0.298$
2 $\Upsilon = 0.5961$

to amplification of the incident wave by an active (for example, semiconducting) film. When we have a corrugated active surface the effect of resonance amplification occurs. Then, unlike the case of a smooth surface, the amplification factor at the working frequency increases greatly and the out-of-band radiation is suppressed essentially.

6.2 Wave diffraction by comb-shaped structures in gyrotropic media

Let us consider the following problem. The extraordinary plane wave

$$H_y^0(x, z) = \exp\{-jk_e(x\cos\vartheta - z\sin\vartheta)\} \qquad (H_x^0 = H_z^0 = 0)$$

is incident on a comb-shaped structure placed in a gyrotropic medium with the tensor

$$\overleftrightarrow{\varepsilon} = \varepsilon_0 \begin{pmatrix} \varepsilon_1 & 0 & -j\varepsilon_2 \\ 0 & \varepsilon_3 & 0 \\ j\varepsilon_2 & 0 & \varepsilon_1 \end{pmatrix} \tag{6.11}$$

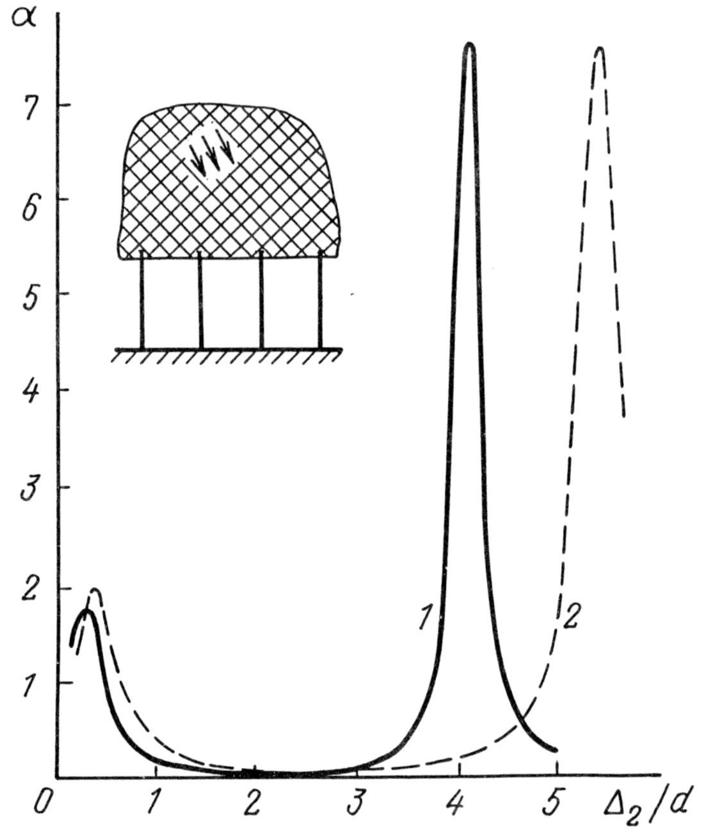

Figure 6.4 *Absorption in comb-shaped structure placed near dielectric halfspace as a function of* Δ_2/d

H polarisation

1 $\Upsilon\sqrt{\varepsilon_1} = 2.5132$
2 $\Upsilon\sqrt{\varepsilon_1} = 1.8849$

Here $k_e = 2\pi\{(\varepsilon_1^2 - \varepsilon_2^2)/\varepsilon_1\}^{1/2}/\lambda$, λ being the wavelength in free space (Figure 6.8). Suppose that the dielectric medium is lossless (ε_1, ε_2, ε_3 real), and the thickness of the ridges Δ is small compared with λ, d, and D, though it exceeds considerably the field-penetration depth into the metal (in this case ridges can be considered as infinitely thin).

The linear (per-period) dissipated power P for a structure of a highly conductive material can be calculated using the perturbation technique (see Section 5.4). In the case considered there is a nonreciprocal energetic effect: $P(\vartheta) \neq P(-\vartheta)$ (note that for reflection from a smooth surface this effect does not exist; there is only a nonreciprocal phase shift). This effect can be characterised by a factor of nonreciprocity

$$\beta_1(\vartheta) = P(\vartheta)/P(-\vartheta) \tag{6.12}$$

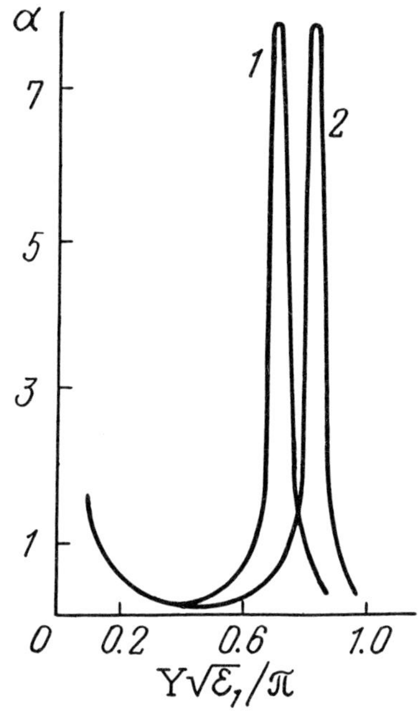

Figure 6.5 *Resonance absorption of H-polarised plane wave in comb-shaped structure placed near dielectric halfspace*

$\vartheta = 60^\circ$, $\varepsilon_2 = 1$, $\Delta_2/d = 4$
1 $n_{12} = 0.387$
2 $n_{12} = 0.331$

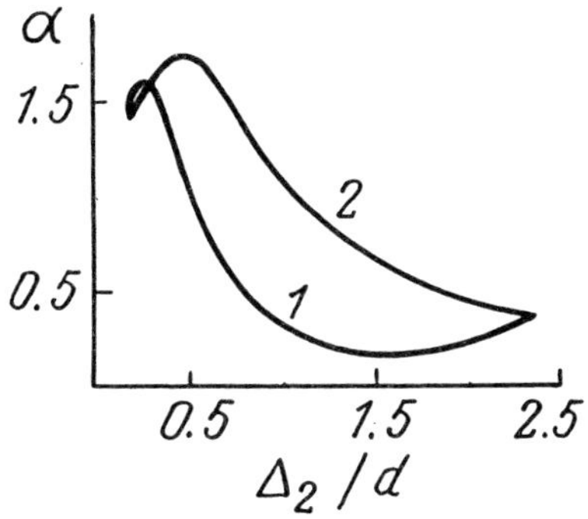

Figure 6.6 *Absorption in comb-shaped structure placed near dielectric halfspace against* Δ_2/d

1 $\Upsilon\sqrt{\varepsilon_1} = 1.8849$
2 $\Upsilon\sqrt{\varepsilon_1} = 0.9424$

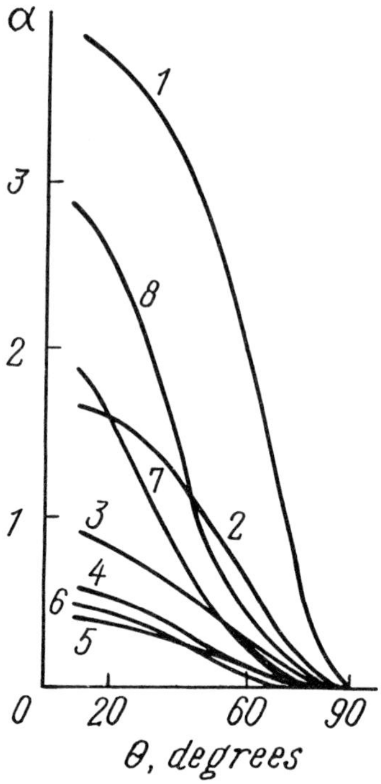

Figure 6.7 *Absorption in comb-shaped structure placed near dielectric halfspace against angle of incidence*

1 $\Upsilon\sqrt{\varepsilon_1} = 0.3167$
2 $\Upsilon\sqrt{\varepsilon_1} = 0.6333$
3 $\Upsilon\sqrt{\varepsilon_1} = 0.95$
4 $\Upsilon\sqrt{\varepsilon_1} = 1.2667$
5 $\Upsilon\sqrt{\varepsilon_1} = 1.5833$
6 $\Upsilon\sqrt{\varepsilon_1} = 2.2167$
7 $\Upsilon\sqrt{\varepsilon_1} = 2.85$
8 $\Upsilon\sqrt{\varepsilon_1} = 3$

Let us suppose that $k_e d < \pi$, i.e. only the fundamental mode can propagate in the grooves. In this case

$$\beta_1(\vartheta) \simeq \frac{|A(\vartheta)|^2}{|A(-\vartheta)|^2} \tag{6.13}$$

where $A(\vartheta)$ is the amplitude of the forward fundamental mode in the grooves.

The field-formation process inside the grooves can be presented as a chain of successive diffractions on structure ridges and reflections from bottoms. Therefore, the higher evanescent modes in this chain can be ignored since for large D (which we consider as such), coming from the point of their excitation, they do not in practice reach the other discontinuity.[40] In this case $A(\vartheta)$, according

[40] However, the higher modes should be taken into consideration when determining the reflection and transmission coefficients.

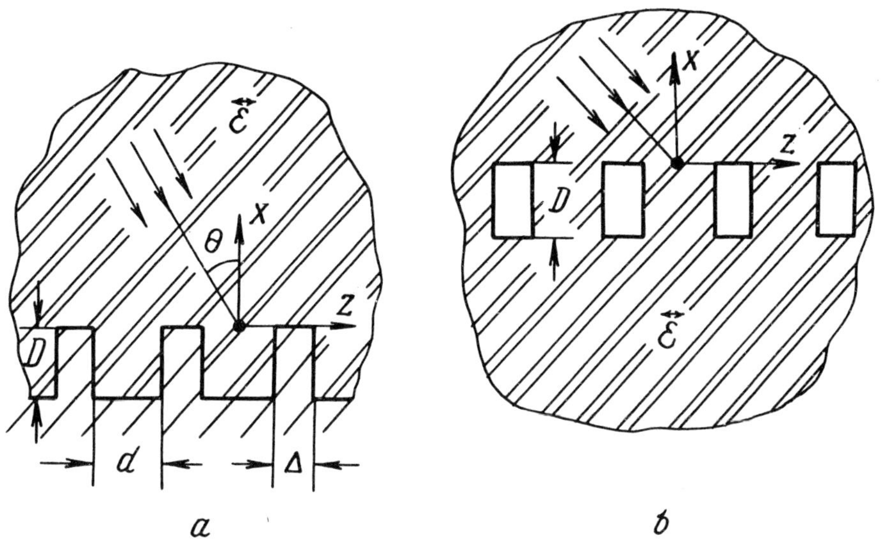

Figure 6.8 *Periodic structures in gyrotropic medium*

a Reflecting grating
b Transmission grating

to the generalised scattering-matrix technique (Mittra and Lee, 1971), is expressed by

$$A(\vartheta) = t_0(\vartheta)\{1 + s_0 R \exp(2j\gamma_0 D) + s_0^2 R^2 \exp(4j\gamma_0 D) + \cdots\}$$
$$= t_0(\vartheta)\{1 - s_0(\vartheta) R \exp(2j\gamma_0 D)\}^{-1} \qquad (6.14)$$

where $t_0(\vartheta)$ and $s_0(\vartheta)$ are the transmission and reflection coefficients of the fundamental mode for the array of semi-infinite parallel-plate waveguides; t_0 corresponds to the case of plane-wave excitation, s_0 corresponds to the periodic excitation by fundamental modes (phase change per unit cell equals $k_e d \sin\vartheta$), R is the reflection coefficient of the fundamental mode from the bottom of the groove,[41] and $\gamma_0 = 2\pi(\sqrt{\varepsilon_1})/\lambda$ is a fundmental wave-phase constant.

Taking into account eqn. 6.14 and also the fact that

$$|t_0(\vartheta)| = |t_0(-\vartheta)|, \qquad |s_0(\vartheta)| = |s_0(-\vartheta)|,$$

we can transform eqn. 6.13 as

$$\beta_1(\vartheta) = \frac{1 + r^2|s_0(\vartheta)|^2 - 2r|s_0(\vartheta)|\cos\{2\gamma_0 D + \phi + \phi_s(-\vartheta)\}}{1 + r^2|s_0(\vartheta)|^2 - 2r|s_0(\vartheta)|\cos\{2\gamma_0 D + \phi + \phi_s(\vartheta)\}} \qquad (6.15)$$

where $r = |R|$ and ϕ and ϕ_s are phases of the coefficients R and s_0 respectively. The values $|s_0|$ and ϕ_s are determined exactly using the Wiener–Hopf technique

[41] We reserve the traditionally used names for R and $s_0(\vartheta)$ (see Wu, 1967), though physically the term 'transformation coefficient' is more correct as, in the presence of gyrotropy, the transverse-field distributions of forward and backward modes are essentially different.

(Wu, 1967); an approximate expression for R was obtained by Popov (1981):

$$r \simeq \exp\{-2\pi d\varepsilon_2/\lambda(\sqrt{\varepsilon_1})\} \qquad \phi = 0$$

The nonreciprocal effect considered is strongest when for the angle of incidence ϑ the resonance condition $\cos\{2\gamma_0 D + \phi_s(\vartheta)\} \simeq 1$ is satisfied, and for the angle of incidence $-\vartheta$ the accuracy of meeting the antiresonance condition $\cos\{2\gamma_0 D + \phi_s(-\vartheta)\} \simeq -1$ is maximised. Then

$$\beta_1^{max}(\vartheta) = \frac{1 + r^2|s_0(\vartheta)|^2 - 2r|s_0(\vartheta)|\cos\{2\Delta\phi_s(\vartheta)\}}{(1 - r|s_0(\vartheta)|)^2} \qquad (6.16)$$

where $\Delta\phi_s(\vartheta) = \{\phi_s(\vartheta) + \phi_s(-\vartheta)\}/2$; according to Wu (1967) we have

$$|s_0(\vartheta)| = \frac{\varepsilon_1 - (\varepsilon_1^2 - \varepsilon_2^2)^{1/2}\cos\vartheta}{\varepsilon_1 + (\varepsilon_1^2 - \varepsilon_2^2)^{1/2}\cos\vartheta} \qquad (6.17)$$

$$\Delta\phi_s(\vartheta) = -2\tan^{-1}[\tan(k_e d\sin\vartheta/2)\coth\{2\pi d\varepsilon_2/\lambda(\sqrt{\varepsilon_1})\}] \qquad (6.18)$$

6.3 Nonreciprocal resonance effects

In Figure 6.9 the dependencies $\beta_1^{max}(\vartheta)$ are shown for different values of $k_e d$, calculated by means of eqns. 6.16–6.18. Analysis of these curves shows that at sufficiently large angles of incidence ($\vartheta \gtrsim 70$–$80°$) a significant nonreciprocal effect occurs. It is important to stress that nonreciprocity, because of a specific resonance character of the effect, can be very strong even at a rather low medium gyrotropy (in our calculations $v = \varepsilon_2/\varepsilon_1 = 0.1$). The effect is reduced when the cutoff condition $k_e d = \pi$ is approached.

As for the physical meaning of this effect, there is a nonreciprocal phase rotation when the wave is reflected from the periodic structure ridges. It is therefore possible to fulfil both the resonance condition at the angle of incidence ϑ, and, at the same time, the antiresonance condition at the angle of incidence $-\vartheta$. This is not enough, however. The phase conditions become essential only when the large-amplitude partial waves interfere. This is why the effect described takes place only at large values of ϑ when the reflection from ridges is substantial.

It would be of interest to study the trend of the dependencies $\beta_1^{max}(\vartheta)$ in the domain of large angles of incidence ($\vartheta > 80°$), absent in Figure 6.9. Analysis of eqns. 6.16–6.18 shows that at $\vartheta \to 90°$ β_1^{max} increases monotonously tending to a certain finite value. At the condition $2\pi d\varepsilon_2/\lambda\sqrt{\varepsilon_1} \ll \tan(k_e d/2)$, which in this case is fulfilled, it is not difficult to obtain the following:

$$\lim_{\vartheta \to \pi/2} \beta_1^{max}(\vartheta) = 1 + 16\cot^2(k_e d/2) \qquad (6.19)$$

The magnitude of the limit in eqn. 6.19 decreases when $k_e d$ increases.

Proceeding from a certain analogy between interference mechanisms of wave formation in the grooves and wave interference in layered media we can also expect similar nonreciprocal effects to occur in layered media. To analyse them consider a plane-wave reflection from a metal plane placed under the interface between two media (Figure 6.10), one of which for simplicity is assumed to be isotropic. This will result in two cases of the problem (Figure 6.10*a* and *b*).

There are exact solutions of the wave-propagation problems in plane-layered

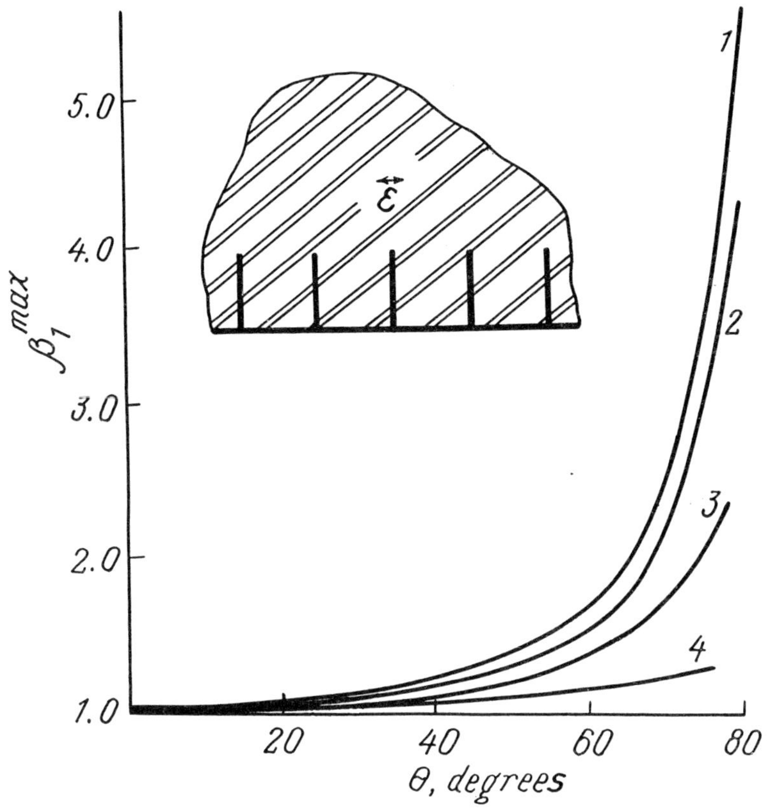

Figure 6.9 *Peak nonreciprocity coefficient for comb-shaped structure placed near dielectric halfspace against angle of incidence*

$\varepsilon_1 = 10$, $\varepsilon_2 = 1$

1 $k_e d = 0.05\pi$
2 $k_e d = 0.2\pi$
3 $k_e d = 0.5\pi$
4 $k_e d = 0.8\pi$

media. The latter are normally obtained by means of fields matching at the interfaces. We prefer, however, to use a method of successive reflections as this form of solving the problem is more convenient to compare with the results obtained for the comb-shaped structures. This is also a rigorous solution, because in layered media there are no higher modes. It is equivalent to the Jones calculus widely used in optics (Yariv and Yeh, 1983).

To begin we consider the structure shown in Figure 6.10*a*. The tensor of permittivity $\overleftrightarrow{\varepsilon}$ for a gyrotropic medium is defined by eqn. 6.11. As in the previous Section, we obtain

$$\beta_2^{max} = \frac{1 + |s_0|^2 - 2|s_0| \cos(2\phi_s)}{(1 - |s_0|)^2} \tag{6.20}$$

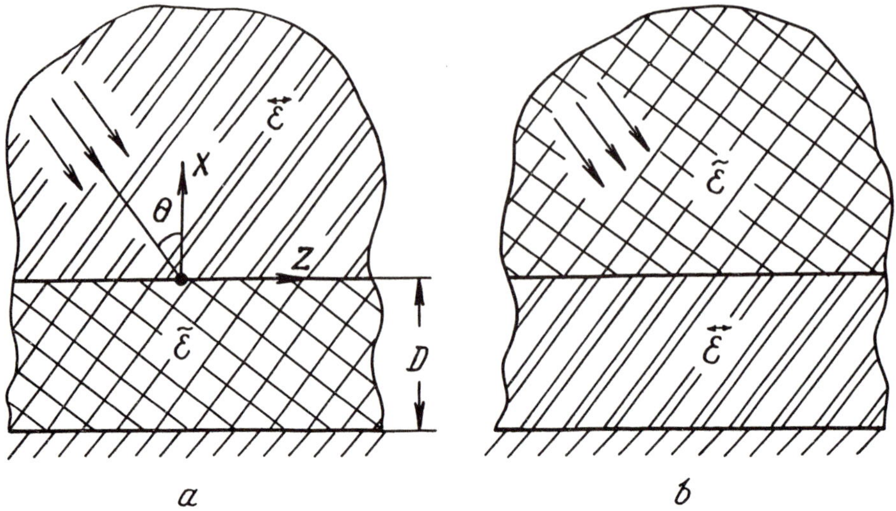

Figure 6.10 *Layered structures with gyrotropic media*

a With gyrotropic halfplane
b With gyrotropic slab

where $s_0 = |s_0| \exp(j\phi_s)$ is the reflection coefficient of the plane wave, coming from an isotropic medium with permittivity $\tilde{\varepsilon}$. Eqn. 6.20 corresponds to the isotropic-layer 'resonance' thickness; the resonance condition has the form

$$\cos\{4\pi D(1 - \sin^2 \vartheta/p^2)^{1/2}/\lambda + \phi_s\} = 1 \tag{6.21}$$

where $p = \{\varepsilon_1 \tilde{\varepsilon}/(\varepsilon_1^2 - \varepsilon_2^2)\}^{1/2}$. For $s_0(\vartheta)$, the following expression is obtained:

$$s_0(\vartheta) = \frac{\cos\phi - p\cos\vartheta + jvp\sin\vartheta}{\cos\phi + p\cos\vartheta + jvp\sin\vartheta} \tag{6.22}$$

where $v = \varepsilon_2/\varepsilon_1$ and $\cos\phi = (1 - \sin^2 \vartheta/p^2)^{1/2}$.

For the structure shown in Figure 6.10*b* eqn. 6.20 is also valid, if $s_0(\vartheta)$ is expressed by the formula

$$s_0(\vartheta) = \frac{p\cos\phi + jvp\sin\phi - \cos\vartheta}{p\cos\phi - jvp\sin\phi + \cos\vartheta} \tag{6.23}$$

where $\cos\phi = (1 - p^2 \sin^2 \vartheta)^{1/2}$, and thereby the resonance condition given by eqn. 6.21 will be slightly changed.

In Figure 6.11 the dependencies $\beta_1^{max}(\vartheta)$ for the structure shown in Figure 6.10*a* are presented for different values of the parameter p. First note that the nonreciprocal effect here is much lower than in periodic structures (with curve 1 as an exception, corresponding to a very high gyrotropy parameter p of 10). In addition, these dependencies have quite a different character from those for periodic structures: they have an extremum. When ϑ tends to 90°, from eqns. 6.20 and 6.22 there follows

$$\lim_{\vartheta \to \pi/2} \beta_1^{max}(\vartheta) = 1 + 4p^2 v^2$$

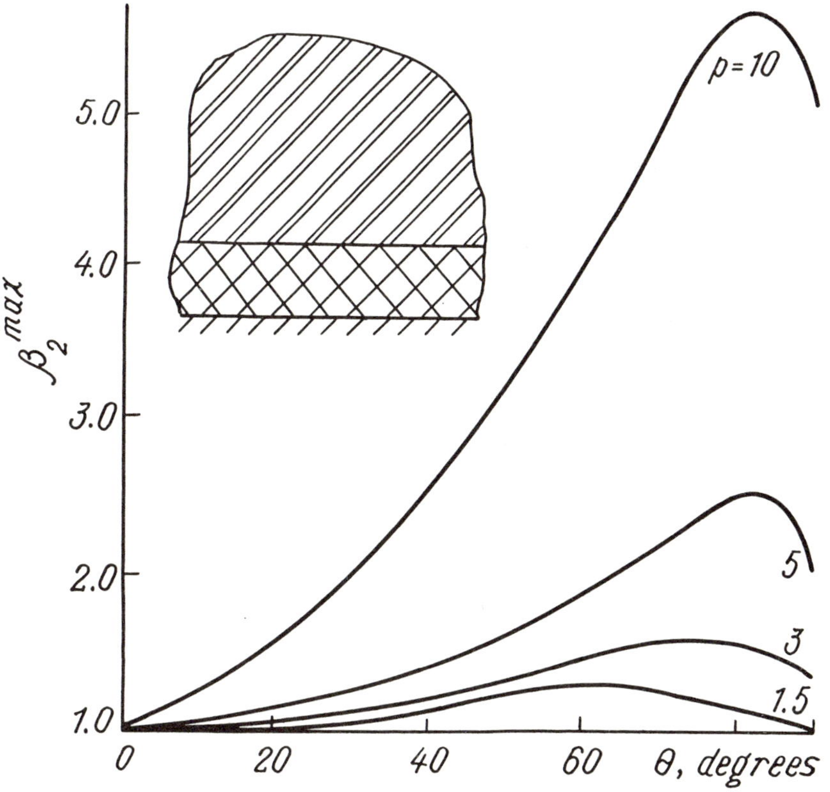

Figure 6.11 *Peak nonreciprocity coefficient of layered structure against angle of incidence* $\nu = 0.1$

Thus, no obvious analogy is observed between nonreciprocal effects in comb-shaped structures and layered media, unlike the resonances in the case of isotropic media. This means that comb-shaped structures cannot be looked upon as layers of artificial dielectrics. This results from the difference between the dependencies of the nonreciprocal phase component on the angle of incidence for periodic and plane-layered structures. Optimal phase and amplitude conditions in the latter (intensive partial-wave reflection from the interface between isotropic and gyrotropic media) cannot be achieved simultaneously—they correspond to different angles of incidence.

The dependencies of the nonreciprocity coefficient on the gyrotropy parameter $\nu = \varepsilon_2/\varepsilon_1$ for corrugated (solid line) and layered (broken line) structures are presented in Figure 6.12. The trends of these curves are different: for comb-shaped structures the curve $\beta(\nu)$ has an extremum, whereas for layered structures it is monotonic. This underlines once again the difference between the nonreciprocity mechanisms in the two cases. For corrugated surfaces the parameter β reaches values of the order of 10 with a much lower gyrotropy than for layered structures.

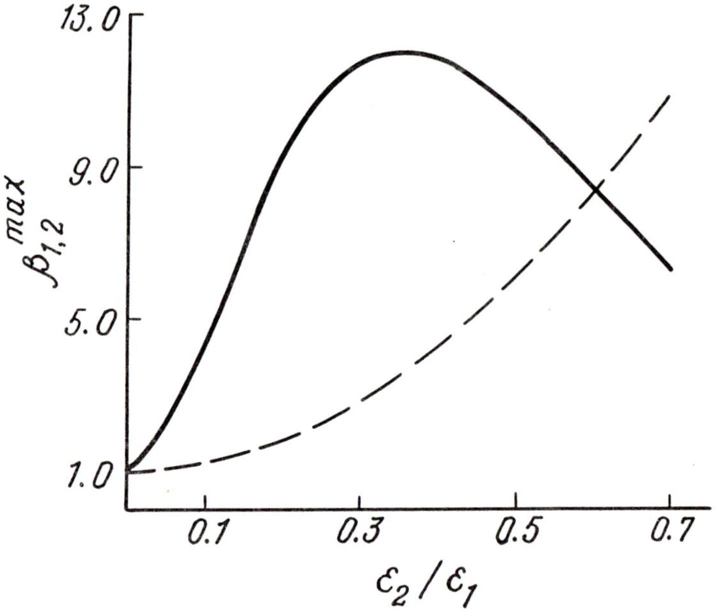

Figure 6.12 *Peak nonreciprocity coefficient as a function of* $\varepsilon_2/\varepsilon_1$

——— comb-shaped structure ($\varepsilon_1 = 10$, $kd = 0.21$)
- - - - layered structure ($p = 2$)

For the structure shown in Figure 6.10*b*, nonreciprocal effects are very weak (as follows from eqns. 6.21 and 6.23); at $\nu \lesssim 0.1$ they can hardly be observed.

Consider now the reflection of a plane wave from a smooth finitely conducting surface in a homogeneous gyrotropic medium. The simplest way to obtain an expression for the reflection coefficient lies in the replacements $\phi \rightarrow \vartheta$, $\vartheta \rightarrow \phi$, $p = 1/\zeta_S w$ in eqn. 6.23. Here ζ_S is the surface impedance of the metal, $w = \{(\varepsilon_1^2 - \varepsilon_2^2)/\varepsilon_1 \mu_0\}^{1/2}$ and μ_0 is the free space permeability (all media are nonmagnetic). We therefore have

$$R_0(\vartheta) = \frac{\cos\vartheta + j\nu\sin\vartheta - \zeta_S w}{\cos\vartheta - j\nu\sin\vartheta + \zeta_S w} \tag{6.24}$$

As $\zeta_S w \ll 1$, we can reduce eqn. 6.24 to a more simple form neglecting the terms above the first order of infinitesimals, which corresponds to the ordinary perturbation technique. Therefore

$$|R_0(\vartheta)|^2 \simeq 1 - 4\zeta_S w \cos\vartheta \tag{6.25}$$

Then, to an accuracy of the terms of the order of $\zeta_S w$, the nonreciprocal energetic effect is absent (only nonreciprocal phase rotation takes place).

Resonance nonreciprocity is also observed for 'reflection–transmission' resonances in transmission periodic structures. The simplest model for this case is shown in Figure 6.8*b*: the extraordinary plane wave falls at angle ϑ on a periodic structure of vertical metal strips. Under certain conditions the transmission coefficient T may depend greatly on the path in which the incident wave arrives (at identical absolute value of ϑ). The nonreciprocity factor in this case can be defined as

$$\beta_3(\vartheta) \simeq \frac{|T(\vartheta)|^2}{|T(-\vartheta)|^2}$$

Calculation of $\beta_3(\vartheta)$ is completely identical to that in Section 6.2; the results were presented by Slepyan (1983). The effect has its maximum when the resonance condition is satisfied, i.e. when $\gamma_e D + \phi_s(\vartheta) \simeq \pi$. For example, at $\varepsilon_1 = 10$, $\varepsilon_2 = 1$, $\vartheta = 80^\circ$, $k_e d = 0.2$ the coefficient β_3^{max} is of the order 5.

Finally, we note that the nonreciprocity of 'reflection–transmission' effects is somewhat lower than that of energy losses in comb-shaped structures at the same parameters of the medium, angles of incidence and structure periods.

The effects described above are promising for the design of nonreciprocal devices in microwave and millimetre-wave bands. They can also be used as a basis for developing equipment for measuring gyrotropic media parameters.

Chapter 7

Eigenmodes in corrugated waveguides and resonators with finitely conducting walls

This Chapter deals with the electromagnetic fields in systems with corrugated surfaces: waveguides, resonators and horns. Special attention is given to the investigation of dissipation characteristics: attenuation coefficients of different modes in corrugated waveguides and Q-factors of eigenoscillations in resonators with corrugated walls. Application of corrugated surfaces of both step shape and smooth is considered. When a period of the structure is small compared with the wavelength, a method of equivalent impedance-boundary conditions is used (in Section 7.2 a generalisation of this method for finite conductivity is presented). Low-loss corrugated waveguides (Sections 7.6–7.8) and millimetre-waveband high-quality resonators with a rare spectrum of eigenoscillations (Section 7.9) based on the effect of abnormally small absorption in periodic structures are described. In Section 7.10 a rigorous method of calculation of electrodynamic characteristics of corrugated horns, which are highly effective feeds for microwave antennae, is considered.

7.1 Eigenmodes in periodic structures

Let us first consider the properties of eigenmodes in periodic structures of a rather general type. We define a periodic structure as an area V, internal or external in relation to a conducting surface Σ, given by the equation $F(x, y, z) = 0$, where F satisfies the condition of periodicity $F(x, y, z) = F(x, y, z + d)$. Another variant of a periodical structure is an inhomogeneous medium with periodically varying permittivity and permeability

$$\varepsilon(x, y, z) = \varepsilon(x, y, z + d) \qquad \mu(x, y, z) = \mu(x, y, z + d)$$

Finally, a combination of both cases is possible: a periodically inhomogeneous medium inside the volume V.

Eigenmodes of periodic structures are nontrivial solutions of Maxwell's equations satisfying Floquet's condition of pseudoperiodicity:

$$\begin{Bmatrix} \boldsymbol{E}(x, y, z+d) \\ \boldsymbol{H}(x, y, z+d) \end{Bmatrix} = \exp(j\gamma d) \begin{Bmatrix} \boldsymbol{E}(x, y, z) \\ \boldsymbol{H}(x, y, z) \end{Bmatrix} \tag{7.1}$$

where γ is the phase change coefficient.

Eqn. 7.1 corresponds to an intuitive idea of waves in periodic structures: from one unit cell to another fields differ only in phase factor. The phase change per unit cell for the periodic structure illuminated by a plane wave is defined by the angle of incidence ϑ, i.e. $\gamma = k \sin \vartheta$, (see Figures 5.1 and 5.12). For eigenmodes γ can be regarded as a spectral parameter.

Using eqn. 7.1, we can represent the eigenmode sought as a space-harmonic expansion

$$\begin{Bmatrix} \boldsymbol{E}(x, y, z) \\ \boldsymbol{H}(x, y, z) \end{Bmatrix} = \sum_{m=-\infty}^{\infty} \begin{Bmatrix} \boldsymbol{e}_m(x, y) \\ \boldsymbol{h}_m(x, y) \end{Bmatrix} \exp\left\{j\left(\gamma + \frac{2\pi m}{d}\right)z\right\} \tag{7.2}$$

where the functions $\boldsymbol{e}_m(x, y)$ and $\boldsymbol{h}_m(x, y)$ are determined by substituting eqn. 7.2 into Maxwell's equations and boundary conditions. Thus we obtain a characteristic equation for γ.

It is noteworthy that each of the space harmonics is not an actually existing wave. Only eigenmodes containing an infinite set of space harmonics are a physical reality. The fact that under certain conditions one of the space harmonics can prevail over others (for example, at $d \ll \lambda$ only zero-order harmonics are essential) is not contradictory to this statement.

The conditions of pseudoperiodicity given by eqn. 7.1 are, to a certain extent, taken *a priori* and need to be rigorously substantiated. First, do such solutions of Maxwell's equations exist at all? Secondly, do all possible solutions satisfy eqn. 7.1? In the 1-dimensional case the answers to these questions are given by the theory of ordinary differential equations with periodical coefficients (Floquet–Bloch theorem). For distributed systems of general type there is no rigorous proof of this theorem analogue.

As an answer to the first question we can consider, for example, pseudoperiodical solutions for specific periodic structures. As for the second question, when solving a dispersion equation, imaginary values of γ can be obtained (we are considering now systems without losses!). Finally, in the work by Beliantsev and Gaponov (1964) complex modes in periodic structures were found. They are characterised by a complex γ even for lossless systems. For such modes the coefficient γd cannot be interpreted as a phase change per unit cell. It can be shown that waves with imaginary γ do not carry energy along the z axis; in this respect they are similar to evanescent modes of ordinary smooth waveguides.[42] Such solutions correspond to stop bands of periodic structures (there is an infinite number of alternating transmission and stop bands). Complex waves are always excited in pairs and form reactive fields (i.e. energy is not carried along the z axis), the spatial distribution of which along the z axis is not periodical.

A formal solution of dispersion equations for systems with dissipation leads to a complex γ ($\gamma = \gamma' + j\gamma''$). In this case it follows from eqn. 7.1 that the field

---

[42] Unlike the case of evanescent modes in smooth waveguides, in this case fields vary along the z axis as $f(z)\exp(-|\gamma|z)$, where $f(z)$ is a periodic function with period d.

diminishes with $\exp(-\gamma'' z)$. A question arises: to what extent is this law general and do nonexponential solutions exist? To answer these we shall not impose the conditions of pseudoperiodicity *a priori*. In this case the periodical structure will be considered as an irregular waveguide, and the theory of such waveguides will have to be applied (Schelkunoff, 1955, Katsenelenbaum, 1961). Now we consider this approach for lossless structures.

The fields sought are written as the expansions

$$\left.\begin{aligned} \boldsymbol{E}(x, y, z) &= \sum_{p=0}^{\infty} c_p(z)\tilde{\boldsymbol{e}}_p(x, y) \\ \boldsymbol{H}(x, y, z) &= \sum_{p=0}^{\infty} b_p(z)\tilde{\boldsymbol{h}}_p(x, y) \end{aligned}\right\} \tag{7.3}$$

where $\{\tilde{\boldsymbol{e}}_p(x, y)\}$, $\{\tilde{\boldsymbol{h}}_p(x, y)\}$ are systems of vector-functions, complete in an arbitrary cross-section of the waveguide, and $c_p(z)$ and $b_p(z)$ are the coefficients sought. Substituting eqn. 7.3 into Maxwell's equations and boundary conditions, we obtain a system of ordinary differential equations for the functions $b_p(z)$ and $c_p(z)$ which can be reduced to the normal Hamilton form (a specific example is given in Section 7.6):

$$\begin{aligned} J\frac{\mathrm{d}\boldsymbol{x}}{\mathrm{d}z} &= H(z)\boldsymbol{x} \\ \boldsymbol{x} &= \begin{pmatrix} b \\ \hline c \end{pmatrix} \\ J &= \left[\begin{array}{c|c} I & 0 \\ \hline 0 & -I \end{array}\right] \end{aligned} \tag{7.4}$$

where $\boldsymbol{x}$ is an unknown vector and $H(z)$ is a periodical matrix Hamiltonian, i.e. $H(z) = H(z + d)$. Eqns. 7.4 have a form similar to Hill's system and the application of general methods of Hill's theory (Yakubovich and Starzhinskii, 1975) to eqns. 7.4 would be promising. However as the system given by eqns. 7.4 is infinite-dimensional, the transition to a finite-dimensional analogue should be substantiated in detail for every specific case. For example, in the paper by Sveshnikov (1963) there is an appropriate proof for circular irregular waveguides. In the papers by Koroza (1970) and Shankin (1974) the possibility of the transition to a finite-dimensional system is postulated with a subsequent use of Hill's theory. It is shown that the solution of eqns. 7.4 satisfies the relationship

$$\boldsymbol{x}(z + d) = \exp(\Lambda)\boldsymbol{x}(z)$$

where Λ is called the monodromy matrix.[43] Eigenvalues ρ_k of the matrix $\exp(\Lambda)$ are called multipliers. They can be expressed by

$$\rho_k = \exp(j\gamma_k d)$$

where γ_k is the propagation coefficient of the kth eigenmode.

If matrix Λ satisfies the conditions of theorems 3.1 and 3.2 (Section 3.3), it

[43] For this method to be used, the monodromy matrix should exist. For some complex periodical structures the monodromy matrix may not exist at all.

can be reduced to a diagonal form (all multipliers in this case are either simple, or D-multiple). Thus, we come to eqns. 7.1 and 7.2. If some of the multipliers are of J-multiplicity, then the monodromy matrix can be reduced to a normal Jordan form (see Section 3.4). In this case, as shown by Shankin (1974), apart from the eigenmodes there are associated modes not satisfying the conditions of pseudoperiodicity. Such modes can not be presented in the form of eqn. 7.2. This can be looked upon as an example of a nonexponential solution of Maxwell's equations for periodic structures. If $\rho_k = \exp(j\gamma_k d)$ is a multiplier, then $\rho_k^* = \exp(-j\gamma_k d)$ is also a multiplier. The numbers ρ_k correspond to forward modes and ρ_k^* to backward modes.

In periodic structures without losses the multipliers are usually located on the circle of unit radius or on the real axis. The first case corresponds to a transmission band, and the second to a stop band; band-edge frequencies satisfy the equations

$$\exp\{j\gamma_k(\omega_{0,1})d\} = \pm 1 \tag{7.5}$$

The general theory of wave transformation developed by Krasnushkin and Fedorov (1972) is applicable to periodic structures. As shown by Krasnushkin (1974) and Shankin (1974), the frequencies $\omega_{0,1}$ determined from eqn. 7.5 are points of J-multiplicity. At these points transformation of two propagating modes into two evanescent ones (and vice versa) occurs and associated modes emerge. We note that the points of J-multiplicity do not necessarily correspond to the values $\rho_k = \pm 1$. They can also be located at an arbitrary point on the unit circle in the plane of the complex variable ρ. This occurs when one of the degenerated modes is characterised by normal dispersion, and the other by anomalous dispersion, which gives rise to complex waves (Krasnushkin, 1974).

Considering the eigenmodes in a periodic structure with losses in the walls, we have to solve Maxwell's homogeneous equations with impedance-boundary conditions. We assume that eigenmodes of the same structure but without losses are known (let us denote them by $\boldsymbol{E}_n$, $\boldsymbol{H}_n$). The method, described below, is based on the expansion of eigenmodes in the system with losses in terms of eigenmodes of the same lossless system and is a generalisation of the technique used in Section 3.3 for periodic structures. Using the Lorentz lemma we easily obtain the following relationship for $\boldsymbol{E}_n$ and $\boldsymbol{H}_n$ (Weinstein, 1988):

$$[1 - \exp\{2j(h_n + h_{n'})d\}] \int_{S_\perp(z)} (\boldsymbol{E}_n \times \boldsymbol{H}_{n'} - \boldsymbol{E}_{n'} \times \boldsymbol{H}_n)\hat{i}_z \, dS_\perp = 0$$

where h_n is a propogation coefficient of the nth mode in the system without losses.

From the latter it follows that either $h_n + h_{n'} = \pi m/d$ (m integer), or the integral over $S_\perp$ equals zero. The first condition at $m \neq 0$ corresponds to the band-edge frequencies. In the case of $m = 0$ we come to a condition $n' = -n$. Thus, we obtain

$$\int_{S_\perp(z)} (\boldsymbol{E}_n \times \boldsymbol{H}_{n'} - \boldsymbol{E}_{n'} \times \boldsymbol{H}_n)\hat{i}_z \, dS_\perp = \delta_{n,-n'} \mathcal{N}_n$$

This is known as the orthogonal property of eigenmodes. From the Lorentz lemma in differential form (see eqn. 3.20) it follows that in lossless structures $\mathcal{N}_n$ does not depend on z.

Now we assume that systems $\{\boldsymbol{E}_n\}$, $\{\boldsymbol{H}_n\}$ are complete in an arbitrary cross-section $S_\perp(z)$.[44] Then eigenmodes of the system with losses can be presented (as in Section 3.3) by

$$\begin{Bmatrix}\boldsymbol{E}\\ \boldsymbol{H}\end{Bmatrix} = \sum_n \left[\Upsilon_n(z) \begin{Bmatrix}\boldsymbol{E}_n\\ \boldsymbol{H}_n\end{Bmatrix} + \Upsilon_{-n}(z) \begin{Bmatrix}\boldsymbol{E}_{-n}\\ \boldsymbol{H}_{-n}\end{Bmatrix} \right] \tag{7.6}$$

where $\Upsilon_{\pm n}(z)$ are functions to be determined.

To determine $\Upsilon_{\pm n}(z)$ we apply eqn. 3.20, substituting the expansion given by eqn. 7.6 as $\boldsymbol{E}_1$, $\boldsymbol{H}_1$ and $\boldsymbol{E}_{-p}$, $\boldsymbol{H}_{-p}$ as $\boldsymbol{E}_2$, $\boldsymbol{H}_2$ ($\boldsymbol{J}^l_{1,2} \equiv 0$, $Z_{S1} = Z_S$, $Z_{S2} = 0$). After some transformations we obtain the following system of differential equations for $\Upsilon_{\pm n}(z)$:

$$\frac{d\Upsilon_p}{dz} = -\frac{Z_S}{N_p} \sum_{q=-\infty}^{\infty} \Upsilon_q(z) \oint_{C(z)} \{(\boldsymbol{n}_0 \times \boldsymbol{H}_q) \times \boldsymbol{H}_{-p}\}\boldsymbol{n}_0 \, dc \tag{7.7}$$

where $C(z)$ is the perimeter of the cross-section $S_\perp(z)$.

Replacing the unknown functions $\Upsilon_n(z)$ by $b_n(z)\exp(-jh_n z)$ in eqn. 7.7, we have

$$\frac{db_p}{dz} - jh_p b_p = -\frac{Z_S}{N_p} \sum_{q=-\infty}^{\infty} b_q B_{qp}(z) \tag{7.8}$$

where $B_{qp}(z)$ are periodical functions with period d, expressed by the formula

$$B_{qp} = \oint_{C(z)} \exp\{j(h_p - h_q)z\} \boldsymbol{H}_{q\tau} \boldsymbol{H}_{-p\tau} \, dc$$

$$= \sum_m \sum_n \exp\left\{\frac{j2\pi(m+n)z}{d}\right\} \oint_{C(z)} \boldsymbol{h}^{(q)}_{m\tau}(x, y) \boldsymbol{h}^{(p)}_{n\tau}(x, y) \, dc \tag{7.9}$$

Let us consider an approximate solution of eqn. 7.8 on the following assumptions:

(i) there are no degenerated modes among the eigenmodes $\boldsymbol{E}_n$, $\boldsymbol{H}_n$ (except trivial multiplicity cases);
(ii) frequencies considered are distanced from the band-edge ones;
(iii) surface impedance of the periodic structure material is small ($|Z_S|/W_0 \ll 1$).

In agreement with the assumptions mentioned above we can reduce the system described by eqn. 7.8, as in Section 3.3, to the equation of the first order; the latter has the form

$$\frac{db_p}{dz} - jh_p b_p = -\frac{Z_S}{N_p} b_p B_{pp}(z) \tag{7.10}$$

Taking into account the periodicity of $B_{pp}(z)$ expressed by eqn. 7.9, the solution

[44] Our case is confined to areas internal as related to the surface Σ. In other cases, the theory of open waveguides should be used (Weinstein, 1969*a*), which takes into account the functions of continuous spectrum.

of eqn. 7.10, in accordance with the Floquet–Bloch theorem, can be written as

$$b_p(z) = \exp(j\gamma_p z) \sum_{r=-\infty}^{\infty} c_{pr} \exp\left\{\frac{j2\pi rz}{d}\right\} \tag{7.11}$$

where c_{pr} are constant factors.

By direct substitution of eqn. 7.11 into eqn. 7.10 we can show that, within the framework of the perturbation theory (Z_S small), the eigenvalue γ_p is expressed by

$$\gamma_p = h_p + j\frac{Z_S}{N_p}\int_0^d B_{pp}(\zeta)\,d\zeta \tag{7.12}$$

For modes in a transmission band[45] we can transform eqn. 7.12 (as in Section 3.7) into the formulae

$$\gamma'_p = h_p - \frac{\operatorname{Im} Z_S \int_{\Sigma_d} |\boldsymbol{H}_{p\tau}|^2\,dS}{2d \operatorname{Re} \int_{S_\perp} (\boldsymbol{E}_p \times \boldsymbol{H}_p^*)\hat{i}_z\,dS_\perp} \tag{7.13}$$

$$\gamma''_p = \frac{\operatorname{Re} Z_S \int_{\Sigma_d} |\boldsymbol{H}_{p\tau}|^2\,dS}{2d \operatorname{Re} \int_{S_\perp} (\boldsymbol{E}_p \times \boldsymbol{H}_p^*)\hat{i}_z\,dS_\perp} \tag{7.14}$$

where Σ_d is the metallised surface of a unit cell of the structure.

Eqn. 7.14 coincides with that obtained by application of the energy-perturbation technique for periodic structures. It is similar to eqn. 3.62 for regular waveguides. Eqn. 7.14 is also valid when areas external to the surface Σ are considered, though its derivation becomes more complicated. For Σ with edges, when the induced current density has singularities, eqn. 7.14 should be modified in a manner identical to that for regular waveguides. The use of eqns. 7.13 and 7.14 is relevant when the integrals in the numerator and denominator have a form convenient for calculation. Otherwise, a direct solution of the original problem with impedance-boundary conditions would be recommended (see Section 3.7).

7.2 Equivalent boundary conditions for finitely conducting comb-shaped structures

The equivalent-boundary-condition technique is widely used for the analysis of electromagnetic fields in periodic structures when the period of the latter is much less than the wavelength. This method consists of the imaginary replacement of the periodical structure by a smooth (plane) surface where special

---

[45] Note that in the presence of absorption the boundaries of transmission and stop bands become slightly washed out in a similar way to cutoff frequencies of waveguides (see Section 3.7).

Weinstein (1969*b*), at $kd \ll 1$ we have

$$\gamma_0 \simeq \frac{2kd \ln 2}{\pi} + \pi$$

Then for $\mathcal{Z}^h$ we obtain the approximate expression

$$\mathcal{Z}^h \simeq \frac{W_0 kd \ln 2}{\pi} \left\{ \frac{\pi v(k)}{2kd \ln 2} + j \right\} \tag{7.24}$$

where $v(k)$ is defined by eqn. 5.35.

Thus, for an arbitrary magnetically polarised field the boundary condition is

$$(E_y + \mathcal{Z}^h H_z)|_{x=0} = 0 \tag{7.25}$$

Combining eqns. 7.15 and 7.25 we obtain an anisotropic impedance-boundary condition for an arbitrary polarised field

$$(\hat{i}_x \times \boldsymbol{E})|_{x=0} = \overleftrightarrow{\mathcal{Z}}\{\hat{i}_x \times (\hat{i}_x \times \boldsymbol{H})\}|_{x=0} \tag{7.26}$$

where $\overleftrightarrow{\mathcal{Z}}$ is an impedance tensor having the form

$$\overleftrightarrow{\mathcal{Z}} = \begin{bmatrix} \mathcal{Z}^e & 0 \\ 0 & \mathcal{Z}^h \end{bmatrix}$$

There can be a different approach to the problem of calculating $\mathcal{Z}^{e,h}$ in eqn. 7.26. It is based on the conformal-mapping method. Depine and Brudny (1990) have studied the problem of wave diffraction by periodic surface $x = g(z)$ $[g(z+d) = g(z)]$, where the Leontovich impedance-boundary condition is imposed. They used the conformal mapping relating $w = z + jx$ with $U = X + jY$ in such a way that the region $x \geq g(z)$ was transformed into the upper halfplane $Y \geq 0$. Then introducing the function

$$F(X, Y) = \begin{cases} H_y\{x(X, Y), z(X, Y)\}, & \text{electric polarisation} \\ E_y\{x(X, Y), z(X, Y)\}, & \text{magnetic polarisation} \end{cases}$$

they wrote the boundary condition on the surface $x = g(z)$ as

$$F(X, 0) = \frac{j\mathcal{Z}_S}{kW_0 s(X)} \frac{\partial F(X, 0)}{\partial Y} \tag{7.27}$$

for magnetic polarisation and

$$\frac{\partial F(X, 0)}{\partial Y} = \frac{k\mathcal{Z}_S}{jW_0} s(X) F(X, 0) \tag{7.28}$$

for electric polarisation. The function $s(X)$ is expressed by

$$s(X) = \left| \frac{dw}{dU} \right| \Bigg|_{Y=0}$$

Eqns. 7.27 and 7.28 are the boundary conditions of the third kind with periodic coefficients which are to be fulfilled at $Y = 0$.

These results, as we shall illustrate later, can be used for the calculation of $\mathcal{Z}^{e,h}$. If the depth and period of the structure are small compared with the wavelength, the field in the grooves is quasistatic. Then, through averaging eqns. 7.27 and 7.28 over the period, we obtain the following formulae for real parts of $\mathcal{Z}^{e,h}$:

$$\operatorname{Re} \mathcal{Z}^e = \frac{\operatorname{Re} \mathcal{Z}_S}{d} \int_0^d s(x)\, dx \tag{7.29}$$

$$\operatorname{Re} \mathcal{Z}^h = \frac{\operatorname{Re} \mathcal{Z}_S}{d} \int_0^d \frac{dx}{s(x)} \tag{7.30}$$

The calculation of the reactive components of $\mathcal{Z}^{e,h}$ is carried out in the framework of the model of perfect conductivity. This problem has been considered by many authors (see, for example, Nefedov and Sivov, 1977), so we do not touch upon it here. It is worth noting that this averaging is in fact equivalent to the use of the two first terms in the perturbation series in $\mathcal{Z}_S$. Eqns. 7.29 and 7.30, unlike eqns. 7.22 and 7.24, are applicable for structures of an arbitrary shape and not very deep ($D \ll \lambda$).

Let us consider the following examples. For the rectangular-groove grating shown in Figure 5.1*b* the conformal mapping function is expressed in the form (Lavrentyev and Shabat, 1987)

$$U(w) = \frac{d}{\pi} \cos^{-1}\left[\frac{\cos(\pi w/d)}{\operatorname{ch}(\pi D/d)}\right] \tag{7.31}$$

Using eqns. 7.29 and 7.31, we obtain

$$\operatorname{Re} \mathcal{Z}^e = \operatorname{Re} \mathcal{Z}_S \left(1 + \frac{2D}{d}\right)$$

which at $kD \ll 1$ coincides with eqn. 7.22. The integral in eqn. 7.30 diverges if the function $U(w)$ is given by eqn. 7.31. This confirms once again the inapplicability of the infinitely thin-surface model (when used directly without modifications) for calculating the power loss in the case of magnetic polarisation.

Depine and Brudny (1990) considered the cycloid-shaped periodic surface described in a parametric form by the equations

$$\begin{aligned} z &= X - \frac{D}{2} \sin\left(\frac{2\pi X}{d}\right) \\ x &= \frac{D}{2}\left\{1 + \cos\left(\frac{2\pi X}{d}\right)\right\} \end{aligned} \tag{7.32}$$

where D is the groove depth.

The conformal-mapping function in this case, according to the work by Depine and Brudny (1990), has the form

$$w = \frac{jD}{2}\left\{1 + \exp\left(\frac{2\pi j U}{d}\right)\right\} + U \tag{7.33}$$

Substituting eqn. 7.33 into eqns. 7.29 and 7.30 we can obtain the following expressions for $\operatorname{Re} \mathcal{Z}^{e,h}$:

$$\operatorname{Re} \mathcal{Z}^e = \frac{2}{\pi} \operatorname{Re} \mathcal{Z}_S \begin{cases} 2E(p) - (1 - p^2)K(p) & (p < 1) \\ p_1 E(2/p_1) & (p > 1) \end{cases} \tag{7.34}$$

$$\operatorname{Re} \mathcal{Z}^h = \frac{2}{\pi} \operatorname{Re} \mathcal{Z}_S \begin{cases} K(p) & (p < 1) \\ K(2/p_1)/p_1 & (p > 1) \end{cases} \tag{7.35}$$

where $p = \pi D/d$, $p_1 = (3 + p^2)^{1/2}$ and $K(q)$ and $E(q)$ are the complete elliptic integrals of the first and second kinds, respectively.

From eqns. 7.34 and 7.35 it follows that at $p \to 1$ $\mathrm{Re}\,\zeta^e \to 4\,\mathrm{Re}\,\zeta_S/\pi$ and $\mathrm{Re}\,\zeta^h \to \infty$. However, it is necessary to point out that at $p \simeq 1$ eqn. 7.35 is invalid, since for $p = 1$ eqn. 7.32 defines a surface with infinitely thin ridges.

Though the boundary condition given by eqn. 7.26 has been obtained for a plane infinitely wide comb-shaped structure, it can also be used for curvilinear comb surfaces, if the main curvature radii significantly exceed the period and depth of the structure, which can then be considered as locally plane. This boundary condition can be also used for finite corrugated surfaces if auxiliary conditions, describing the field behaviour at the ends of the structure, are introduced. These conditions should be formulated so that fictitious field sources would not emerge on edges.

The situation described is rather typical when using equivalent boundary conditions of different types: these conditions are derived rigorously for a certain key problem, simplified to the maximum. Then, on the basis of physical sense and intuition, we extrapolate the equivalent boundary conditions to a certain class of much more complicated problems to fundamentally simplify their mathematical formulation. In this case such small-scale details of the field structure are not considered. However, we must underline that they do not, to any notable degree, affect the field-integral characteristics (e.g. phase and attenuation coefficients etc.).

In our further considerations, we shall describe a simple analytical theory of eigenmodes in corrugated waveguides and resonators, based on approximate boundary conditions developed.

7.3 Surface waves in finitely conducting comb-shaped structures

A metal comb-shaped structure (Figure 5.1*b*) is the simplest guiding system capable of supporting a surface wave. Dispersion properties of such slow-wave structures in an approximation of perfect conductivity have been examined quite extensively; the most rigorous theory for this case had been developed by Weinstein (1969*b*). It is of interest to develop such a theory which would allow us to take into consideration energy losses in the metal.

Let us consider a 2-dimensional (infinite in the direction of the y axis) comb-shaped structure, the ridges of which are considered infinitely thin. Suppose also that $kd \ll 1$. For our purposes only electrically polarised fields are of interest: there are no surface modes of opposite polarisation. The field components are expressed by

$$\left.\begin{aligned} H_y &= \psi(x, z) \\ E_x &= \frac{1}{j\omega\varepsilon_0}\frac{\partial\psi}{\partial z} \\ E_x &= \frac{j}{\omega\varepsilon_0}\frac{\partial\psi}{\partial x} \end{aligned}\right\} \qquad (7.36)$$

From eqns. 7.15 and 7.36 we can obtain the following boundary condition for the function $\psi(x, z)$:

$$\left.\frac{\partial\psi}{\partial x}\right|_{x=0} = -jk\frac{Z^e}{W_0}\psi\bigg|_{x=0} \qquad (7.37)$$

The surface-wave field can be written as

$$\psi(x, z) = \exp(-\chi x + j\gamma z) \qquad (7.38)$$

where $\gamma = \gamma' + j\gamma'' = (k^2 + \chi^2)^{1/2}$, χ being an unknown transverse complex wave number ($\lim_{Z_S \to 0} \chi^2 < 0$). Substituting eqn. 7.38 into eqn. 7.37 we come to the formula

$$\chi = jkZ^e / W_0$$

Taking into consideration eqn. 7.22, we have

$$\chi = -\frac{jk \operatorname{Re} Z_S}{\cos^2(kD)}\left\{1 + \frac{D}{d}\left(1 + \frac{\sin 2kD}{2kD}\right)\right\} + k\tan(kD) \qquad (7.39)$$

It is easy to determine the surface-wave attenuation coefficient γ'' from eqn. 7.39. For convenience, let us introduce the dimensionless slowing factor $N = v_{ph}/c$ (v_{ph} being the surface-wave phase velocity). According to the elementary theory by Weinstein (1969*a*), in an approximation of perfect conductivity $N = \cos kD$. This makes it possible to connect the attenuation coefficient γ'' directly with the slowing factor N. Neglecting the terms above the first order of $\operatorname{Re} Z_S$, we obtain (Slepyan, 1981)

$$\frac{\gamma''}{k} = \operatorname{Re} Z_S \frac{p}{W_0 N^2}\{1 + \frac{1}{kd}(pN + \cos^{-1} N)\} \qquad (7.40)$$

where $p = (1 - N^2)^{1/2}$.

Using eqn. 7.40, we can show that the surface-wave attenuation grows with the increase of retardation. This can be explained by the fact that while N is increasing the wave is 'pressed' to a large degree to the surface, and this leads to an increase in induced currents. On approaching the lower boundary of the transmission band ($N \to 1$) losses in walls, according to eqn. 7.40, disappear ($\gamma'' \to 0$). This is a result of the Rayleigh diffractional anomaly dealt with in Section 5.5.

While approaching the upper boundary of the transmission band ($N \to 0$) the losses, as follows from eqn. 7.40, increase infinitely. However, in a certain narrow neighbourhood of the band-edge frequency, corresponding to the condition $N = 0$, the perturbation technique, and eqn. 7.40 based on it, are inapplicable. More accurate analysis shows that at $N \to 0$ γ'' increases sharply while remaining finite.

7.4 TM modes in plane comb-shaped waveguides

Let us examine propagation and attenuation properties of TM modes in 2-dimensional corrugated waveguides (Figure 7.1). The theory described is based on equivalent impedance boundary conditions obtained in Section 7.2. The assumptions made are similar to those in Sections 3.4 and 7.2. The mathematical formulation of the problem is to find a solution of the 2-dimensional Helmholtz equation satisfying the boundary condition given by eqn. 7.37 at $x = \pm\Delta$.

We present the eigenmode as

$$\psi(x, z) = \begin{Bmatrix} \cos \chi x \\ \sin \chi x \end{Bmatrix} \exp(j\gamma z) \tag{7.41}$$

where $\chi = \chi' + j\chi''$ is the unknown transverse wave number and $\gamma = (k^2 - \chi^2)^{1/2}$, the upper line in eqn. 7.41 corresponding to the modes symmetrical in x, and the lower line to modes antisymmetrical in x. The field components are determined from eqn. 7.36.

Substituting eqn. 7.41 into eqn. 7.37, we obtain the following dispersion equations for χ:

$$\begin{Bmatrix} -\chi\Delta \tan(\chi\Delta) \\ \chi\Delta \cot(\chi\Delta) \end{Bmatrix} = \zeta' + j\zeta'' \tag{7.42}$$

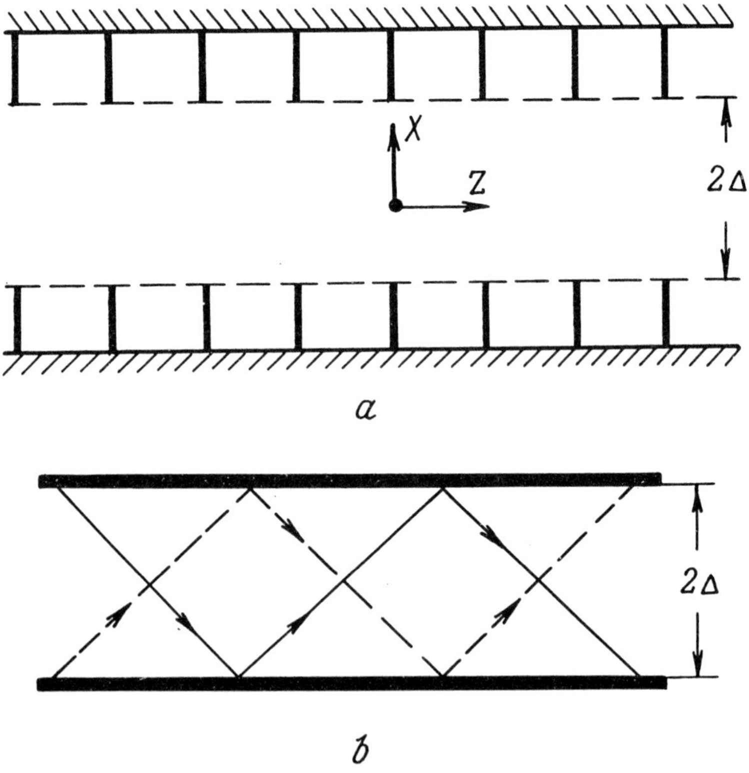

Figure 7.1 *Geometry of plane comb-shaped waveguide*

a Original structure
b Impedance model (arrows indicate the formation of an eigenmode according to Brillouin's concept)

where $\zeta' + j\zeta'' = jk\Delta \mathcal{Z}^e / W_0$. The roots of eqn. 7.42 located near the real axis correspond to fast (spatial) modes, the roots located near the imaginary axis to slow (surface) ones.

Eqn. 7.42 for fast modes has a concrete physical sense. It corresponds to the Brillouin concept according to which fast eigenmodes are formed as a result of consecutive reflections of two plane partial waves from the upper and lower comb-shaped surfaces.[47] In this case we do not take into consideration the interaction of corrugated structures through higher space harmonics, excited at every reflection of partial plane waves and distorted by an opposite reflecting surface. These effects, however, are essential only for closely located comb structures,[48] which are not of great interest, but show an approximate character of the equivalent boundary condition given by eqn. 7.15. Thus a theory, developed here, is applied at considerably large Δ.

Considering the term ζ'' describing the losses as a small perturbation, we can simplify eqn. 7.42. Let $\chi \simeq \chi_{(0)} + j\delta$ where $\chi_{(0)}$ is the transverse wave number of a corresponding waveguide without losses. Then δ is expressed by the formula

$$\delta = \zeta'' \chi_{(0)} \{\zeta' - \zeta'^2 - \chi_{(0)}^2 \Delta^2\}^{-1} \tag{7.43}$$

For $\chi_{(0)}$ we have real transcendental equations

$$\left\{ \begin{array}{l} -\chi_{(0)}\Delta \tan\{\chi_{(0)}\Delta\} \\ \;\;\chi_{(0)}\Delta \cot\{\chi_{(0)}\Delta\} \end{array} \right\} = \zeta' \tag{7.44}$$

an approximate solution of which does not appear to be too difficult to obtain.

Classification of modes in the waveguide under consideration can be given in an approximation of perfect conductivity on the basis of eqn. 7.44. The qualitative trend of dispersion curves for symmetrical and antisymmetrical modes is shown in Fig. 7.2*a* and *b*. The cutoff frequencies determined by the condition $\gamma(k_{c1}) = 0$ can be found from eqn. 7.44, assuming that $\chi_{(0)} = k_{c1}$. Then it is possible to solve the equations exactly; the roots of these equations are expressed by

$$k_{c1} = \begin{cases} n\pi/(\Delta + D) & \text{for symmetrical modes} \\ (n - \frac{1}{2})\pi/(\Delta + D) & \text{for antisymmetrical modes} \end{cases}$$

where n is the mode index ($n = 1, 2, \ldots$).

At $k > k_{c1}$ the mode is a propagating one, and at $k < k_{c1}$ an evanescent one.

In the frequency range determined by the condition

$$k_{c1} < k < k_t$$

the given mode is a fast one; at $k = k_t$ it transforms into a slow (surface) one. Transformation frequencies of fast waves into slow waves are determined from the condition $\lim_{\zeta' \to 0} \gamma(k_t) = k_t$ and can be obtained from eqn. 7.44, assuming that $\chi_{(0)} = 0$. For antisymmetrical modes we obtain the transcendental equation

$$k_t \Delta \tan(k_t D) = 1$$

---

[47] In the presence of losses, Brillouin's angle ϑ_n, formally introduced by the relationship $\vartheta_n = \cos^{-1}(\chi_n/k)$, is a complex number and does not have a direct sense of the angle between the x axis and the normal to the front of a partial plane wave.

[48] For example, such as $\exp[-\{(2\pi/d - k\sin\vartheta)^2 - k^2\}^{1/2}\Delta] \simeq 1$.

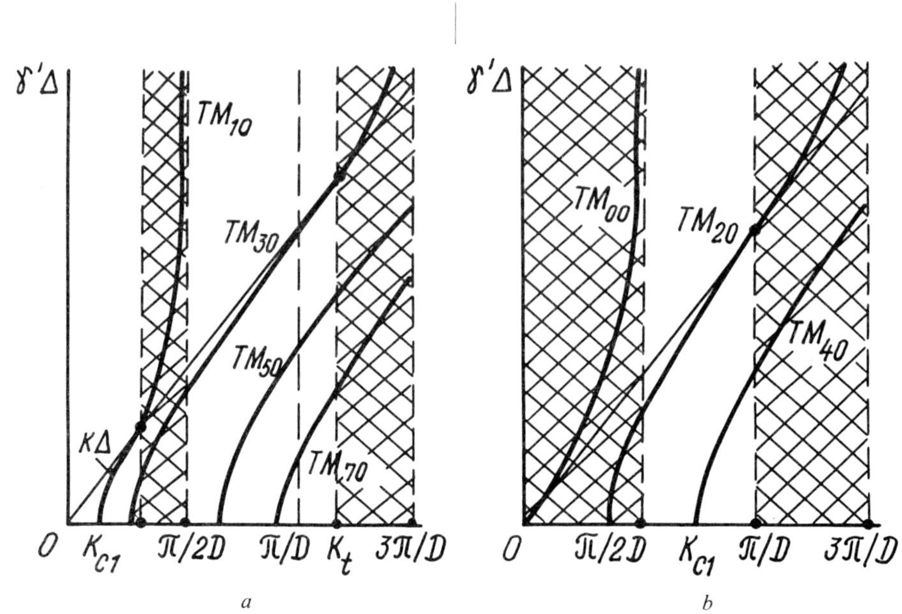

Figure 7.2 *Dispersion diagrams for plane comb-shaped waveguide*

a Odd modes
b Even modes

For symmetrical modes the corresponding equation can be solved exactly. The roots of the latter are

$$k_t = n\pi/D, n = 0, 1, 2, \ldots$$

At $k_t < k < k_{c2}$ there exist surface waves in a waveguide. When k increases these waves are more 'pressed' to the surface, and, at $k_{c2} = \pi(n + \frac{1}{2})/D$, undergo a cutoff. For antisymmetrical modes the mth frequency band δk in which a surface wave exists grows when Δ increases at a constant k. If Δ is constant, δk grows when n increases, tending asymptotically to the value $\delta k = \pi/2D$. For symmetrical modes $\delta k = \pi/2D$ independent of Δ and a band number.

Among symmetrical modes one mode is a slow one throughout the range of its existence; there are no antisymmetrical modes of such type.

Simple approximate relationships for the attenuation coefficients of TM modes in a plane corrugated waveguide can be given at $|\tan(kD)| \gg 1$ ($|\mathrm{Re}\, Z^e| \ll |\mathrm{Im}\, Z^e|$). This case, as is shown by calculations made for the key problem (see Section 5.5), correspond exactly to minimal attenuation and, hence, are of the greatest practical interest.

To begin, let us consider a transverse resonance for the TM_{20} mode[49] on the cutoff frequency. Supposing that in eqn. 7.42 $\chi = k$, we obtain a characteristic

[49] When the attenuation is minimal ($k \simeq \pi/2D$), the TM_{00} mode is a surface one ($k < \pi/D$) or does not exist at all ($k > \pi/D$). That is why at $k \simeq \pi/2$ the TM_{20} mode is the lowest symmetrical spatial mode of E-type.

equation for the wave number $k = k' - jk''$, which, owing to the presence of losses, becomes a complex one. Solving this equation and taking into consideration the smallness of losses, we obtain for the Q-factor of transverse resonance the formula

$$Q^{-1} = \frac{2k''}{k'} = Q_{(0)}^{-1}\left(1 + \frac{D}{d}\right) \tag{7.45}$$

where $Q_{(0)} = \pi W_0/2 \operatorname{Re} Z_S$ is the Q-factor of a similar resonance in a parallel-plate waveguide at the same frequency. From eqn. 7.45 it follows that the Q-factor of a plane corrugated waveguide on a cutoff frequency is less than that in a parallel-plate one.

For the attenuation coefficient of the TM_{20} mode in a corrugated waveguide it is possible to obtain from eqns. 7.43 and 7.44 the formula

$$\gamma'' \simeq \gamma''_{(0)} \frac{\pi^2 (d + D)}{2d (k\Delta)^2} \tag{7.46}$$

where $\gamma''_{(0)}$ is the attenuation coefficient of the E_{20} mode in a corresponding smooth-wall waveguide (at the same propagation coefficient) expressed as

$$\gamma''_{(0)} = \frac{\operatorname{Re} Z_S k}{W_0 \{(2k\Delta)^2 - \pi^2\}^{1/2} \pi}$$

From eqn. 7.46 it follows that TM modes in a corrugated waveguide have lower attenuation than in a smooth-walled waveguide. The larger values of the ratio $\gamma''/\gamma''_{(0)}$ correspond to the larger values of $k\Delta$ (to larger distances between the operational frequency and the cutoff frequencies).

We wish to emphasise that eqn. 7.46 is suitable only for fast modes ($k \lesssim k_t$). However, it cannot be used for a fast wave in close vicinity to the frequency of transformation, when $\chi' \simeq \chi''$. In this case, as mentioned in Section 7.2, the simplified formula for the equivalent surface impedance Z^e given by eqn. 7.22 cannot be used.

We can give a clear physical interpretation to the results obtained in the light of the Brillouin concept. In fact, at a cutoff frequency partial Brillouin waves are incident normal to the comb-shaped structure, and their absorption (in accordance with the results of Section 5.5) is larger than that in a smooth surface. Far from cutoff frequencies, the incidence of Brillouin waves is close to a grazing one and this leads to an opposite situation: energy losses of partial waves in every single reflection act are considerably smaller than in the case of a smooth surface. Note that the angle of incidence ϑ is determined by $\vartheta = \cos^{-1}\{\pi/(2k\Delta)\}$. For our assumptions the absorption coefficient α introduced in Section 5.5 can be expressed as

$$\alpha(\vartheta) \simeq \left(1 + \frac{D}{d}\right) \cos^2 \vartheta$$

which means that the ratio $\gamma''(\vartheta)/\gamma''_{(0)}(\vartheta)$ is directly determined by the value of α:

$$\gamma''(\vartheta) \simeq 2\alpha(\vartheta)\gamma''_{(0)}(\vartheta) \tag{7.47}$$

Minimal attenuation takes place when k is somewhat less than $\pi/2D$, i.e. in the region where the surface wave exists. In this case, on various irregularities of a waveguide (which are always present in real waveguides), a transformation of a spatial wave into a surface one is possible. In this connection attenuation characteristics can become worse as a surface wave attenuates rather strongly (see Section 7.3). To avoid this phenomenon it would be reasonable to choose the parameters of a corrugated waveguide so that $k \gtrsim \pi/2D$, when there is no surface wave.

7.5 Attenuation in waveguides with azimuthal corrugation

Figure 7.3 shows the cross-section of the waveguide under study and defines the notation. Supposing that $ka \gg 1$, it is posible to apply the impedance boundary condition given by eqns. 7.22, 7.24 and 7.26 to curvilinear surfaces. The other assumptions correspond to those accepted in Sections 5.5 and 7.2.

The impedance boundary conditions at $r = a$ are

$$\begin{aligned} E_z &= Z^h H_\phi \\ E_\phi &= -Z^e H_z \end{aligned} \tag{7.48}$$

Let the dependence of the modes on z be $\exp(j\gamma z)$. Then the azimuthal compo-

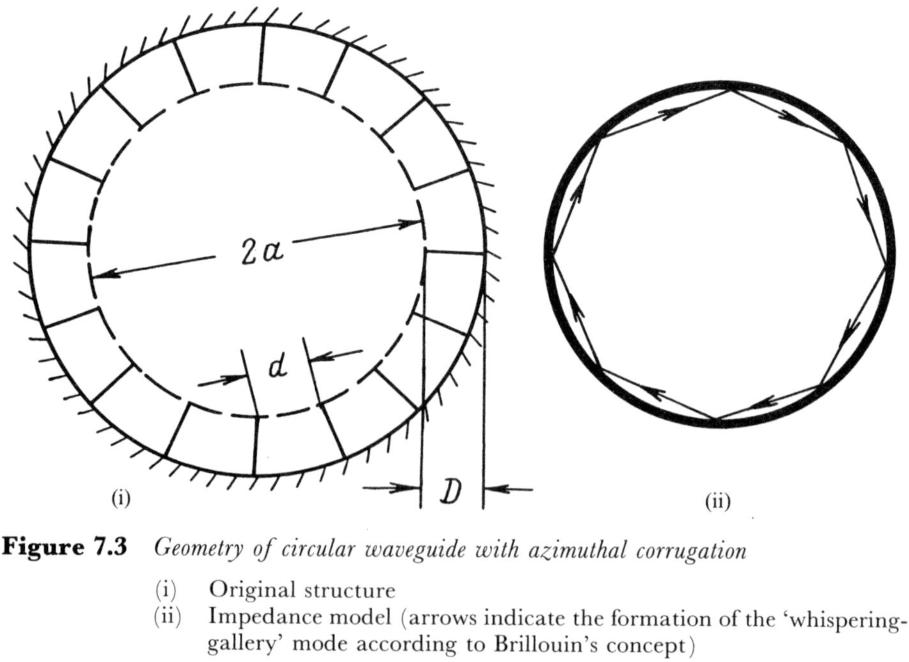

Figure 7.3 *Geometry of circular waveguide with azimuthal corrugation*

(i) Original structure
(ii) Impedance model (arrows indicate the formation of the 'whispering-gallery' mode according to Brillouin's concept)

nents of field E_ϕ and H_ϕ can be expressed in terms of the longitudinal ones as

$$\left.\begin{aligned} E_\phi &= \frac{j\gamma}{r\chi^2}\frac{\partial E_z}{\partial \phi} + \frac{j\omega\mu_0}{\chi^2}\frac{\partial H_z}{\partial r} \\ H_\phi &= -\frac{j\omega\varepsilon_0}{\chi^2}\frac{\partial E_z}{\partial r} + \frac{j\gamma}{r\chi^2}\frac{\partial H_z}{\partial \phi} \end{aligned}\right\} \tag{7.49}$$

where $\gamma = (k^2 - \chi^2)^{1/2}$, χ being the transverse-wave number. The problem considered, unlike those in Sections 7.3 and 7.4, cannot be reduced to a scalar one: the angularly dependent TM and TE modes do not exist separately in such a waveguide.

The longitudinal components of the field can be written as

$$\left.\begin{aligned} E_z &= C^e \mathcal{J}_m(\chi r)\exp(jm\phi) \\ H_z &= C^m \mathcal{J}_m(\chi r)\exp(jm\phi) \end{aligned}\right\} \tag{7.50}$$

where C^e and C^m are unknown amplitude coefficients. Substituting eqn. 7.50 into eqn. 7.49 and using eqn. 7.48 we obtain for C^e and C^m a homogeneous system of two linear equations. Equating the determinant with zero, we have a characteristic equation for χ, which can be written in the form

$$\{M_m(\beta) - \zeta\}\{M_m(\beta) + \xi^{-1}\} = p_m(\beta) \tag{7.51}$$

where $\beta = \chi a$, $M_m(\beta) = \mathcal{J}'_m(\beta)/\beta\mathcal{J}_m(\beta)$ (a prime stands for Bessel's-function derivative with respect to the argument), $p_m(\beta) = m^2\{(ka)^2 - \beta^2\}/(ka)^2\beta^4$, $\zeta = j\mathcal{Z}^e/(W_0 ka)$ and $\xi = j\mathcal{Z}^h ka/W_0$.

We are most interested in the EH_{mn} modes,[50] for which the effect of abnormally small attenuation may occur. In this case eqn. 7.51 can be presented in a more convenient form:

$$M_m(\beta) - \zeta' = \frac{\xi p_m(\beta)}{1 + \xi M_m(\beta)} + j\zeta'' \tag{7.52}$$

where $\zeta' + j\zeta'' = \zeta$.

To solve eqn. 7.52 we apply the perturbation method, considering the right-hand side of the latter as a small perturbation, and the equation

$$M_m(\beta) = \zeta' \tag{7.53}$$

to be of a zero-order approximation.

It is essential that eqn. 7.53 should be solved not in a complex plane but on the real axis β. For this many standard techniques of numerical solution can be used; one of the most effective is the method of differentiation with respect to the parameter (Modenov and Slepyan, 1984). Let us present the solution of eqn. 7.52 as $\beta = \beta_{(0)} + j\beta^{(1)}$, where $\beta_{(0)}$ is the solution of eqn. 7.53 and $\beta^{(1)}$ is a small correction. We are interested in the real part of $\beta^{(1)}$, which determines the

[50] When $D \to 0$, the EH_{mn} and HE_{mn} modes are transformed into the H_{mn} modes and E_{mn} modes of a smooth-walled waveguide, respectively.

eigenmode attenuation:

$$\beta'' = \operatorname{Re} \beta^{(1)} = \frac{1}{\tilde{M}_m(\beta_{(0)})}\left[\zeta'' + \frac{\xi'' p_m\{\beta_{(0)}\}}{(1+\zeta'\xi')^2}\right] \tag{7.54}$$

where $\tilde{M}_m(\beta) = \mathrm{d}M_m(\beta)/\mathrm{d}\beta$ and $\xi' + j\xi'' = \xi$.

The attenuation coefficient is expressed from eqn. 7.54 by

$$\gamma'' \simeq \frac{\beta_{(0)}\beta''}{a\{(ka)^2 - \beta_{(0)}^2\}^{1/2}}$$

Using the anisotropy of the impedance of a corrugated surface it is easy to specify the contribution of field components with different polarisations in the solution obtained: the augend in eqn. 7.54 corresponds to the electrically polarised component, and the addend to the magnetically polarised component.

This form of the perturbation technique is applicable when the following condition is fulfilled:

$$\left(\frac{\zeta''\zeta'}{1+\xi'\zeta'}\right)^2 \ll 1 \tag{7.55}$$

When the condition expressed by eqn. 7.55 is not true, the absolute value of the second factor on the left-hand side of eqn. 7.51 can be small and, as a consequence, the right-hand side of eqn. 7.52 will not be a small perturbation. The latter is indicative of the proximity of β sought to a D-multiplicity point of the EH_{mn} and HE_{mn} modes (see Section 3.7). Thus, for convenience, eqn. 7.51 can be rewritten as

$$\left\{\frac{1}{M_m(\beta)} - \frac{1}{\zeta}\right\}\left\{\frac{1}{M_m(\beta)} + \xi\right\} = -\frac{\xi p_m(\beta)}{\zeta M_m^2(\beta)} \tag{7.56}$$

and now the right-hand side of this equation can be considered as a small perturbation. Let $\beta = \beta_{(0)} + j\beta^{(1)}$. Neglecting the terms above the first order of $\beta^{(1)}$ in both factors on the left-hand side of eqn. 7.56, we obtain a quadratic equation for $\beta^{(1)}$. The solution of the latter results in

$$\beta_{1,2}^{(1)} = \frac{\zeta'' + q \pm [(\zeta'' - q)^2 - 4\xi\zeta' p_m\{\beta_{(0)}\}]^{1/2}}{2\tilde{M}_m(\beta_{(0)})} \tag{7.57}$$

where $q = -j(\xi\zeta' + 1)\zeta'$ and the plus sign corresponds to the EH_{mn} modes, the minus sign to HE_{mn} modes. Eqn. 7.57 is similar to eqn. 3.67 obtained by means of another method for a waveguide with an arbitrary shape of cross-section. However, there is the following difference between them: for cases considered in Section 3.7 the coupling of the D-multiplicity modes results from the finite conductivity of the walls, whereas in our case it is from azimuthal periodicity of the surface (when, even neglecting the losses, $\xi, \zeta' \neq 0$ and the coupling of the D-multiplicity modes remains).

Let us consider now eigenoscillations in a 2-dimensional resonator which is a cross-section of the waveguide under study. Equations for complex eigenfrequencies can easily be obtained from eqn. 7.51, assuming that $\beta = ka$. Then we have

$$\frac{J'_m(ka)}{J_m(ka)} = \frac{jZ^e}{W_0} \tag{7.58}$$

$$\frac{J_m(ka)}{J'_m(ka)} = -\frac{jZ^h}{W_0} \tag{7.59}$$

Note that in this system separate TE_{mn} and TM_{mn} oscillations do exist. The TE_{mn} oscillations are described by eqn. 7.58 and the TM_{mn} oscillations by eqn. 7.59; we shall be dealing below with the TE_{mn} oscillations alone. For our analysis we can apply the perturbation method in a form similar to that used above. Let us suppose that $k = k' - jk''$ and, taking into account the fact that $k'' \ll k'$, from eqn. 7.58 we obtain

$$\frac{J'_m(k'a)}{J_m(k'a)} = \tan(k'D) \tag{7.60}$$

and

$$k'' \simeq \frac{\operatorname{Re} Z_S \left\{1 + \frac{D}{d}\left(1 + \frac{\sin 2k'D}{2k'D}\right)\right\}}{W_0 a \left\{1 + \frac{D}{a}\left(1 + \frac{\sin 2k'D}{2k'D}\right) - \left(\frac{m}{k'a}\right)^2 \cos^2(k'D)\right\}} \tag{7.61}$$

Eqns. 7.60 and 7.61 are valid under the condition $\operatorname{Re} Z^e \ll |\operatorname{Im} Z^e|$. Now assuming that $|\tan(k'D)| \gg 1$, which does not contradict the previous condition due to the smallness of $\operatorname{Re} Z^e/W_0$, we can obtain from eqn. 7.60,[51] in a zero-order approximation,

$$k'a \cong \nu_{mn} \tag{7.62}$$

Combining eqn. 7.61 with eqn. 7.62, we have

$$Q^{mn} = \frac{k'}{2k''} = \frac{W_0(\nu_{mn} + \pi/2)}{2 \operatorname{Re} Z_S(1 + D/d)} \tag{7.63}$$

It is of interest to compare the obtained value of Q^{mn} with the Q-factor of corresponding resonance $Q^{mn}_{(0)}$ in a 2-dimensional smooth-walled circular resonator for the same resonance frequency. The relative Q-factor is

$$\frac{Q^{mn}_{(0)}}{Q^{mn}} = \frac{\mu_{mn}(1 + D/d)}{(\nu_{mn} + \pi/2)}\left(1 - \frac{m^2}{\mu^2_{mn}}\right) \tag{7.64}$$

From eqn. 7.64 it follows that the 'whispering gallery' modes ($n = 1$, $m \gg 1$) are characterised by the highest relative Q-factor. The ratio $Q^{mn}_{(0)}/Q^{mn}$ increases with increasing m and diminishes when n increases at fixed m. At $m = 0$ the Q-factor of a corrugated system is lower than that of a smooth-walled one. Thus, an additional rarefaction of the eigenmode spectrum occurs which is of a purely dissipative nature and is not connected with the radiation effects.

The physical meaning of the results obtained becomes quite clear when using the Brillouin interpretation of this class of fields. In fact, the TE_{mn} mode in a circular waveguide can be presented as a superposition of travelling (in azimuth direction) plane waves undergoing $2m$ reflections on the path $l = 2\pi a$. The angle

[51] The quantity k' determined by eqn. 7.60 is a cutoff wave number of the waveguide under consideration.

of incidence ϑ is determined by the formula (Weinstein, 1969*a*)

$$\vartheta_{mn} = \cos^{-1}\{(1 - m^2/\mu_{mn}^2)^{1/2}\} \tag{7.65}$$

Owing to the curvature of the guiding surface the reflected waves do not abandon the latter, returning to it to undergo a new reflection etc. [see Figure 7.3(ii)]. This interpretation, being essentially geometro-optical, is ever more precise for larger m and n. Qualitatively, however, it also suits the lower modes as well.

It is easy to see from eqn. 7.65 that 'whispering gallery' modes are characterised by angles of incidence close to grazing angles; that is why the relative Q-factor is especially high for them ($[Q^{nm}]^{-1} \sim \cos^2\vartheta_{mn}$). For the angular symmetric modes ($m = 0$) $\vartheta_{n0} = 0$, which corresponds to normal incidence.

To reach the maximum Q-factor it is necessary to set $kd \simeq \pi/2$, i.e. $d/a \simeq \pi/2\nu_{mn}$. Thus, the optimum number of combs is determined by the formula $N_{opt} = 4\{\nu_{mn}\}$, where $\{\alpha\}$ denotes the greatest integer in α.

A more precise determination of k' and Q^{mn} can be achieved by proceeding to a numerical solution of eqn. 7.60. In this case the method of differentiation with respect to the parameter, described by Modenov and Slepyan (1984), is the most effective one. Figure 7.4 shows χ_{mn} as a function of D/a with m and n as parameters. It can be seen that $\chi_{m1} < m$. This means that for any values of D/a the first root of eqn. 7.60 describes the surface mode (in the radial direction). This mode is not described by the approximate solution of eqn. 7.60, given by eqn. 7.62. The roots with $n = 2$ and 3 correspond to 'whispering gallery' modes for which approximate eqns. 7.62 and 7.63 are valid (in this case, for $D \to 0$, $\chi_{m+1,n} \to \mu_{mn}$). With increasing of D/a spatial modes are transformed into radially surface ones and this transformation takes place at $\chi_{mn} = m$.

It is noteworthy that the optima in Q resonance frequencies determined by eqn. 7.62 are located near the multiplicity points of the TM_{mn} and TE_{mn} modes. However, here the multiplicity is trivial. The reason for this lies in a different polarisation structure of these modes: the TE_{mn} modes have only field components H_z, E_r and E_ϕ, the TM_{mn} modes the components E_z, H_r and H_ϕ.

7.6 Projection method for calculation of propagation and attenuation coefficients of corrugated waveguides with arbitrary shapes of cross-section and corrugation

The structure under consideration in this Section is a good model of a flexible corrugated waveguide with smoothly shaped slots. Owing to low attenuation of the HE_{11} mode and also high operating characteristics they are widely used as feeders in stationary and mobile radio-relay, space and tropospheric-scatter communication facilities.

Designing such waveguides is impossible without an accurate estimation of their electrodynamic characteristics. This is a rather complicated problem because for such structures the method of separation of variables and other analytical techniques cannot be applied. Angular dependent modes in such waveguides are hybrid ones (i.e. having all six nonzero components of the electromagnetic field), owing to which a solution of full vector problem for Maxwell's equations is needed. The considerable depths of the slots are of the

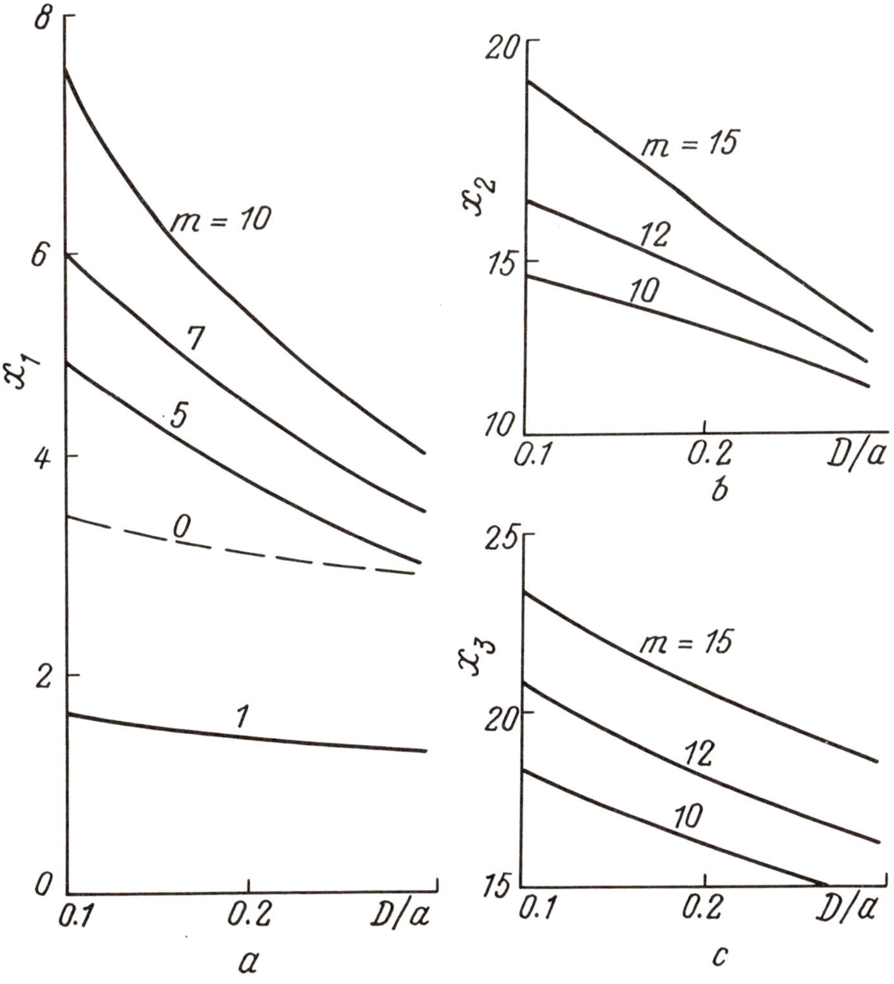

Figure 7.4 *Cutoff characteristics for various modes in circular waveguide with azimuthal corrugation*

a $n = 1$
b $n = 2$
c $n = 3$

most practical interest (it is just when the effect of abnormally low attenuation is observed). Therefore, methods considering corrugation as a small perturbation of the smooth-wall waveguide surface are inefficient. The best approach seems to be the projection method with co-ordinate transformation suggested by Ilyinsky and Sveshnikov (1968). Based on the latter, a general numerical method suitable for calculation of propagation and attenuation characteristics of corrugated waveguides with arbitrary shapes of cross-section and corrugation has

Table 7.2 *Results of calculation of HE_{11}-mode cutoff wave number obtained for various sets of basis functions*

N_1	N_2	k_0 (cm^{-1})
3	3	2.0649
5	5	2.0657
8	7	2.0658
8	8	2.0658

From these Tables we see fast inner convergence for the HE_{11} mode and slower convergence for the EH_{11} mode. These facts about the convergence are general in nature as far as this calculation technique is concerned. Tables 7.3 and 7.4 show the results of calculating the cutoff wave numbers of the first transmission band for EH_{11} and HE_{11} modes at $N_1 = N_2 = 3$ and various sets of space harmonics.

Figure 7.7 illustrates the trend of the convergence with respect to the number of basis functions for the EH_{11} mode at $M = 1$. The even values of N correspond to equal numbers of 'electric' and 'magnetic' functions ($N_1 = N_2$), whereas the odd ones correspond to numbers of 'magnetic' functions exceeding those of 'electric' functions by one. In Figure 7.8 are shown the theoretical and experimental dispersion characteristics of EH_{11} and HE_{11} modes for frequencies near the cutoff frequency.

The results discussed above were obtained for a waveguide with a shallow corrugation, though, as mentioned previously, the projection method described in Section 7.6 has no limitations in q.

In Figure 7.9 the values of $f_0 = k_0 c/2\pi$ for various modes are shown as functions of the corrugation depth. These data were obtained at $a = 28.925$ mm and $d = 20$ mm; 21 basis vector functions and seven space harmonics were taken into account (Alkhovsky *et al.*, 1979).

Figure 7.10 shows the dispersion characteristics for a waveguide with deep sine corrugation calculated by means of the projection method. The waveguide parameters are $a = 28.925$ mm, $d = 20$ mm and $l = 8.925$ mm. The curves in Figure 7.10 are obtained for the same sets of basic functions and space harmonics as those in Figure 7.9. The experimental data in the paper by Alkhovsky *et al.* (1979) are marked with circles. As can be seen from Figure 7.10, there is good

Table 7.3 *Results of calculation of EH_{11}-mode cutoff wave number obtained for various sets of space harmonics*

M	1	2	3	4
k_0 (cm^{-1})	1.0126	1.0109	1.0105	1.0105

Table 7.4 *Results of calculation of HE_{11}-mode cutoff wave number obtained for various sets of space harmonics*

M	1	2	3
k_0 (cm^{-1})	2.0667	2.0649	2.0649

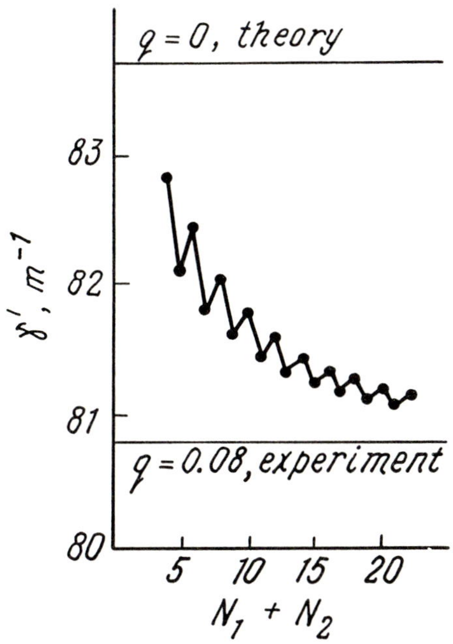

Figure 7.7 *Propagation coefficient of EH_{11} mode against number of basis functions*
$a = 18.512$ mm, $d = 8$ mm, $k = 1.3$ cm^{-1}, $M = 1$

agreement between the theoretical and experimental results. A sharp rise in the dispersion curves for the TM_{01}, EH_{11} and EH_{21} modes is due to their transformation into surface modes (compare with Section 7.4).

In Figure 7.11 the dispersion characteristics of a circular waveguide with a step corrugation calculated by Clarricoats *et al.* (1975*a*) using the partial-regions method, are shown for comparison. Their shapes correspond to those in Figure 7.10.

We confine our discussion to the most typical examples of the calculation of propagation characteristics, the main goal of this book being absorption analysis. More complete numerical results and their applications for the design of reflector-antenna feeds are described by Clarricoats and Olver (1984), and those for flexible waveguides by Alkhovsky *et al.* (1986).

7.8 Attenuation characteristics of circular corrugated waveguides

A detailed study of the attenuation characteristics of circular corrugated waveguides with step corrugation has been carried out by Clarricoats *et al.* (1975*a*, 1975*b*). In this Section the results obtained for waveguides with smooth corrugation are presented and compared with those obtained by Clarricoats *et al.* The experimental data are also given. The computational algorithm is based on the projection method described in Section 7.6.

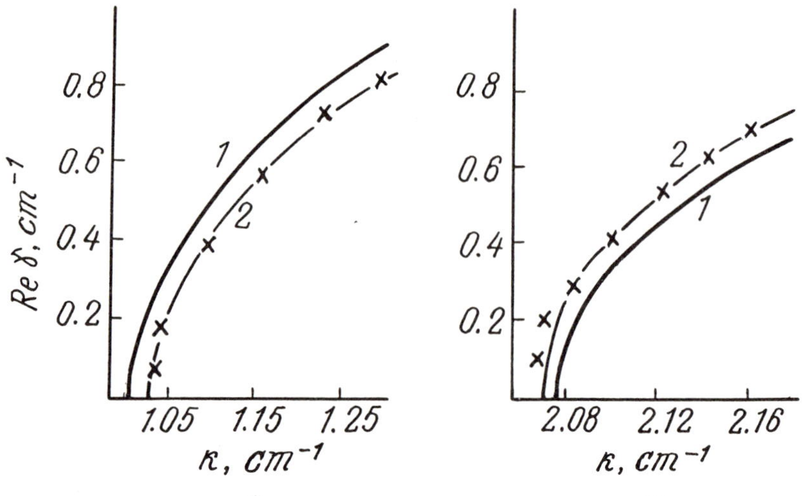

Figure 7.8 *Dispersion characteristics of EH_{11} and HE_{11} modes*

$a = 18.512$ mm, $d = 8$ mm
1 $q = 0$
2 $q = 0.08062$
——— theory
× × × experiment
(i) EH_{11}
(ii) HE_{11}

In Figure 7.12 the attenuation coefficient of the EH_{11} mode in a circular waveguide with sine corrugation[54] is presented as a function of the number of basis functions. In this Figure the value of the attenuation coefficient of the TE_{11} mode in a smooth-wall circular waveguide with radius equal to the mean radius of the corrugated waveguide (all other parameters being the same) is shown by a horizontal line for comparison. Figure 7.13 illustrates the inner convergence of the projection method with respect to the number of space harmonics.

In Figure 7.14 the frequency dependence of the attenuation coefficient of the HE_{11} mode in a sine-corrugated circular waveguide (solid line) and similar dependence for the TE_{11} mode in a smooth-walled circular waveguide with a radius equal to the mean radius of corrugated waveguide (broken line) are shown. The experimental results from the work by Alkhovsky *et al.* (1978) are shown by crosses. The conductivity of the wall material is determined experimentally by measuring the *Q*-factor of a cylindrical resonator manufactured from the same copper band as the waveguide under examination. The agreement between theoretical and experimental results confirms the high accuracy of the projection method.

Figures 7.12–7.14 give the results for waveguides with shallow corrugations ($q \ll 1$). In such waveguides the attenuation is higher than in smooth-walled

[54] The equation of the surface of a sine-corrugated waveguide was given by eqns. 7.66 and 7.74.

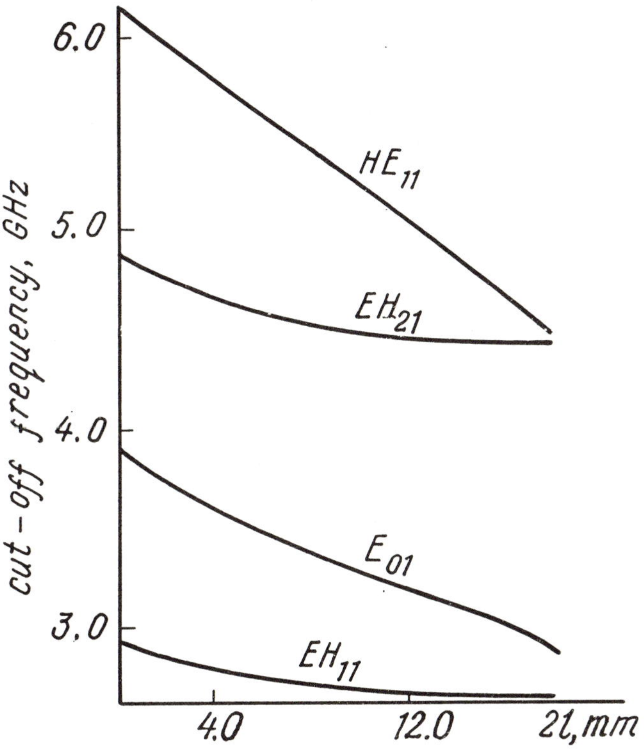

Figure 7.9 *Cutoff frequencies of various modes as functions of the slot depth*

ones, which corresponds to the results obtained for the model problem in Section 5.5. From the physical point of view this can be explained by an increase in the surface current path while the resonance effect of the field 'being forced out' from the periodic structure (see Section 5.5) is not observed at such small depths of corrugation.

The results of calculations for deep corrugations (Alkhovsky *et al.*, 1979) are shown in Figure 7.15. The circles represent the experimental data. The waveguide was made by means of the electroforming technique. Measurements were carried out through the resonance method. Both theory and experiment point to the fact that there can be an abnormally small attenuation of the HE_{11} mode.

Physically, this phenomenon can be understood using the Brillouin concept which, while being more complicated in this case than for a plane waveguide, is still valid. The HE_{11} mode can be expressed as a spectrum of plane waves of both polarisations, the reflection of which from the round walls of the waveguide at each point is similar to that from the plane corrugated surface. The waves of electric polarisation dominate in this spectrum of which the resonance character of the dependence of γ'' on the corrugation depth is indicative. In the angular spectrum of this mode the near-grazing plane waves are prevailing when its frequency is distant from cutoff. The absorption of such waves at corresponding d and l in the corrugated surface is very low. At $f \to f_0$ the maximum of the

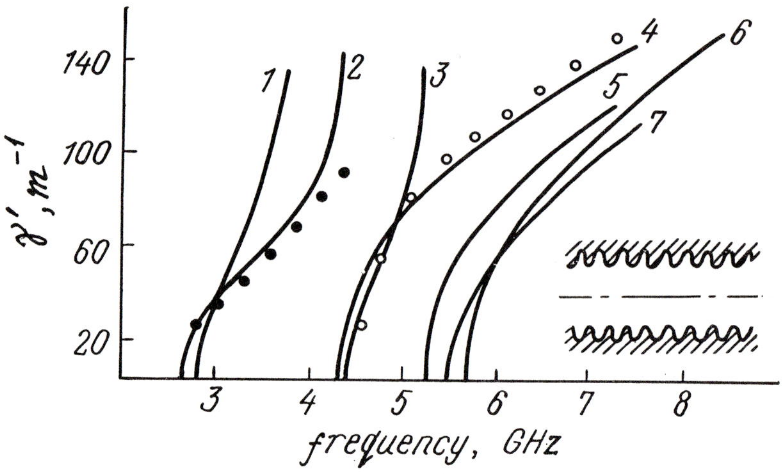

Figure 7.10 *Dispersion characteristics of corrugated waveguide*

1 TM_{01} 2 EH_{11} 3 EH_{21} 4 HE_{11} 5 TE_{01}
6 TM_{02} 7 HE_{21}
○ ○ ○ experimental results

angular spectrum shifts towards the region of smaller angles of incidence and the loss at each reflection increases considerably.

Attenuation in circular corrugated waveguides of a non-sine shape of corrugations has also been investigated on the basis of the projection method described in Section 7.6. The equation of the waveguide surface was given in the form (Alkhovsky *et al.*, 1987)

$$R(z) = R_0 - l|\cos(\pi z/d)|^v \tag{7.75}$$

where

$$v = -\ln 2/\ln\{\cos(\pi s/2)\} \tag{7.76}$$

s being a preset numerical parameter.

Figure 7.16*a* shows the shapes of the slots described by eqns. 7.75 and 7.76 for different values of s. In Figure 7.16*b* the attenuation coefficient of the HE_{11} mode as a function of s is presented for $R_0 = 47.85$ mm, $l = 17.85$ mm, $d = 20$ mm, $f = 7.5$ GHz and $\sigma_0 = 5 \times 10^7$ S/m. As can be seen from Figure 7.16*b* there is an opportunity for a considerable decrease in attenuation through optimisation of the corrugation shape. Table 7.5 compares the theoretical and experimental data on attenuation for circular waveguides with non-sine-shape corrugations. There is a good agreement between calculations and experiments. The data in Table 7.5 correspond to frequencies where the attenuation of the HE_{11} mode is minimal.

Figures 7.17 and 7.18 show the calculation results for a circular waveguide with a step corrugation obtained by Clarricoats *et al.* (1975*a*). These results are similar to those mentioned previously.

In Figure 7.19 the frequency dependence of attenuation of the TE_{01} mode in

Table 7.5 *Comparison of theoretical and experimental results for corrugated waveguides with non-sine-shaped corrugations*

f (GHz)	R_0 (mm)	l (mm)	d (mm)	s	Theory γ'' (dB/m)	Measurement γ'' (dB/m)
7.5	47.85	17.85	20.00	0.5	0.0149	0.0127
8.6	49.97	15.22	16.98	0.21	0.0071	0.0062

corrugated waveguide with sine corrugation is shown (Alkhovsky and Ilyinsky, 1979). Broken lines correspond to those of smooth-walled waveguides with $R = R_1$ and $R = R_2$ ($R_{1,2} = a \pm l$, the inner and outer radii of the corrugated waveguide respectively).

From Figure 7.19 one can see that the attenuation of the TE_{01} mode in a corrugated waveguide is higher than that in a smooth-walled waveguide of comparable size. From the physical point of view this result can be explained using the Brillouin concept. In the expansion of the TE_{01} mode in a spectrum of plane waves there are only those of magnetic polarisation ($\boldsymbol{E}$ is parallel to the structure ridges). For such waves the losses in a corrugated surface exceed those in a smooth one (compare with Section 5.5), which results in the growth of attenuation.

We have described only some of the numerical results illustrating the application of the projection method to study the phenomenon of abnormally low attenuation of the HE_{11} mode in corrugated waveguides. Important information on attenuation characteristics of elliptical, rectangular and double-ridge corrugated waveguides is given in the book by Alkhovsky *et al.* (1986). Physical peculiarities of corrugated structures were used to develop flexible corrugated waveguides, which are used as highly effective microwave-band transmission lines manufactured in many countries.

7.9 Millimetre-waveband high-quality corrugated resonators

In this Section we describe how the effect of abnormally low dissipation in periodic structures is used to develop high-quality resonators at millimetre waveband. The material presented is based on work by Luk *et al.* (1988), Rodionova and Slepyan (1989) and Rodionova *et al.* (1990). In addition to their high quality, resonators of this type are characterised by a wide range of frequency retuning, small overall dimensions and weight, and a high level of parasitic-mode suppression. Such resonators can be used for frequency stabilisation of millimetre-waveband generators, material parameters measurement, phase detection etc.

In Figure 7.20 a photograph of this resonator in knock-down form is shown. Figure 7.21 shows a sketch of an experimental model. The resonator is a hollow cylinder on the inner side surface of which there is a periodical structure (1). End covers (2, 3) are smooth and can be either plane or spherical with a radius

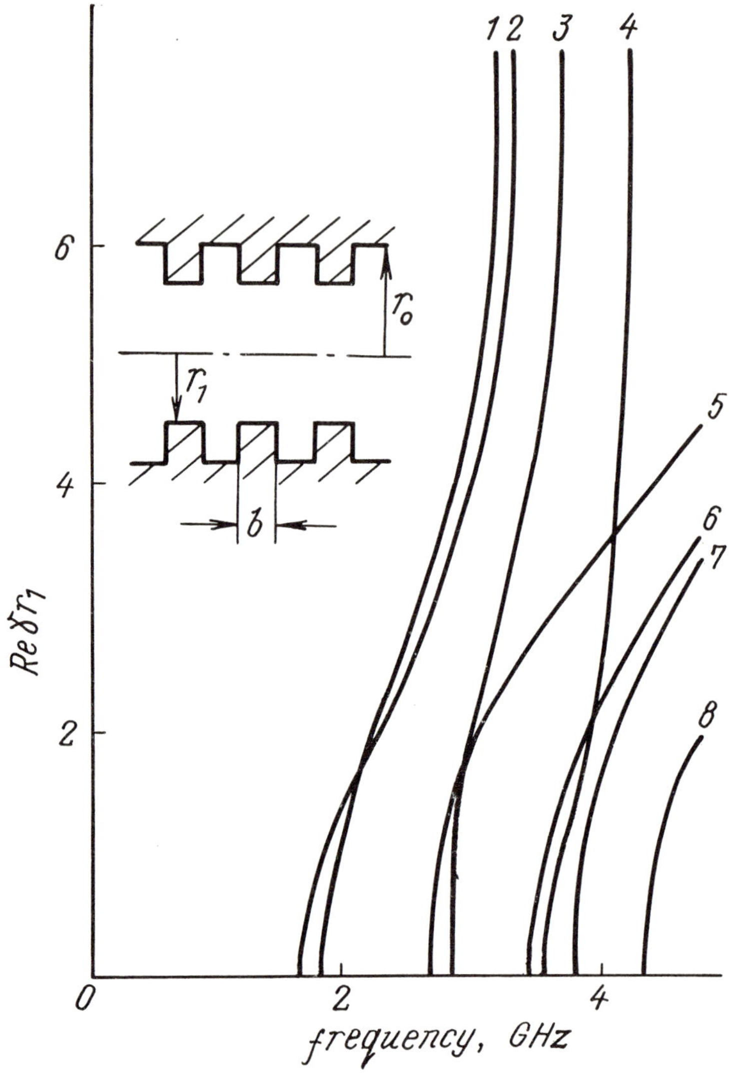

Figure 7.11 *Dispersion characteristics of circular waveguide with step corrugation (Clarricoats et al., 1975a)*

$r_1 = 25$ mm, $d = 10$ mm, $b = 2$ mm, $r_1/r_0 = 0.714$

1 EH_{11} 2 TM_{01} 3 HE_{11} 4 EH_{21} 5 TE_{01}
6 TM_{02} 7 HE_{31} 8 EH_{31}

of curvature of about $3b$ (one is made mobile for retuning over a frequency range). The excitation of the resonator was implemented with a thin plug (4) which was fed from a standard rectangular waveguide through a waveguide-to-coaxial transition.

To provide thermal stability, the resonator body was made of superinvar; the

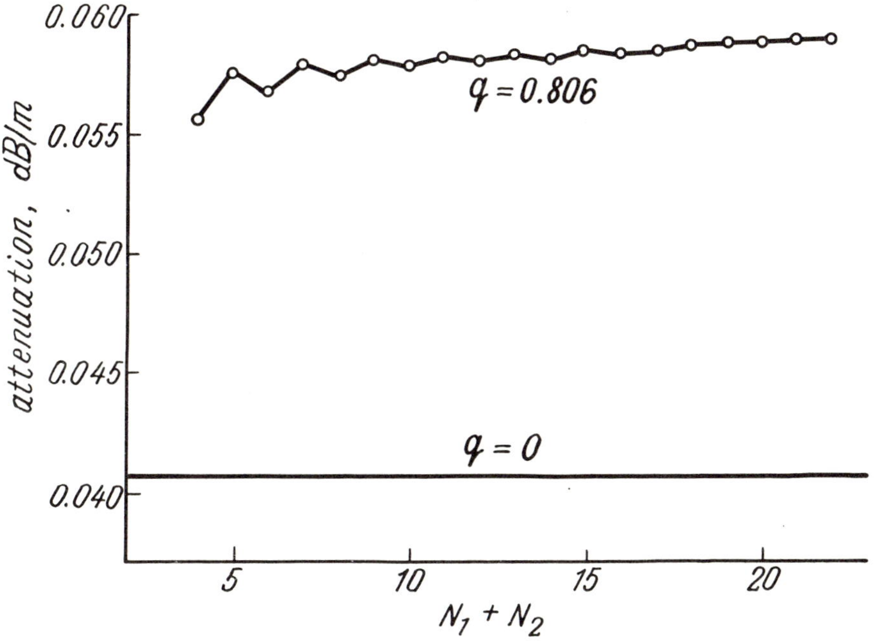

Figure 7.12 *Attenuation coefficient of EH_{11} mode in corrugated waveguide against number of basis functions*

$a = 18.512$ mm, $d = 8$ mm, $k = 1.3\ \mathrm{cm}^{-1}$, $M = 1$, $\sigma_0 = 5 \times 10^7$ S/m

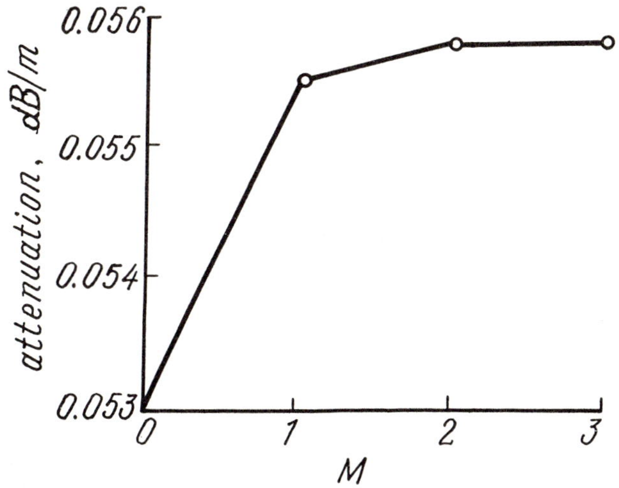

Figure 7.13 *Attenuation coefficient of EH_{11} mode in corrugated waveguide against number of space harmonics*

$a = 18.512$ mm, $d = 8$ mm, $k = 1.3\ \mathrm{cm}^{-1}$, $N_1 = N_2 = 2$, $\sigma_0 = 5 \times 10^7$ S/m

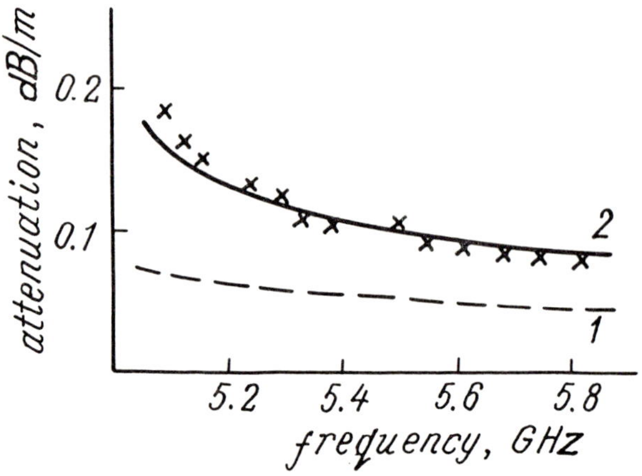

Figure 7.14 *Attenuation characteristics of EH_{11} mode in corrugated waveguide*

$N_1 = 11$, $N_2 = 10$, $M = 1$
$a = 19.1$ mm, $d = 7.19$ mm, $\sigma_0 = 5.05 \times 10^7$ S/m
1 $q = 0$
2 $q = 0.1063$
× × × experiment

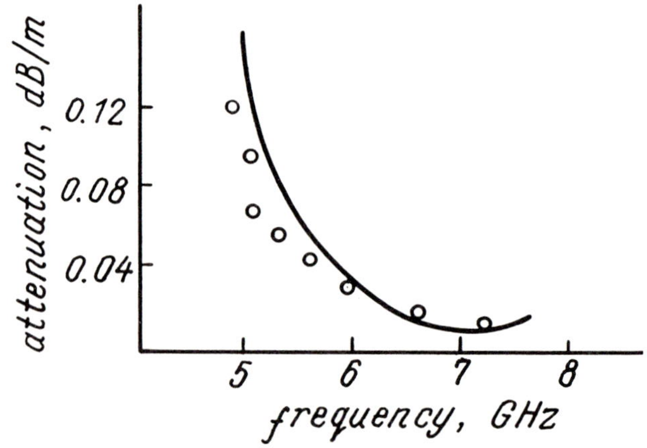

Figure 7.15 *Attenuation characteristics of HE_{11} mode in corrugated waveguide*

$N_1 = N_2 = 5$, $M = 2$
$a = 38.925$ mm, $d = 20$ mm, $l = 8.925$ mm
——— theory
○ ○ ○ ○ experiment

inner surface was coated with a layer of copper with a layer of silver on top of it by electroforming. After compacting, the silver coating was polished to 6-7 surface finish class for the comb-shaped surface and to 10-12 surface finish class for the smooth one.

The TM_{0mn} oscillations are the working ones. According to the information

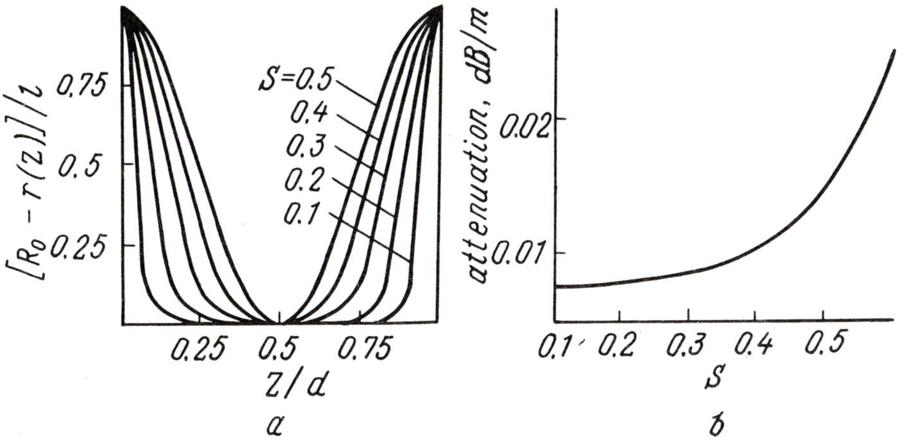

Figure 7.16 *HE_{11}-mode attenuation in circular corrugated waveguide with non-sine-shape corrugations*

a Shapes for corrugations
b Attenuation coefficient against *s*

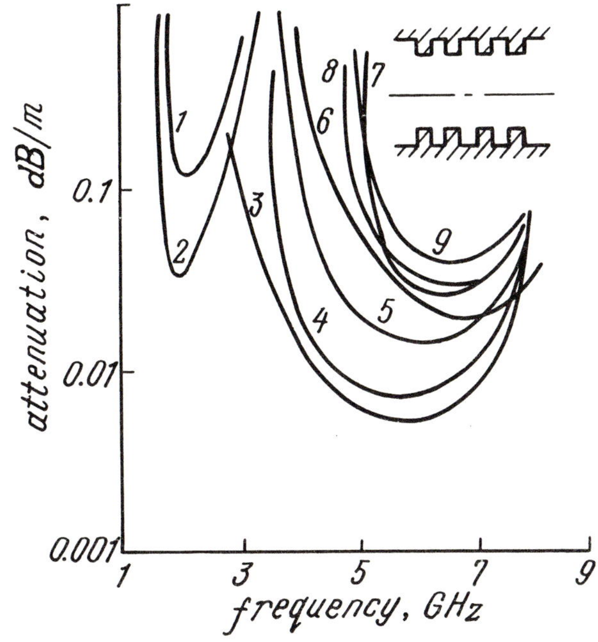

Figure 7.17 *Attenuation characteristics of circular waveguide with step corrugation (Clarricoats et al., 1975a)*

$r_1 = 25$ mm, $d = 10$ mm, $b = 2$ mm, $r_1/r_0 = 0.714$, $\sigma_0 = 5.8 \times 10^7$ S/m

1	TM_{01}	2	EH_{11}	3	HE_{11}	4	TE_{01}	5	HE_{21}
6	TM_{02}	7	HE_{12}	8	HE_{31}	9	EH_{12}		

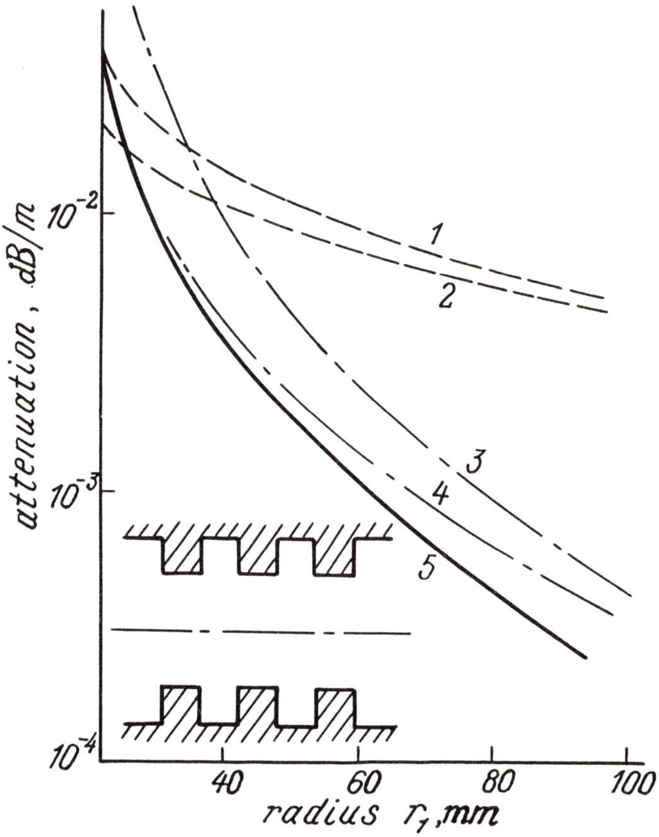

Figure 7.18 *Attenuation in circular waveguide as a function of radius (Clarricoats et al., 1975a)*

Smooth-wall waveguides	Corrugated waveguide
1 HE_{11} $(R=r_1)$	5 HE_{11}
2 TE_{11} $(R=r_0)$	
3 TE_{01} $(R=r_1)$	
4 TE_{01} $(R=r_0)$	

given in Section 7.4, a considerable decrease in energy absorption in a comb-shaped surface may occur at small Brillouin angles, compared with a smooth surface, i.e. at $n \gg m$.[55] To obtain a higher Q-factor due to lower losses in the side surface, their contribution to the total losses should be dominant. This will be the case when $L \gtrsim b$ (in our case $L/b \simeq 2$–4, $b = 26.35$ mm).

Filtration of parasitic modes is carried out by means of a narrow ($\Delta \sim 1$ mm) slot in the diametrical plane of the immobile end wall [Figure 7.21 (8)). For

[55] As is shown in Section 7.8, the Brillouin concept for circularly symmetric structures can still be used. In this case the Brillouin angle is determined by the relation $\vartheta_m = \cos^{-1}(\chi_m/h_m)$, where χ_m and h_m are transverse and longitudinal wave numbers of the TM_{0m} modes in a circular waveguide of radius b.

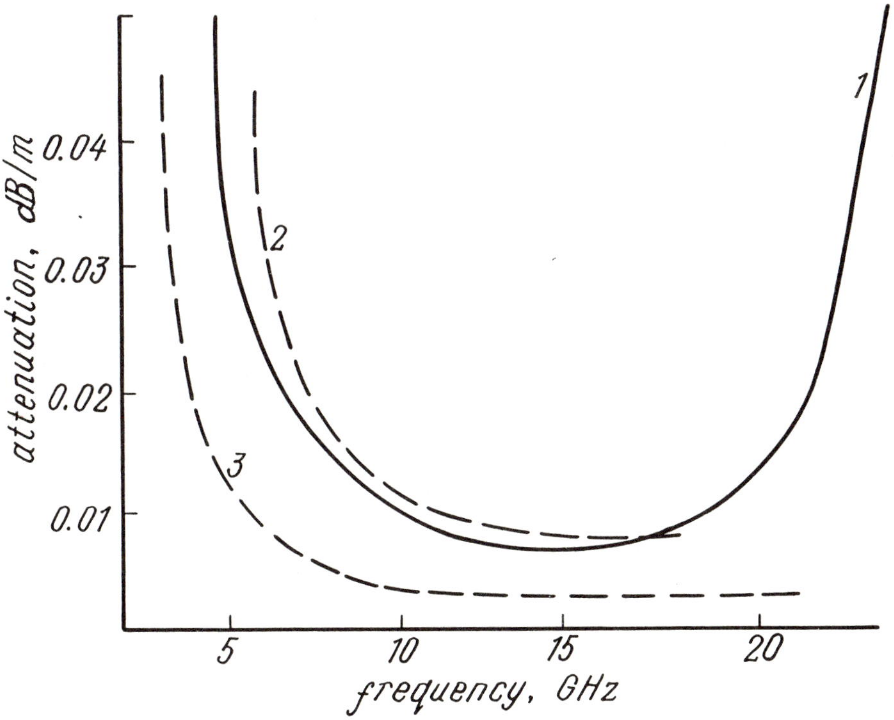

Figure 7.19 *Attenuation characteristics of TE_{01} mode in circular waveguide*

$a = 34.9$ mm, $d = 5.5$ mm, $q = 0.0552$

1 Corrugated waveguide
2 Smooth-wall waveguide ($R = R_1$)
3 Smooth-wall waveguide ($R = R_2$)

the TM_{0mn} oscillations, such a slot is practically nonradiating while for the other types of oscillation the slot is transparent (as for azimuthal slots in circular waveguides for TE_{0n} modes). The radioabsorbing elements are inserted into the post-plunger space to suppress parasitic resonances. The method of resonator excitation hindering the generation of all modes except TM_{0mn} and the effect of a comb structure giving rise to stop bands and increasing absorption of modes with large radial indices are also essential to filtrate spurious oscillations.

The general structure of the eigenoscillation spectrum was examined through the use of an automatic SWR and attenuation meter. The spectral-line width at a level of 3 dB and SWR in resonance was evaluated in the manual retuning operation. A more accurate spectral-line-width measurement was performed by means of an electronic frequency meter and SWR measurement, with a precision measuring line as recommended by Valitov (1963).

Figure 7.22 shows the resonator spectral characteristics (solid line for a resonator with a slot, broken line for a resonator without a slot). As can be seen from Figure 7.22, the slot considerably enhances the selection of the oscillations, only the TM_{0mn} oscillations having a high Q-factor. Their resonance frequencies

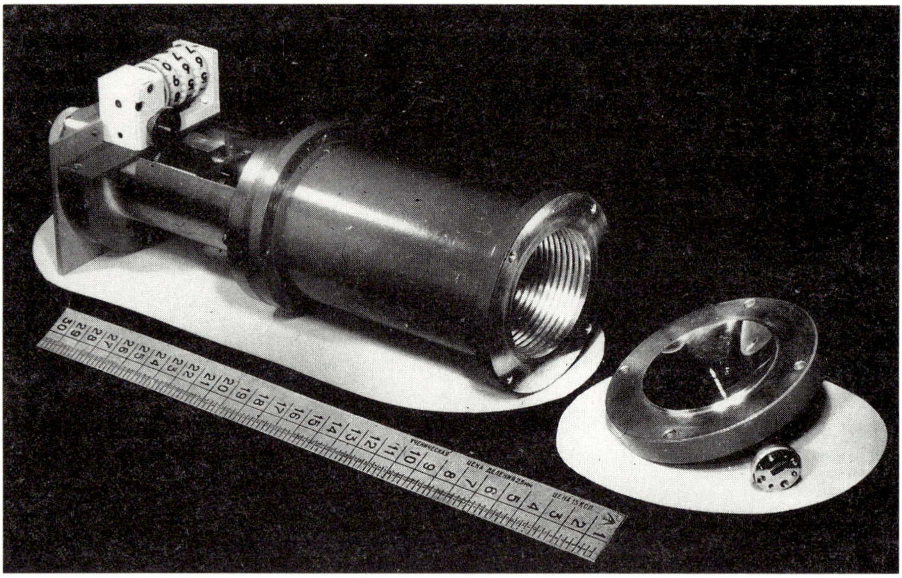

Figure 7.20 *Cylindrical corrugated resonator in 'knock-down' form*

are slightly shifted due to the slot; the widths of spectral lines remain the same despite the presence of the latter. This is confirmed by the fact that in one of the experimental models a slit was shifted 1 mm from the diameter line parallel to it, which did not change the eigenoscillation-spectrum structure (solid line in Figure 7.22), increasing the widths of spectral lines 2–2.5 times.

The widths of spectral lines, shown in Figure 7.22 (solid line) are within the limits of 0.5–1.2 MHz which corresponds to unloaded Q-factors of the order of 10^5 (taking into account the differences in values of SWR). Thus, the resonator described has a high unloaded Q-factor and, at the same time, a rarefied eigenoscillation spectrum. The spectrum density and Q-factor of the corrugated resonator without a slot are approximately the same as for an open resonator with spherical mirrors having a volume an order of magnitude greater. The resonator with a slot can compete, as far as spectrum rarefaction is concerned, with resonators operating at 'near-boundary' oscillations (Nefedov and Rossiisky, 1978) and has a much higher Q-factor.

The resonator has such characteristics in a wider frequency range than that in Figure 7.22 (up to $f = 37.5$ GHz). Within these limits, the resonator can be retuned by moving one of the end walls. When changing the operational mode, coupling readjustment (by moving the excitation plug) is not needed.

Assuming the end walls to be plane, we can estimate the resonator characteristics. The effect of the comb-shaped surface can be described mathematically by impedance-boundary conditions generalised for the case of finite conductivity (see Section 7.2). In this approximation a resonance frequency for the TM_{0mn} oscillations is determined as the mth root of the equation

$$p\frac{J_0(p)}{J_1(p)} = -\frac{2\pi f b}{c}\tan\left(\frac{2\pi f D}{c}\right) \qquad (7.77)$$

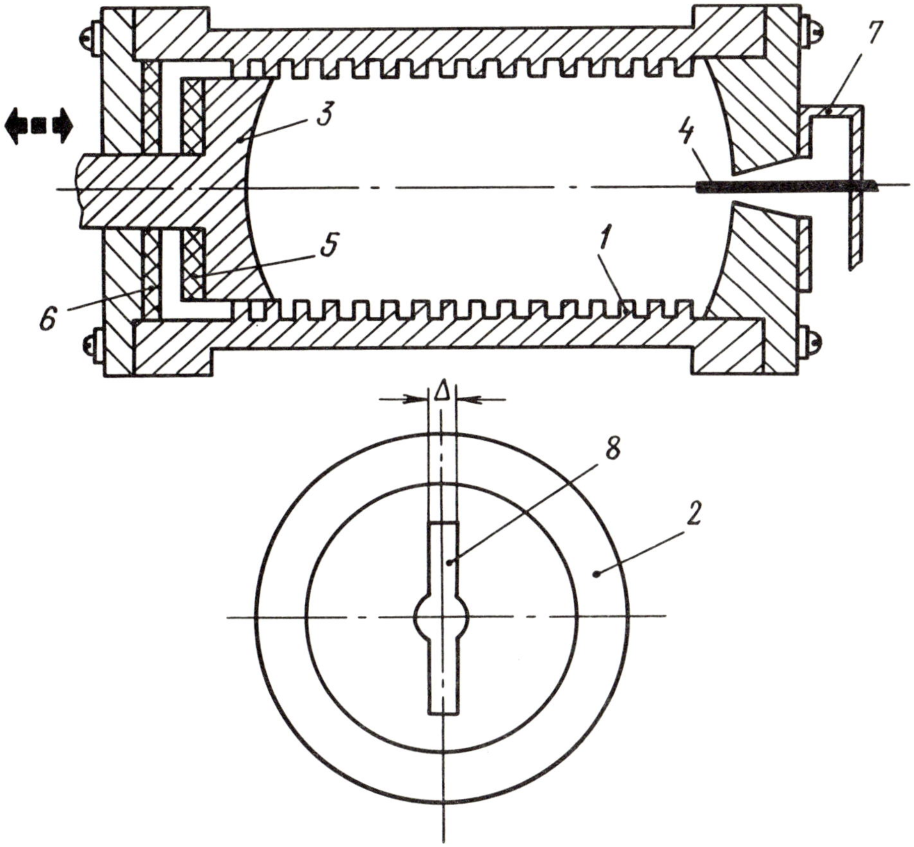

Figure 7.21 *Sketch of cylindrical corrugated resonator*

1 Periodic structure
2 Fixed end cover
3 Movable end cover
4 Thin plug
5 and 6 Radioabsorbing elements
7 Waveguide-to-coaxial transition
8 Filtering slot

where $p = \{(2\pi fb/c)^2 - (n\pi b/L)^2\}^{1/2}$, c being the velocity of light. Eqn. 7.77 can easily be solved numerically, for example, by the method of differentiation with respect to the parameter (Modenov and Slepyan, 1984).

In this case, we obtain the following formula for the Q-factor:

$$Q^{-1} = \frac{2c \operatorname{Re} Z_S}{W_0 \pi f L}\left[1 + \left(1 + \frac{D}{d}\right)\frac{\beta nc}{2bf}\left\{1 - \left(\frac{nc}{2Lf}\right)^2\right\}\right] \tag{7.78}$$

where $\beta = \operatorname{Re} Z_{S1}/\operatorname{Re} Z_{S2}$, Z_{S1} and Z_{S2} are surface impedances of the end and side surface materials, and D and d are the depth and period of the comb-shaped structure.

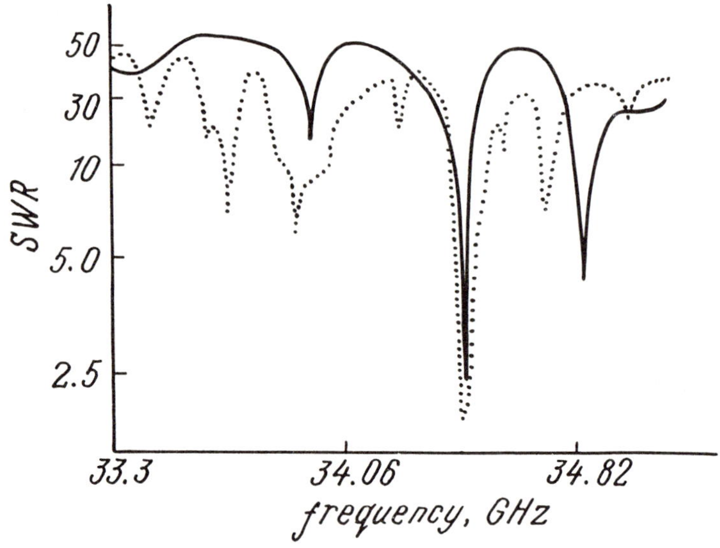

Figure 7.22 *Experimental frequency characteristics of corrugated resonator*

——— with slot
········ without slot

Introducing different surface impedances Z_{S1}, Z_{S2} (then $\beta \neq 1$) we can approximately take into account the difference in the surface finish class of polishing the comb and smooth surfaces.

The results obtained through this technique and the experimental data at $m = 1, 2, n = 15 \ldots 25$ are of the same order of magnitude.

7.10 Radiation from a corrugated horn

The scope for using corrugated surfaces is not confined to low-loss waveguides and high-quality resonators. Another important field of application is antenna engineering, in particular horn corrugated feeds for reflector antennae (Clarricoats and Olver, 1984). The approximate methods are based on the Huygens–Kirchhoff principle commonly used for the calculation of such feeds. The accuracy of these methods, however, cannot easily be estimated. Apart from this they are not applicable for calculating near-field characteristics. The development of rigorous methods, based on a complete solution of electrodynamic problems, is therefore of great interest. To formulate one such method we use Galerkin's incomplete method with semi-inversion in boundary conditions. Originally this technique was applied to solve the problem of diffraction on a corrugated surface with complex shaped grooves (Kopenkin *et al.*, 1987, 1988).

In this Section we consider the problem of wave radiation from an open-ended plane irregular waveguide [Figure 7.23(i) and (ii)] having a profile described by an arbitrary twice differentiable even function $a(z)$. This general formulation of the diffraction problem permits calculation of a wide class of horns, in particular profiled corrugated horns.

Consider the field in the E-plane of a horn, assuming the metal surfaces to be perfectly conducting (extension of this approach to include the case of finitely conducting surfaces would not be difficult). Much of the material presented is based on the work by Kopenkin *et al.* (1989).

Let the waveguide be excited by an arbitrary superposition of TM modes. Then an original diffraction problem is reduced to the following boundary value one:

$$\nabla_{xz}^2 \psi + k^2 \psi = 0 \qquad (7.79)$$

$$\left.\frac{\partial \psi}{\partial \boldsymbol{n}}\right|_{\Sigma} = 0 \qquad (7.80)$$

where Σ is the total surface of conducting bodies.

The solution of eqns. 7.79 and 7.80 is sought in a class of functions satisfying the condition of the finiteness of the energy integral

$$\iint (k^2 |\psi|^2 - |\text{grad}_{xz} \psi|^2)\, dx\, dz$$

over any finite region in the plane xOz including an edge.

The incident field can be presented as $(z < 0)$

$$\psi^0(x, z) = \sum_n{}' S_n \exp\{-\gamma_n(0) z\} \cos\left[\frac{n\pi}{2a(0)} \{x - a(0)\}\right]$$

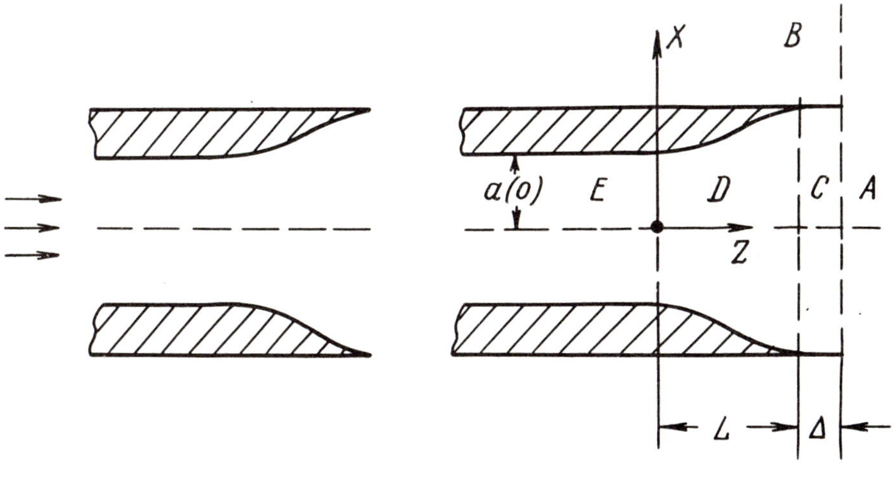

(i) (ii)

Figure 7.23 *Geometry for wave radiation from open-ended plane irregular waveguide*

(i) Original structure
(ii) Modified structure

where S_n are given coefficients, $\gamma_n(0) = [\{n\pi/2a(0)\}^2 - k^2]^{1/2}$, $\operatorname{Im}\gamma_n \leq 0$ and $\operatorname{Re}\gamma_n \geq 0$, a prime here and later meaning that the summation is taken with respect to odd values of n. The scattered field $\psi - \psi^0$ should satisfy the radiation condition at $r = (x^2 + z^2)^{1/2} \rightarrow \infty$.

It follows from symmetry that $\psi(-x, z) = -\psi(x, z)$. After the function $\psi(x, z)$ satisfying the conditions mentioned above has been found, the electromagnetic field can be expressed by

$$\boldsymbol{E} = \frac{j}{\omega\varepsilon_0}\operatorname{curl}(\hat{i}_y\psi) \qquad \boldsymbol{H} = \hat{i}_y\psi$$

As in Section 5.8, let us consider a modified structure shown in Figure 7.23(ii), passing to the limit $\Delta \rightarrow 0$ only in the final equations. At the initial stage the method of partial regions is used. Division into partial regions is shown in Figure 7.23(ii). The peculiarity of this problem is in the fact that the structure under study includes the open regions $\mathscr{A}$, $\mathscr{B}$, where the field cannot be described in the form of series but should be written as Fourier's integrals. The procedure of field matching for open regions was developed by Mittra and Lee (1971); later we shall show how it can be combined with Galerkin's incomplete method applied to express the field in the irregular partial region $\mathscr{D}$ (in the regions $\mathscr{C}$ and $\mathscr{E}$ the field is expanded in series in terms of eigenmodes of corresponding regular waveguides).

Therefore, in the regions $\mathscr{A}$ and $\mathscr{B}$ the field sought can be presented in the following form (Mittra and Lee, 1971):

Region $\mathscr{A}$: $z > L + \Delta \qquad (-\infty < x < \infty)$

$$\psi(x, z) = \int_0^\infty A(\alpha)\sin\alpha x\exp\{-(\alpha^2 - k^2)^{1/2}(z - L - \Delta)\}\,\mathrm{d}\alpha \tag{7.81}$$

Region $\mathscr{B}$: $|x| > a(L) \qquad (z < L + \Delta)$

$$\psi(x, z) = \int_0^\infty B(\beta)\cos\{\beta(x - a(L))\}\exp\{(\beta^2 - k^2)^{1/2}(z - L - \Delta)\}\,\mathrm{d}\beta \tag{7.82}$$

where $A(\alpha)$ and $B(\beta)$ are unknown functions. In the course of integration over the real halfaxis in eqns. 7.81 and 7.82 we get round the branching point α, $\beta = k$ from below.

As to the rest of the partial regions, the field there is expressed in a manner similar to that in Section 5.8:

Region $\mathscr{C}$: $|x| < a(L) \qquad (L < z < L + \Delta)$

$$\psi(x, z) = \sum_n{}' \psi_n(x, L)[T_n\exp\{\gamma_n(L)(z - \Delta - L)\} + Q_n\exp\{-\gamma_n(L)(z - \Delta - L)\}] \tag{7.83}$$

Region $\mathscr{E}$: $|x| \leq a(0) \qquad (z < 0)$

$$\psi(x, z) = \psi^0(x, z) + \sum_n{}' \psi_n(x, 0)R_n\exp\{\gamma_n(0)z\} \tag{7.84}$$

Region $\mathscr{D}$: $|x| \leq a(z) \qquad (0 \leq z \leq L)$

$$\psi(x, z) = \sum_n{}' P_n(z)\psi_n(x, z) \tag{7.85}$$

In eqns. 7.83–7.85 T_n, Q_n and R_n are unknown coefficients, $P_n(z)$ are unknown functions and $\psi_n(x, z) = \cos[\pi n\{x - a(z)\}/2a(z)]$, $\gamma_n(L)$ being obtained from the expression for $\gamma_n(0)$ through the replacement: $a(0) \to a(L)$.

According to Galerkin's incomplete method we substitute eqns. 7.79 and 7.80 in the region $\mathscr{D}$ by a set of equivalent projection relationships

$$\int_0^{a(z)} (\nabla_{xz}^2\psi + k^2\psi)\psi_m(x, z)\,dx = \left.\frac{\partial\psi}{\partial \boldsymbol{n}}\right|_{x=a(z)} \left\{1 + \left(\frac{da}{dz}\right)^2\right\}^{1/2} \tag{7.86}$$

where $m = 1, 3, \ldots$.

Substituting eqn. 7.85 into eqn. 7.86 we obtain an infinite system of ordinary differential equations for $\boldsymbol{P} = \{P_n(z)\}$

$$\frac{d^2\boldsymbol{P}(z)}{dz^2} + \tilde{H}(z)\frac{d\boldsymbol{P}(z)}{dz} + \tilde{B}(z)\boldsymbol{P}(z) = 0 \tag{7.87}$$

where $\tilde{H}(z)$ and $\tilde{B}(z)$ are known matrix functions, elements of which are calculated in a manner similar to that in Section 5.8.

The boundary conditions for the system described by eqn. 7.87 are obtained as a result of matching of ψ and $\partial\psi/\partial z$ at the interfaces between the partial regions ($z = 0$, $z = L$, $z = L + \Delta$). The continuity of the tangential-field components across $z = 0$ implies

$$S_n + R_n = P_n(0) \tag{7.88}$$

$$(R_n - S_n)\gamma_n(0) = \frac{dP_n(0)}{dz} + \sum_m{}' \xi_{nm}(0)P_m(0) \tag{7.89}$$

where the elements of the matrix $\xi(0)$ are

$$\xi_{nm}(0) = -(4 - 3\delta_{mn})\frac{(n^2\delta_{mn} - m^2)a'(0)}{(n^2 - m^2)2a(0)}$$

$a'(0) = (da/dz)|_{z=+0}$. Excluding R_n from eqn. 7.89 we obtain a boundary condition for eqn. 7.87 at $z = 0$:

$$\frac{d\boldsymbol{P}(0)}{dz} + \{\xi(0) - \Gamma(0)\}\boldsymbol{P}(0) = -2\Gamma(0)\boldsymbol{S} \tag{7.90}$$

where $\boldsymbol{S} = \{S_n\}$ is a given vector and $\Gamma(0)$ is a matrix with the elements $\Gamma_{nm}(0) = \delta_{mn}\gamma_n(0)$.[56]

Similarly, at $z = L$ we obtain

$$T_n \exp\{-\gamma_n(L)\Delta\} + Q_n \exp\{\gamma_n(L)\Delta\} = P_n(L) \tag{7.91}$$

$$\gamma_n(L)[T_n \exp\{-\gamma_n(L)\Delta\} - Q_n \exp\{\gamma_n(L)\Delta\}] = \frac{dP_n(L)}{dz} + \sum_m{}' \xi_{nm}(L)P_m(L) \tag{7.92}$$

[56] In fact, eqn. 7.90 is the radiation condition and shows that there are no incoming (from $z = -\infty$) waves in the scattered field.

The expressions for the matrix elements $\xi_{nm}(L)$ are easily obtained from the expressions for $\xi_{nm}(0)$ through the replacements: $a(0) \to a(L)$, $a'(0) \to a'(L)$ and $a'(L) = \mathrm{d}a/\mathrm{d}z|_{z=L-0}$.

The field matching at $z = L + \Delta$, as our next stage, is performed using the analytical solution of the problem of radiation from an open-ended plane regular waveguide (Mittra and Lee, 1971). Let us present the coefficients T_n and functions $A(\alpha)$ and $B(\beta)$ in the form of the series

$$T_n = \sum_m{}' Q_m \Lambda_{nm} \tag{7.93}$$

$$\begin{Bmatrix} A(\alpha) \\ B(\beta) \end{Bmatrix} = \sum_m{}' Q_m \begin{Bmatrix} A_m(\alpha) \\ B_m(\beta) \end{Bmatrix} \tag{7.94}$$

where Λ_{nm} is a coefficient of transformation of the mth incident mode into the nth reflected one for the open end of a regular waveguide with a width of $2a(L)$ and $A_m(\alpha)$, $B_m(\beta)$ are spectral amplitude functions in eqns. 7.81 and 7.82 corresponding to the case of excitation of an open-ended regular waveguide with a width of $2a(L)$ by the mth eigenmode of unit amplitude. The coefficients Λ_{nm} and functions $A_m(\alpha)$ and $B_m(\beta)$ can be determined in a closed form using the factorisation method (Weinstein, 1969*b*) or a variant of the residue-calculus technique modified for the case of open structures (Mittra and Lee, 1971). Both methods, naturally, produce identical results:

$$\Lambda_{nm} = \frac{j[\{k + j\gamma_m(L)\}\{k + j\gamma_n(L)\}]^{1/2}}{a(L)\gamma_n(L)\{\gamma_n(L) + \gamma_m(L)\}} M_+\{j\gamma_n(L)\} M_+\{j\gamma_m(L)\} \tag{7.95}$$

$$A_m(\alpha) = \frac{j[\{k + j\gamma_m(L)\}(\alpha + k)]^{1/2}}{(2\pi)^{1/2}\{\alpha + j\gamma_m(L)\}\gamma} \exp\{\gamma a(L)\} M_+(\alpha) M_+\{j\gamma_m(L)\} \tag{7.96}$$

where $\gamma = (\alpha^2 - k^2)^{1/2}$. The special function $M_+(\alpha)$, obtained as a result of factorisation, is expressed in terms of infinite products:

$$M_+(\alpha) = [\cos\{ka(L)\}]^{1/2} \exp\left(\frac{ja(L)\gamma}{\pi} \ln\left(\frac{\alpha - \gamma}{k}\right) + \frac{ja(L)\alpha}{\pi}\left[1 - C + \ln\left\{\frac{\pi}{2ka(L)}\right\} + j\frac{\pi}{2}\right]\right) \times \prod_{n=1,3,\ldots}^{\infty} \left\{1 + \frac{\alpha}{j\gamma_n(L)}\right\} \exp\left\{\frac{2j\alpha a(L)}{\pi n}\right\}$$

where C is Euler's constant. The function $B_m(\beta)$ is expressed similarly to eqn. 7.96.

Use of the expressions for T_n, $A(\alpha)$ and $B(\beta)$ given by eqns. 7.95 and 7.96 is equivalent to the fulfilment of the continuity conditions for ψ and $\partial\psi/\partial z$ at the $z = L + \Delta$ plane. Note that this variant of field matching at the interfaces between the partial regions $\mathscr{A}$, $\mathscr{B}$ and $\mathscr{A}$, $\mathscr{C}$ requires an extraction and analytical inversion of the operator, corresponding to the problem of radiation from the open-ended plane regular waveguide of a width of $2a(L)$. In fact, this is a procedure to regularise the 2-point boundary-value problem obtained similarly to that described in Section 5.8.

To obtain the boundary condition for eqn. 7.87 at $z = L$, it is necessary to exclude from eqn. 7.92 the unknown coefficients T_n and Q_n using eqns. 7.91 and 7.93. At the limit $\Delta \to 0$, we obtain the required boundary condition in the form

$$\frac{d\boldsymbol{P}(L)}{dz} + \{\xi(L) - \Gamma(L)\}\boldsymbol{P}(L) = -2\Gamma(L)(I+\Lambda)^{-1}\boldsymbol{P}(L) \tag{7.97}$$

where $\Gamma(L)$ is a matrix with the elements $\Gamma_{nm}(L) = \delta_{nm}\gamma_n(L)$, and Λ is a matrix whose elements are expressed by eqn. 7.95.

Thus the original diffraction problem has been reduced to a 2-point linear boundary-value problem for the system of ordinary differential equations of infinite order (eqns. 7.87, 7.90 and 7.97). To find a numerical solution, we should reduce this problem to that of the finite order N. We can then use one of the shooting-method modifications (see Appendix 1). The convergence of an approximate solution to the accurate one at $N \to \infty$ can be proved rigorously (as was done by Kopenkin *et al.*, 1987).

For the field in the far zone we can obtain the following expression using the saddle-point method:

$$\lim_{r\to\infty} \psi(r, \vartheta) \simeq \frac{j \sin \vartheta F(\vartheta)}{(2\pi kr)^{1/2}|\sin \vartheta|} \exp\left\{j\left(kr - \frac{\pi}{4} + ka(L)\right)\right\}$$

where the radiation pattern $F(\vartheta)$ is expressed by

$$F(\vartheta) = (1 - \cos \vartheta)^{1/2} M_+(-k \cos \vartheta)$$
$$\times \sum_m{}' \frac{j\gamma_m(L) + k}{j\gamma_m(L) - k \cos \vartheta} Q_m M_+\{j\gamma_m(L)\} \tag{7.98}$$

The vector of coefficients $\boldsymbol{Q} = \{Q_m\}$ in eqn. 7.91 is expressed in terms of the solution of eqns. 7.87, 7.90 and 7.97 in the form

$$\boldsymbol{Q} = (I+\Lambda)^{-1}\boldsymbol{P}(L)$$

As in other cases when this method is used, the number N, sufficient for the inner convergence, should exceed the number of the eigenmodes propagating in a plane waveguide of the maximum width by 2–3 units.

To monitor the accuracy of the approximate solution, we keep checking on how the active power-balance and reciprocity relationships are fulfilled. One of the forms of the reciprocity relationships can be written as

$$R_p^{(q)}\gamma_p(0) = R_q^{(p)}\gamma_q(0) \tag{7.99}$$

where $R_p^{(q)}$ is the complex amplitude of the pth reflected mode at the incidence of the qth mode ($S_q = 1$, $S_n = 0$, $n \neq q$). The numerical experiments have shown that for such N the relative deviation from these relationships does not, on average, exceed 10^{-4}–10^{-5}.

To apply the technique described in this Section for examining corrugated horns we used the following function for the horn profile:

$$a(z) = a(0) + \alpha_0 z + \beta_0 \sin(2\pi z/l) \tag{7.100}$$

where α_0, β_0 and l are constant values.

In Figure 7.24 the normalised radiation patterns are shown when the horn is excited by the TM_{01} mode ($S_1 = 1$, $S_n = 0$, $n \neq 1$). The lengths of irregular sections and sizes of the apertures are the same for all curves, the only difference being in the number of the slots defined by the ratio l/L. The radiation patterns, as we note, are of a complicated, sharply oscillating nature. The peaks are sharper for larger ratios L/l. Physically this means that the higher modes are predominant in the process of the formation of the radiation pattern. These higher modes are intensively excited due to fundamental-mode transformation which occurs with the horn profile nonmonotonic.

When the period of corrugation is small compared with the wavelength, use of the surface-impedance model of a corrugated horn appears to be more suitable (Clarricoats and Olver, 1984). A rigorous solution of the corresponding diffraction problem, not based on the physical optics approximations, can easily be obtained within the framework of the approach described above.

To conclude with let us note that Galerkin's incomplete method with semi-inversion in the boundary conditions is a universal and highly effective new method applicable to various problems of mathematical physics. The advantages of this technique arise from its being a combination of powerful analytical

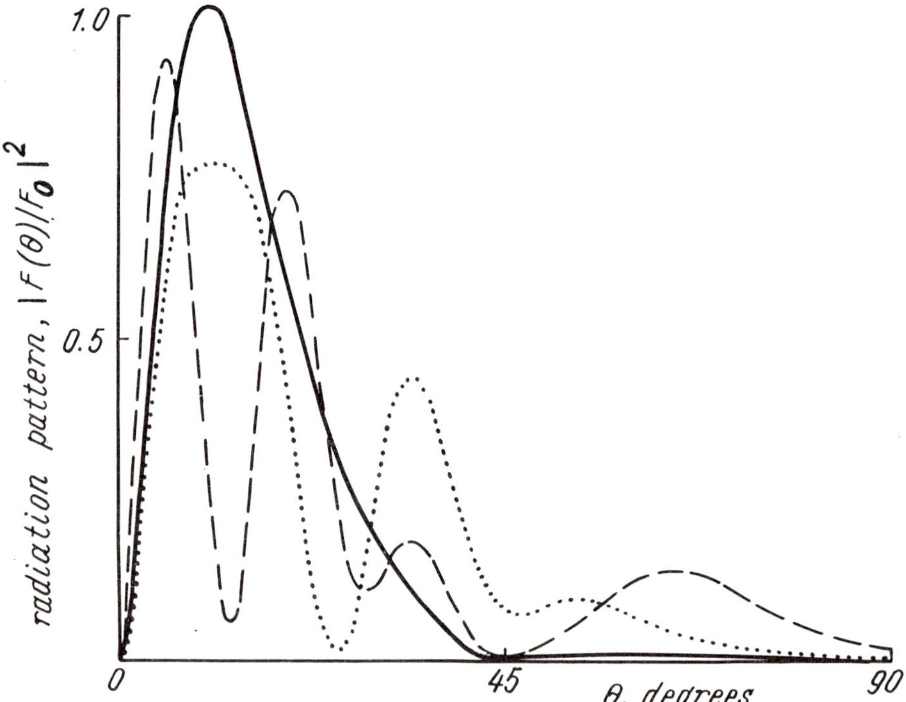

Figure 7.24 *Normalised radiation patterns for corrugated horns*

$F_0 = \max_{\vartheta} \{F(\vartheta)\}$ for $L/l = 2.25$

$a(0) = 0.5\lambda$, $\alpha_0 = 0.4$, $\beta_0 = 0.1\lambda$

——— $L/l = 2.25$

- - - - $L/l = 4.25$

· · · · · · $L/l = 6.25$

methods (for example, the Wiener–Hopf method) and the general methods of the theory of smoothly inhomogeneous media and smoothly irregular structures. Thus, this method can be applied, for example, to the problems of wave radiation from open-ended plane and circular waveguides which are irregular not because of their variable cross-section but because of the inhomogeneity of the medium inside. In some special cases (for example, when a waveguide is 'slowly' irregular) this method allows us to obtain approximate analytical solutions using the WKB technique or Langer's method (Felsen and Marcuvitz, 1973, Jones, 1986). An example of this can be found in the work by Ilyinsky *et al.* (1987).

References

ABRIKOSOV, A.A. (1987): 'Principles of the theory of metals' (Moscow, Nauka)

ABRIKOSOV, A.A., GORKOV, L.P., and KHALATNIKOV, I.M. (1958): 'Superconductor in a high frequency field', *Zhurnal Experimentalnoi i Teoreticheskoi Fiziki*, **35**, pp. 265–275

ABRIKOSOV, A.A., GORKOV, L.P., and KHALATNIKOV, I.M. (1959): 'Analysis of experimental data relating to the surface impedence of superconductors', *Zhurnal Experimentalnoi i Teoreticheskoi Fiziki*, **37**, pp. 187–191

AGRANOVICH, V.M., and GINZBURG, V.L. (1979): 'Crystalooptics considering spatial dispersion and exciton theory' (Moscow, Nauka)

ALKHOVSKY, E.A., ASAFYEV, V.I., ILYINSKY, A.S., and TSIBIZOV, N.K. (1987): 'Investigation of the dependence of attenuation in low-loss corrugated waveguides on the corrugation parameters', *Radiotekhnika i Elektronika*, **34**, pp. 1103–1105

ALKHOVSKY, E.A., GOLOVCHENKO, G.S., ILYINSKY, A.S., TROSHIN, G.I., and HYDUYAKOVA, V.A. (1986): 'Flexible waveguides in microwave engineering' (Moscow, Radio i Svyaz)

ALKHOVSKY, E.A., and ILYINSKY, A.S. (1979): 'Numerical investigation of propagation constants of symmetrical magnetic waves in a circular corrugated waveguide', *Radiotekhnika i Elektronika*, **24**, pp. 1684–1685

ALKHOVSKY, E.A., ILYINSKY, A.S., and TROSHIN, G.I. (1975): 'Calculation of phase and attenuation constants for electromagnetic waves in a circular corrugated waveguide', *Radiotekhnika i Elektronika*, **20**, pp. 2250–2255

ALKHOVSKY, E.A., ILYINSKY, A.S., and TROSHIN, G.I. (1978): Remarks on the paper 'Calculation of phase and attenuation constants for electromagnetic waves in a circular corrugated waveguide', *Radiotekhnika i Elektronika*, **23**, pp. 1109–1110

ALKHOVSKY, E.A., ILYINSKY, A.S., and HARLANOV, U.YA. (1979): 'Results of numerical and experimental study of circular deep-corrugated waveguides' in 'Numerical Methods of Electrodynamics' (Moscow, MGU)

AL – HAKKAK, M.J. (1978): 'Residue-calculus solution of the TM_{10}-mode problem in circular corrugated waveguides', *Proc. IEE*, **125**, pp. 787–792

AMITAY, N., GALINDO, V., and WU, C.P. (1972): 'Theory and analysis of phased array antennas' (Wiley-Interscience)

ANDRENKO, S.D., and SHESTOPALOV, V.P. (1979): 'Diffraction logical elements of an optical computer', *Doklady AN SSSR*, **248**, pp. 1070–1073

ARKADAKSKY, S.S., and TSYKIN, B.G. (1975*a*): 'On the theory of impedance waveguides excitation by an electron beam. Pt. I. Coupled-mode form of the equations', *Radiotekhnika i Elektronika*, **20**, pp. 2113–2120

ARKADAKSKY, S.S., and TSYKIN, B.G. (1975*b*): 'On the theory of impedance waveguides excitation by electron flow. Pt. II. Excitation equations of normal waves', *Radiotekhnika i Elektronika*, **20**, pp. 2328–2335

ARKADAKSKY, S.S., and TSYKIN, B.G. (1976): 'The equations of homogeneous waveguides excitation at cut-off frequencies', *Radiotekhnika i Elektronika*, **21**, pp. 608–611

BARSOTTI, E.L., KUESTER, E.F., and DUNN, J.M. (1991): 'A simple method to account for edge shape in the conductor loss in microstrip', *IEEE Trans.*, **MTT-39**, pp. 98–106

BARYSHNIKOV, I.V., DATSKOVSKY, V.A., and UPOLOVNEV, A.V. (1988): 'Influence of surface roughness of guiding systems on RF losses', *Radiotekhnika i Elektronika*, **33**, pp. 2029–2035

BASS, F.G., and FUKS, I.M. (1979): 'Scattering of waves from statistically irregular surfaces' (Pergamon Press)

BELIANTSEV, A.M., and GAPONOV, A.V. (1964): 'Waves with complex constants of propagation in coupled transmission lines without energy dissipation', *Radiotekhnika i Elektronika*, **9**, pp. 1188–1197

BICHYTSKAYA, T.I., and NOVIKOV, V.V. (1979): 'Interaction of normal waves in a plane irregular waveguide in the presence of a degeneration point', *Izvestiya VUZov—Radiofizika*, **22**, pp. 860–870

BICHYTSKAYA, T.I. (1990): 'An effect of mode interaction on the field in the waveguide under the conditions of local degeneration', *Izvestiya VUZov—Radiofizika*, **33**, pp. 758–761

BRAGINSKY, V.B., MITROFANOV, V.P., and PANOV, V.I. (1981): 'Systems with small dissipation' (Moscow, Nauka)

BRAGINSKY, V.B., and VYATCHANIN, S.I. (1980): 'High-quality dielectric resonators', *Doklady AN SSSR*, **252**, pp. 584–585

BRAVER, I.M., FRIDBERG, P.SH., GARB, KH.L., and YAKOVER, I.M. (1988): 'The behaviour of the electromagnetic field near the edge of a resistive half-plane', *IEEE Trans.*, **AP-36**, pp. 1760–1768

BREWS, J.R. (1987): 'Characteristic impedance of microstrip lines', *IEEE Trans.*, **MTT-35**, pp. 30–34

BYKOV, A.A., and ILYINSKY, A.S. (1979): 'The directed orthogonalisation method for solving linear systems of ordinary differential equations', *Zhurnal Vychislitelnoi Matematiki i Matematicheskoi Fiziki*, **19**, pp. 631–639

BYKOV, A.A., and ILYINSKY, A.S. (1982): 'Numerical analysis of dielectric resonances in waveguides', *Radiotekhnika i Elektronika*, **27**, pp. 1830–1833

CARALLO, A., DOANE, J., and CUTLER, R. (1990): 'Low-loss broadband multimode corrugated waveguide performance', *Rev. Science Instrum.*, **61**, pp. 2396–2400

CHANG, K. (Ed.) (1989): 'Handbook of microwave and optical components. Microwave passive and antenna components Vol. 1' (Wiley)

CLARRICOATS, P.J.B., and OLVER, A.D. (1984): 'Corrugated horns for microwave antennas' (Peter Peregrinus Ltd)

CLARRICOATS, P.J.B., OLVER, A.D., and SHONG, S.L. (1975*a*): 'Attenuation in corrugated circular waveguides. Pt. I. Theory', *Proc. IEE*, **122**, pp. 1173–1179

CLARRICOATS, P.J.B., OLVER, A.D., and SHONG, S.L. (1975*b*): 'Attenuation in corrugated circular waveguides. Pt. II. Experiment', *Proc. IEE*, **122**, pp. 1173–1186

CLARRICOATS, P.J.B., and SAHA, P.K. (1970): 'Attenuation in corrugated circular waveguide', *Electron. Lett.*, **12**, pp. 370–372

CLARRICOATS, P.J.B., and TAYLOR, B.C. (1964): 'Evanescent and propagating modes of dielectric loaded waveguide', *Proc. IEE*, **111**, pp. 1951–1956

COLTON, D., and KRESS, R. (1983): 'Integral equation methods in scattering theory' (John Wiley & Sons)

DAS, N.K., and POZAR, D.M. (1991): 'Full wave spectral domain computation of material, radiation and guided wave losses in infinite multilayered printed transmission lines', *IEEE Trans.*, **MTT-39**, pp. 54–64

DAYWITT, W.C. (1990): 'First order symmetric modes for a slightly lossy coaxial transmission line', *IEEE Trans.*, **MTT-38**, pp. 1644–1650

DEPINE, R.A., and BRUDNY, V.L. (1990): 'The surface impedance of metallic objects: rigorous calculations for imperfectly conducting diffraction gratings', *IEEE Trans.*, **AP-38**, pp. 1290–1294

DE SANTO, J.A., and BROWN, G.S. (1986): 'Analytical techniques for multiple scattering from rough surfaces' *in* 'Progress in Optics. Vol. 23' (Elsevier)

EKHOLM, E.B., and McKNIGHT, S. (1990): 'Attenuation and dispersion for high-T_c superconducting microstrip lines', *IEEE Trans.*, **MTT-38**, pp. 387–395
EVGRAFOV, M.A. (1965): 'Analytical functions' (Moscow, Nauka)
FADEEV, D.K., and FADEEVA, V.N. (1963): 'Computational methods of linear algebra' (Dover)
FEL, S.S., FRIDBERG, P.SH., and LEVINSON, I.B. (1962): 'On the theory of broad band nonretunable echo-cavities of a spherical form', *Radiotekhnika i Elektronika*, **6**, pp. 1125–1130
FELSEN, L.B., and MARCUVITZ, N. (1973): 'Radiation and scattering of waves' (Prentice-Hall)
GALISHNIKOVA, T.N., and ILYINSKY, A.S. (1987): 'Numerical methods in diffraction problems' (Moscow, MGU)
GAPLEVSKII, V.V., and KONTOROVICH, V.M. (1971): 'Application of the symmetry theory to study degenerate mode splitting in magnetron resonators', *Izvestiya VUZov—Radiofizika*, **14**, pp. 1906–1917
GAVRIELIDES, A., and PETERSON, P. (1979): 'Power losses in lamellar gratings', *Applied Optics*, **18**, pp. 4168–4172
GENT, A.W. (1959): 'The attenuation and propagation factor of spaced-disc circular waveguides', *Proc. IEE*, **106**, pp. 366–372
GERSHMAN, B.N. (1955): 'On the theory of normal waves in homogeneous plasma in a magnetic field' *in* 'Pamyati A.A. Andronova' (Moscow, AN SSSR)
GINSBERG, D.M. (1966): 'Calculation of the electromagnetic absorption edge in superconducting alloys', *Phys. Rev.*, **151**, pp. 241–245
GODUNOV, S.K. (1961): 'Numerical solution techniques for boundary value problems of linear systems of ordinary differential equations', *Uspekhi Matematicheskikh Nauk*, **16**, pp. 171–174
GOPINATH, A., HORTON, R., and EASTER, B. (1970): 'Microstrip loss calculations', *Electron. Lett.*, **6**, pp. 40–41
GRINBERG, YA.R. (1981): 'Boundary conditions for an electromagnetic field in the presence of thin metal surfaces', *Radiotekhnika i Elektronika*, **26**, pp. 2492–2500
GUBSKY, D.S., LYAPIN, V.P., and SINYAVSKY, G.P. (1982): 'Study of circular resonators retuned by coaxial rods', *Radiotekhnika i Elektronika*, **27**, pp. 1294–1300
GUREEV, A.V. (1990*a*): 'The velocity of energy transmission by electromagnetic waves in a regular screened waveguide', *Zhurnal Tekhnicheskoi Fiziki*, **60**, pp. 23–28
GUREEV, A.V. (1990*b*): 'The normal wave properties in a screened inhomogeneous waveguide', *Izvestiya VUZov—Radiofizika*, **33**, pp. 954–964
HARGREAVES, T.A., FISHER, R.P., MCCOWAN, R.B., and FLIFLET, A.W. (1991): 'Ohmic effects in quasioptical resonators', *Int. J. Infrared and Millimeter Waves*, **12**, pp. 9–21
HARRINGTON, R.F. (1968): 'Field computation by moment methods' (Macmillan)
HEITKAMPER, P., and HEINRICH, W. (1991): 'On the calculation of conductor loss on planar transmission lines assuming zero strip thickness', *IEEE Trans.*, **MTT-39**, pp. 586–588
ILLARIONOV, YU.A., RAYEVSKY, S.B., and SMORGONSKY, V.YA. (1980): 'Numerical analysis of corrugated and partially filled waveguides' (Moscow, Sovetskoe Radio)
ILYINSKY, A.S. (1973): 'Direct analysis method for periodic systems', *Zhurnal Vychislitelnoi Matematiki i Matematicheskoi Fiziki*, **13**, pp. 119–126
ILYINSKY, A.S. (1981): 'The substantiation of the method for calculating eigenmodes in microstrip transmission lines', *Differentsialnye Uravneniya*, **17**, pp. 1868–1874
ILYINSKY, A.S., and BYKOV, A.A. (1980): 'Numerical investigation of the model of an absorbing system in a waveguide', *Izvestiya VUZov—Radiofizika*, **28**, pp. 821–832
ILYINSKY, A.S., KOPENKIN, A.D., KURAEV, A.A., SLEPYAN, A.YA., SLEPYAN, G.YA., and CHEREPENIN, V.A. (1989): 'Galerkin's incomplete method with semi-inversion for diffraction problems in waveguides and gratings' Proceedings of URSI International Symposium on Electromagnetic Theory, Stockholm, pp. 633–635
ILYINSKY, A.S., KURAEV, A.A, SLEPYAN, G.YA., and SLEPYAN, A.YA. (1987):

'Semi-inversion method for problems of wave diffraction in bifurcations of irregular plane waveguides', *Doklady AN SSSR*, **294**, pp. 1345–1349

ILYINSKY, A.S., and SLEPYAN, G.YA. (1983): 'Oscillations and waves in electrodynamic systems with dissipation' (Moscow, MGU)

ILYINSKY, A.S., and SLEPYAN, G.YA. (1986): 'On the calculation of Joule losses in electrodynamic systems with thin unclosed metal surfaces', *Radiotekhnika i Elektronika*, **31**, pp. 670–675

ILYINSKY, A.S., and SLEPYAN, G.YA. (1990): 'Impedance boundary conditions and their application for numerical analysis of the electromagnetic waves absorption in conducting media', *Radiotekhnika i Elektronika*, **35**, pp. 1121–1139

ILYINSKY, A.S., and SVESHNIKOV, A.G. (1968): 'Methods of study of irregular waveguides', *Zhurnal Vychislitelnoi Matematiki i Matematicheskoi Fiziki*, **8**, pp. 363–373

ILYINSKY, A.S., and SVESHNIKOV, A.G. (1969): 'Direct methods for investigation of waveguide systems' *in* 'Vychislitelnye Metody i Programmirovanie. Vol. 13' (Moscow, MGU)

ILYINSKY, A.S., and TUPIKOV, M.V. (1989): 'A new scheme of projection method for investigation of circular corrugated waveguides', *Zhurnal Vychislitelnoi Matematiki i Matematicheskoi Fiziki*, **29**, pp. 1012–1022

ILYINSKY, A.S., and ZARUBANOV, V.V. (1980): 'Galerkin's method application to calculate and investigate current distributions of fundamental and higher modes in a nonsymmetrical microstrip line', *Radiotekhnika i Elektronika*, **25**, pp. 1844–1850

JONES, D.S. (1986): 'Acoustic and electromagnetic waves' (Clarendon Press)

KALHOR, H.A. (1977); 'Approximate analysis of electromagnetic scattering from corrugated surfaces by surface impedance modeling', *IEEE Trans.*, **AP-25**, pp. 721–722

KANTOROVICH, L.V., and AKILOV, G.P. (1977): 'Functional analysis' (Moscow, Nauka)

KANTOROVICH, L.V., and KRYLOV, V.I. (1958): 'Approximate methods of higher analysis' (Interscience Publishers, Inc)

KATO, T. (1966): 'Perturbation theory for linear operators' (Springer)

KATSENELENBAUM, B.Z. (1959): 'Attenuation of H_{0n}-modes in a helix waveguide', *Radiotekhnika i Elektronika*, **4**, pp. 428–432

KATSENELENBAUM, B.Z. (1961): 'Theory of slowly varying waveguides' (Moscow, AN SSSR)

KAZANTSEV, YU.N., MANENKOV, A.B., and KHARLASHKIN, O.A. (1974): 'Hollow-dielectric and metal-dielectric waveguides for fast H-wave transmission', *Izvestiya VUZov—Radiofizika*, **17**, pp. 1529–1538

KAZANTSEV, YU.N., and KRAFTMAKHER, G.A. (1979): 'Hollow infrared transmissive metal-dielectric waveguide', *Radiotekhnika i Elektronika*, **24**, pp. 1084–1088

KELDYSH, M.V. (1951): 'On the eigenvalues and eigenfunctions of some types of non-self-adjoint equations', *Doklady AN SSSR*, **77**, pp. 11–14

KIANG, J.F. (1991): 'Integral equation solution to the skin-effect problem in conductor strips of finite thickness', *IEEE Trans.*, **MTT-39**, pp. 452–460

KISUN'KO, G.V. (1948): 'Electrodynamics of cavity systems' (Leningrad, VKAS)

KITTEL, C. (1963): 'Quantum theory of solids' (John Wiley & Sons)

KLEEV, A.I., and MANENKOV, A.B. (1982): 'Calculation of the fields in coaxial resonators', *Radiotekhnika i Elektronika*, **27**, pp. 1888–1896

KLEEV, A.I., and MANENKOV, A.B. (1984): 'The coaxial resonator with a thin internal wire', *Radiotekhnika i Elektronika*, **29**, pp. 2293–2301

KLIMENKO, V.A., and FIALKOVSKY, A.T. (1990): 'Diffraction of E-electromagnetic wave by a semi-infinite lossy dielectric slab', *Radiotekhnika i Elektronika*, **35**, pp. 479–487

KOPENKIN, A.D., KURAEV, A.A., SLEPYAN, A.YA., SLEPYAN, G.YA., and CHEREPENIN, V.A. (1987): 'Solution technique for two-dimensional problems of wave diffraction by gratings', *Zhurnal Vychislitelnoi Matematiki i Matematicheskoi Fiziki*, **27**, pp. 1894–1900

KOPENKIN, A.D., KURAEV, A.A., SLEPYAN, A.YA., SLEPYAN, G.YA., and CHEREPENIN, V.A. (1988): 'An analytical-numerical method of studying the diffraction of waves by the periodic structures with a non-coordinate shape of the period', *Sov. J. Communications Technology and Electronics*, **33**, pp. 19–25

KOPENKIN, A.D., KURAEV, A.A., SLEPYAN, A.YA., SLEPYAN, G.YA., and CHEREPENIN, V.A. (1989): 'Radiation of waves from the open end of a smoothly-irregular waveguide', *Sov. J. Communications Technology and Electronics*, **34**, pp. 45–50
KORN. G., and KORN, T. (1968): 'Mathematical handbook for scientists and engineers' (McGraw-Hill)
KOROZA, V.I. (1970): 'Parametric resonance in the case of electromagnetic wave propagation in a periodic slowing structure', *Radiotekhnika i Elektronika*, **15**, pp. 450–455
KOTAYAMA, S., FITATSUGI, H., ABE, K., and YAMASAKI, M. (1979): 'Double ridge corrugated waveguide characteristics', *Trans. Inst. Electron. Comm. Eng. Jap.*, **B-62**, pp. 329–336
KOVALENKO, A.N. (1980): 'The analysis of the spectrum of a screened strip line', *Izvestiya VUZov—Radiofizika*, **23**, pp. 1388-1390
KRASNUSHKIN, P.E., and FEDOROV, E.N. (1972): 'Multiplicity of the wave numbers of normal waves in laminated media', *Radiotekhnika i Elektronika*, **17**, pp. 1129–1140
KRASNUSHKIN, P.E. (1974): 'Normal waves transformation in periodical and smooth waveguides without losses', *Radiotekhnika i Elektronika*, **19**, pp. 1345–1358
KURAEV, A.A., NATAROV, M.P., RODIONOVA, V.N., SKRESANOV, V.N., SLEPYAN, A.YA. and SLEPYAN, G.YA. (1991): 'Coupling of an open resonator and a rectangular waveguide through a smooth waveguide transition', *Int. J. Electronics*, **70**, pp. 1005–1014
KURAEV, A.A., and SLEPYAN, A.YA. (1990): 'Iterative algorithm of numerical implementation of Galerkin's incomplete method for problems of wave diffraction in dielectric structures', *Doklady AN SSSR*, **34**, pp. 418–421
LANDAU, L.D., LIFSHITZ, E.M., and PITAEVSKII, L.P. (1984): 'Electrodynamics of continuous media' (Pergamon Press)
LAVRENTYEV, M.A., and SHABAT, B.V. (1987): 'Methods in the theory of functions of complex variables' (Moscow, Nauka)
LERER, A.M., and MIKHALEVSKY, V.S. (1983): 'Dispersion characteristics of microstrips on anisotropic substrata', *Radiotekhnika i Elektronika*, **28**, pp. 36–43
LEVIN, L. 'A method of avoiding the edge current divergence in perturbation loss calculations', *IEEE Trans.*, **MTT-32**, pp. 717–719
LIFSHITZ, E.M., and PITAEVSKII, L.P. (1979): 'Physical kinetics' (Moscow, Nauka)
LITVINENKO, L.N. (1972): 'A method of successive refinements to solve infinite systems of linear algebraic equations', *Doklady AN SSSR*, **203**, pp. 64–67
LUCHKA, A.YU. (1980): 'Projection-iterative methods for solving differential and integral equations' (Kiev, Naukova Dumka)
LUK, L.N., RODIONOVA, V.N., and SLEPYAN, G.YA. (1988): 'Study of millimetre-wave band corrugated resonators with a high Q-factor', *Zhurnal Tekhnicheskoi Fiziki*, **58**, pp. 594–596
MARKOV, G.T., and VASILYEV, E.N. (1970): 'Mathematical methods of applied electrodynamics' (Moscow, Sovetskoe Radio)
MASHKOVTSEV, B.M. TSIBISOV, K.N., and EMELIN, V.F. (1966): 'Theory of waveguides' (Moscow–Leningrad, AN SSSR)
MATTIS, D.CH., and BARDEEN, J. (1958): 'Theory of anomalous skin-effect in normal and superconducting metals', *Phys. Rev.*, **111**, pp. 412–417
MEISSNER, J. (1990): 'Rigorous calculation of losses of single and coupled grooveguides', *Electron. Lett.*, **26**, pp. 41-42
MEIXNER, J. (1972): 'The behaviour of electromagnetic fields at edges', *IEEE Trans.*, **AP-20**, pp. 442–446
MENDE, F.F., and SPITSYN, A.I. (1985): 'Surface impedance of superconductors' (Kiev, Naukova Dumka)
MIKHLIN, S.G. (1966): 'Numerical implementation of variational methods' (Moscow, Nauka)
MIKHLIN, S.G. (1970): 'Variational methods in mathematical physics' (Moscow, Nauka)
MILLER, M.A. (1952): 'Electromagnetic waves propagation over a plane surface with anisotropic impedance boundary conditions', *Doklady AN SSSR*, **87**, pp. 571–574
MIRSHEKAR–SYAHKAI, D., and DAVIES, J.B. (1982): 'An accurate, unified

solution to various fin-line structures, of phase constant, characteristic impedance and attenuation', *IEEE Trans.*, **MTT-30**, pp. 1854–1861

MITTRA, R., and LEE, S.W. (1971): 'Analytical techniques in the theory of guided waves' (Macmillan)

MITZNER, K.M. (1967): 'An integral equation approach to scattering from a body of finite conductivity', *Radio Science*, **2**, pp. 1459–1470

MODENOV, V.P., and SLEPYAN, G.YA. (1984): 'Analysis of eigenmodes in a circular azimuth-corrugated waveguide with impedance walls', *Radiotekhnika i Elektronika*, **29**, pp. 1829–1832

NEFEDOV, E.I., and SIVOV, A.N. (1977): 'Electrodynamics of periodic structures' (Moscow, Nauka)

NEFEDOV, E.I., and ROSSIISKY, I.M. (1978): 'An open coaxial resonator of a new type', *Doklady AN SSSR*, **241**, pp. 812–814

NEIMARK, M.A. (1954): 'Linear differential operators' (Moscow, Gostekhizdat)

NIKOLSKY, V.V., and KOZLOV, A.YU. (1986): 'Losses in microstrip lines: nonsimplified problem formulation considering the technological factors', *Izvestiya VUZov—Radiofizika*, **29**, pp. 566–576

NIKOLSKY, V.V., and KOZLOV, A.YU. (1987*a*): 'Electrodynamic analysis of microstrip structures with resistive films', *Radiotekhnika*, **6**, pp. 55–60

NIKOLSKY, V.V., and KOZLOV, A.YU. (1987*b*): 'On the problem of losses in microstrip transmission lines', *Radiotekhnika i Elektronika*, **32**, pp. 1562–1563

NOBLE, B. (1958): 'Methods based on the Wiener-Hopf technique' (Pergamon Press)

NOSICH, A.I., and SHESTOPALOV, V.P. (1980): 'Ohmic losses in transmission lines with thin conductors', *Soviet Physics Doklades*, **25**, pp. 127–128

OKRESS, E.C. (Ed.) (1968): 'Microwave power engineering. Vols. 1 and 2' (Academic Press)

PARINI, C.G., CLARRICOATS, P.J.B., and OLVER, A.D. (1977): 'Single-mode operation of low-attenuation corrugated waveguide', *Electron. Lett.*, **13**, pp. 16–17

PARTON, V.Z., and PERLIN, P.I. (1977): 'Integral equations in elasticity theory' (Moscow, Nauka)

POGGIO, A.J., and MILLER, E.K. (1973): 'Integral equation solutions of three-dimensional scattering problems' *in* MITTRA, R. (Ed.): 'Computer techniques for electromagnetics' (Pergamon Press)

POPOV, V.P. (1981): 'Reflection from open-circuited and short-circuited ends of a nonsymmetrical ferrite-filled strip line', *Radiotekhnika i Elektronika*, **26,** pp. 269–276

PREGLA, R. (1980): 'Determination of conductor losses in planar waveguide structures', *IEEE Trans.*, **MTT-28**, pp. 433–434

RAWAL, S., and JACKSON, D.R. (1991): 'An exact TEM-calculation of loss in a strip line of arbitrary dimensions', *IEEE Trans.*, **MTT-39**, pp. 694–699

ROBERTS, R.M., and SHIPMAN, J.S. (1972): 'Two-point boundary value problems: shooting methods' (American Elsevier)

RODIONOVA, V.N., and NEKHAY, A.K. (1981): 'Investigation of ways of designing the high-quality resonators', *Tekhnika sredstv svyazi—Radioizmeritelnaya tekhnika*, **5**, pp. 39–44

RODIONOVA, V.N., and SLEPYAN, G.YA. (1986): '*Q*-factor of a cylindrical cavity with a coaxial insert', *Radiotekhnika i Elektronika*, **31**, pp. 1915–1921

RODIONOVA, V.N., and SLEPYAN, G.YA. (1989): 'Investigation of high-quality comb-shaped resonators with rarefied spectrum of eigenoscillations', *Radiotekhnika i Elektronika*, **34**, pp. 1767–1769

RODIONOVA, V.N., SLEPYAN, A.YA., and SLEPYAN, G.YA. (1990): 'Set of millimeter-wave band cavities with a high *Q*-factor and rarefied spectrum of eigenoscillations' 23rd General Assembly of URSI, Prague, Abstracts, Vol. 2, p. 344

ROZZI, T., MOGLIE, F., MORINI, A., MARCHIONNA,, E., and POTITI, M. (1990): 'Hybrid modes, substrate leakage, and losses of slotline at millimeter wave frequencies', *IEEE Trans.*, **MTT-38**, pp. 1069–1078

SAPOGOVA, N.I., and KONTOROVICH, V.M. (1971): 'Application of the group theory for the study of removal of mode degeneration in a spherical resonator', *Izvestiya VUZov—Radiofizika*, **14**, pp. 1869–1877

SCHELKUNOFF, S.A. (1955): 'Conversion of Maxwell equations into generalised telegraphist's equations'. *Bell Syst. Tech. J.*, **34**, pp. 995-1001
SCHIMERT, T.R., BROUNS, A.J., CHAN, C.H., and MITTRA, R. (1991): 'Investigation of millimeter-wave scattering from frequency selective surfaces', *IEEE Trans.*, **MTT-39**, pp. 315–322
SEMAKOV, V.L., and TERESCHENKO, A.P. (1975): 'Angular-symmetric eigenoscillations in spheroidal resonators', *Radiotekhnika*, **32**, pp. 72–79 (Kharkov, Vischa Shkola)
SHANKIN, YU.P. (1974): 'Associated waves in periodic waveguides', *Radiotekhnika i Elektronika*, **19**, pp. 1590–1594
SHESTOPALOV, V.P. (1970): 'Method of the Riemann-Hilbert problem in the theory of diffraction and propagation of electromagnetic waves' (Khar'kov, KhGU)
SHESTOPALOV, V.P. (1976): 'Diffraction electronics' (Khar'kov, KhGU)
SHESTOPALOV, V.P., KIRILENKO, A.A., and MASALOV, S.A. (1984): 'Convolution-type matrix equations in diffraction theory' (Kiev, Naukova Dumka)
SHESTOPALOV, V.P., KIRILENKO, A.A., MASALOV, S.A., and SIRENKO, YU.K. (1986): 'Resonance scattering of waves. Vol. 1. Diffraction gratings' (Kiev, Naukova Dumka)
SHESTOPALOV, YU.V. (1980): 'Properties of the spectrum of one class of non-self-adjoint boundary-value problems for sets of Helmholtz equations', *Doklady AN SSSR*, **252**, pp. 1108–1111
SHESTOPALOV, YU.V. (1983): 'Existence of a discrete spectrum of normal waves of microstrip transmission lines with layered dielectric filling', *Doklady AN SSSR*, **273**, pp. 592–596
SHTEINSHLEYGER, V.B. (1955): 'Mode interaction effects in electromagnetic resonators' (Moscow, Oborongiz)
SIRENKO, YU.K. (1983): 'On the substantiation of the method of semi-inversion of matrix operators in wave diffraction problems', *Zhurnal Vychislitelnoi Matematiki i Matematicheskoi Fiziki*, **23**, pp. 1381–1391
SLEPYAN, A.YA. (1984): 'Reflection of electrically polarised plane wave from inclined finitely conducting comb-shaped structure', *Doklady AN BSSR*, **28**, pp. 419–422
SLEPYAN, A.YA. (1990): 'Electromagnetic wave diffraction by an open-ended plane flanged irregular waveguide', *Izvestiya VUZov—Radiofizika*, **33**, pp. 334-341
SLEPYAN, G.YA. (1977): 'On the calculation of eigenmodes in cavities with rotational symmetry', *Zhurnal Vychislitelnoi Matematiki i Matematicheskoi Fiziki*, **17**, pp. 776–780
SLEPYAN, G.YA. (1980): 'Rigorous solution of the problem of plane wave diffraction by the periodic structure with impedance boundary conditions', *Zhurnal Vychislitelnoi Matematiki i Matematicheskoi Fiziki*, **20**, pp. 461–470
SLEPYAN, G.YA. (1981): 'Normal waves in corrugated waveguides with finitely conducting walls', *Radiotekhnika i Elektronika*, **26**, pp. 1833–1839
SLEPYAN, G.YA. (1983): 'Resonance nonreciprocal effects in layered gyrotropic media and periodical structures with a gyrotropic filling', *Radiotekhnika i Elektronika*, **28**, pp. 749–759
SLEPYAN, G.YA. (1984): 'On the boundary conditions for non-ideally conducting bodies', *Zhurnal Tekhnicheskoi Fiziki*, **54**, pp. 403–405
SLEPYAN, G.YA. (1988): 'More precise impedance conditions for calculating the heat losses in thin open metal screen', *Sov. J. Communications Technology and Electronics*, **33**, pp. 172–175
SLEPYAN, G.YA., and SLEPYAN, A.YA. (1979): 'Energy absorbtion in the case of electrically polarised plane wave reflection from a non-ideally conducting comb-shaped structure', *Pis'ma v Zhurnal Tekhnicheskoi Fiziki*, **5**, pp. 1316–1319
SLEPYAN, G.YA., and SLEPYAN, A.YA. (1980*a*): 'Diffraction of a plane wave by non-ideally conducting comb-shaped structure', *Izvestiya VUZov—Radiofizika*, **23**, pp. 1330–1341
SLEPYAN, G.YA., and SLEPYAN, A.YA. (1980*b*): 'Resonance absorbtion of electrically polarised plane wave by a comb-shaped structure in a layered dielectric', *Pis'ma v Zhurnal Tekhnicheskoi Fiziki*, **6**, pp. 415–418
SLEPYAN, G.YA., and SLEPYAN, A.YA. (1981): 'Absorbtion of electrically polarised

plane wave by a corrugated structure in layered dielectric medium', *Radiotekhnika i Elektronika*, **26**, pp. 689–694

SLEPYAN, G.YA., and SLEPYAN, A.YA. (1986): 'Reflection of H-polarised plane wave from non-ideally conducting inclined comb-shaped structures and echelettes', *Radiotekhnika i Elektronika*, **31**, pp. 1015–1017

SMYTHE, W.R. (1950): 'Static and dynamic electricity' (McGraw-Hill)

SURATTEAN, J.Y., CADILHAC, M., and PETIT, R. (1985): 'The perfectly conducting wire grating; computation of the diffracted field from Maxwell's equations and Hamilton's canonical system', *IEEE Trans.*, **AP-33**, pp. 404–408

SVESHNIKOV, A.G. (1963): 'On the substantiation of a calculation method for electromagnetic oscillations propagation in irregular waveguides', *Zhurnal Vychislitelnoi Matematiki i Matematicheskoi Fiziki*, **3**, pp. 314–326

SVESHNIKOV, A.G. (1969): 'Diffraction by a bounded body', *Doklady AN SSSR*, **184**, pp. 63–65

SVESHNIKOV, A.G., ILYINSKY, A.S., and KOTIK, I.P. (1965): 'Propagation of modes in irregular waveguides with a complicated form of lateral surface' *in* 'Vychislitelnye Metody i Programmirovanie. Vol. 3' (Moscow, MGU)

TAMM, I.E. (1976): 'Principles of electricity' (Moscow, Nauka)

TIKHONOV, A.N., and ARSENIN, V.YA. (1977): 'Solution of ill-posed problems' (Wiley)

TIKHONOV, A.N., and SAMARSKY, A.A. (1977): 'Equations of mathematical physics' (Moscow, Nauka)

TISHER, F.J. (1974): 'Anomalous skin effect of single-crystal copper in the millimeter-wave region at room temperature', *Phys. Lett.*, **47A**, pp. 231–232

TISHER, F.J. (1976): 'Excess conduction losses at millimeter wavelengths', *IEEE Trans.*, **MTT-24**, pp. 853–858

TISHER, F.J. (1978): 'Surface characteristics of metals and waveguide attenuation at millimeter wave frequencies'. Proceedings of 8th European Microwave Conference, Paris, pp. 524–527

TISHER, F.J. (1979): 'Experimental attenuation of rectangular waveguides at millimeter wavelengths', *IEEE Trans.*, **MTT-27**, pp. 31–37

VAINIKKO, G.M., and KARMA, O.O. (1974): 'On the rapidity of convergence of approximate methods for eigenvalue problems with nonlinear dependence on a parameter', *Zhurnal Vychislitelnoi Matematiki i Matematicheskoi Fiziki*, **14**, pp. 1393–1408

VAKANAS, L., CANGELLARIS, A., and PRINCE, J. (1990): 'A parametric study of the attenuation constant of lossy microstrip lines', *IEEE Trans.*, **MTT-38**, pp. 1136–1139

VALITOV, R.A. (1963): 'Radioengineering measurements' (Moscow, Sovetskoe Radio)

VESELOV, G.I., and LYUBIMOV, L.A. (1963): 'On the theory of a two-layer dielectrical waveguide with a cylindrical screen', *Radiotekhnika i Elektronika*, **8**, pp. 1530–1541

VOROVICH, I.I., and BABESHKO, V.A. (1979): 'Dynamic mixed problems of elasticity theory for nonclassical regions' (Moscow, Nauka)

VYMOROKOV, N.V., KLIMATCHEV, P.P., KOVTUNOVA, Z.D., and SILIN, R.A. (1987): 'Analysis of the contribution of various parts of microstrip line cross-section to the attenuation coefficient', *Elektronnaya Tekhnika — Elektronika SVCh*, **3(397)**, pp. 40–46

WANG, Y.C. (1978): 'The screening potential theory of excess conduction loss at millimeter and submillimeter wavelengths', *IEEE Trans.*, **MTT-26**, pp. 858–861

WEINSTEIN, L.A. (1969*a*): 'Open resonators and open waveguides' (Golem Press)

WEINSTEIN, L.A. (1969*b*): 'The theory of diffraction and the factorization method' (Golem Press)

WEINSTEIN, L.A. (1988): 'Electromagnetic waves' (Moscow, Radio i svyaz)

WEINSTEIN, L.A., and SOLNTSEV, V.A. (1973): 'Lectures in microwave electronics' (Moscow, Sovetskoe Radio)

WEINSTEIN, L.A., and VAKMAN, D.I. (1983): 'Frequency separation method in the oscillation and wave theory' (Moscow, Nauka)

WEINSTEIN, L.A., ZHURAV, S.M., and SUKOV, A.I. (1986): 'On the calculation

$\tilde{\varepsilon}$ being the relative permittivity of the cylinder material. Boundary conditions for eqn. A1.12 have the form

$$\left.\begin{aligned}\frac{dC_n(0)}{dt} + \Gamma_n C_n(0) = q_n \\ \frac{dC_n(-1)}{dt} - \Gamma_n C_n(-1) = 0\end{aligned}\right\} \qquad (A1.13)$$

$-N \le n \le N$, $q_n = 2\delta_{n0}\Gamma_0$.

Problem (i) is a mathematical model of a waveguide high-power matched load. The use of water ($\tilde{\varepsilon}_3 = 74.0 + j8.14$) for absorbing the electromagnetic energy requires a special choice of $\tilde{\varepsilon}_2$, α and d to reduce SWR (in the calculations below $\tilde{\varepsilon}_2 = 6.8 + j0.0068$). The difficulties in solving this problem result from a sharp jump in permittivity and considerable absorption in water.

DOM, when applied to the problems described, as shown by numerical investigation, is stable against numerical errors at arbitrary length of the irregular domain and the order of the system of ordinary differential equations. For illustration, Table A1.1 presents the magnitude and phase of the reflection coefficient R_1 for different values of the discretisation error δ at the step (the 4th-order Runge–Kutta method was used). The order of the system $m = 2N$ and all ather parameters of the problem are fixed ($\alpha = 45°$, $d/a = 0.1$, $\chi = ka/2\pi = 0.75$, $P = 18$, $N = 10$).

When using OSM we have to sum large numbers in eqn. A1.4, which leads to considerable roundoff errors. They are especially significant when determining the coefficients R_n^N with large n-indices, as the values of the corresponding components g_m^n are large. This is why near the solvability boundary (see Figure A1.1) OSM results in erroneous values starting with $R_N^N, R_{N-1}^N, R_{N-2}^N, \ldots$ etc. When using DOM, the coefficient C_1 is obtained close to 1 and thus the vector $\boldsymbol{g}_1$ is the main part of the vector $\boldsymbol{y}$ (other vectors $\boldsymbol{g}_i$ play the role of small corrections). This reveals a specific feature of DOM, when each basis vector of the system is responsible only for one growing component of the solution to be found.

Table A1.1 *The influence of discretisation error δ on magnitude and phase of reflection coefficient R_1*

δ	Magnitude of R_1	Phase of R_1 (deg)
10^{-1}	0.3142412640	57.764294
3×10^{-2}	0.3141518220	57.799514
10^{-2}	0.3141462330	57.800933
10^{-3}	0.3141457931	57.803196
10^{-4}	0.3141454484	57.803415
10^{-5}	0.3141454191	57.803419
10^{-6}	0.3141454130	57.803429
10^{-7}	0.3141454129	57.803429
10^{-8}	0.3141454129	57.803430

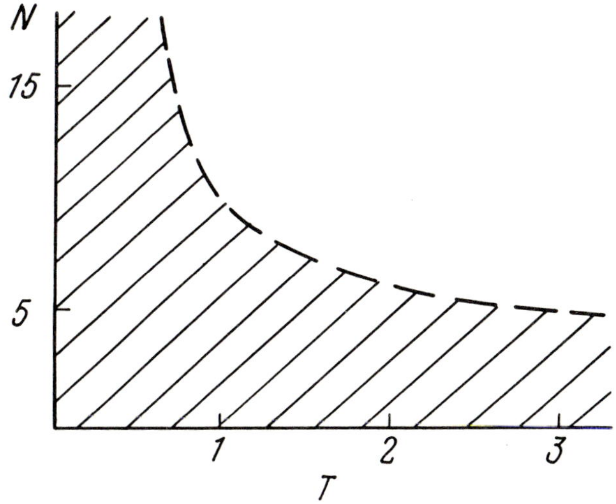

Figure A1.1 *Approximate borderline of the stability zone of the orthogonal shooting method*

To obtain a high accuracy the number of modes N under consideration should be chosen sufficiently large. For example, at $\chi < 1$ we choose N such that among the normal modes of the third region there would be 1–2 modes greatly attenuating at distances of the order of 0.1 *ka*. At $\chi > 1$, when the length of irregularity increases, N should be chosen with a large reserve (3–5 evanescent modes depending on *ka* and $T = L/a$).

As regards the problem under consideration, OSM does not have such stability. Stability against errors is found in the shared region in Fig. A1.1. Outside this region stable values of R_1 and other coefficients were not obtained through OSM by means of step reduction in the Runge–Kutta method or increasing the orthogonalisation process number P. Certainly, the diagram in Fig. A1.1 is only sketchy: it is influenced by floating-point-number system of the computer used and peculiarities of solving linear algebraic systems given by eqns. A1.7.

Table A1.2 presents the values of $|R_1|$ obtained by means of various versions of the shooting method for different N at $\delta = 0.01$. In fact the system given by eqns. A1.10 and A1.11 was solved at $N \leq 28$, $\Lambda \geq 0.025$, $\Lambda = 2\pi/\mathrm{Re}\,\Gamma_1^{(3)}$.

Figure A1.2 shows the frequency dependencies of $|R_1|$ for problem (i), calculated through DOM. Note that the curves have peaks near the points of higher waves transition from the evanescent to the propagating. As for Wood's anomalies in periodic structures, they can only be found using the rigorous approach.

Let us dwell on another modification of the shooting method, effective for a wide range of diffraction problems (Kuraev and Slepyan, 1990). It is based on a synthesis of the shooting method and that of iteration, and can be looked upon as a particular case of the general projection–iteration technique (Luchka, 1980). A similar idea was used by Litvinenko (1972) for the solution of infinite systems of linear algebraic equations and Fredholm integral equations of the

Table A1.2 *Comparison of $|R_1|$ calculated by means of various versions of the shooting method*

N	DOM	NM	OSM
1	0.327667	0.327665	0.327664
2	0.354577	0.354572	0.354573
3	0.383516	0.383503	0.383504
4	0.375138	0.375222	0.375135
5	0.343115	0.343112	0.343111
6	0.297915	0.297909	0.297912
7	0.255860	0.255856	0.255854
8	0.246915	0.246923	0.246759
9	0.280635	0.280652	0.280812
10	0.314146	0.314210	0.313400
11	0.331323	0.335315	0.341123
12	0.339395	0.369691	0.961558
13	0.343420	0.452289	0.465650
14	0.345513	0.372816	0.218452
15	0.346765	–	–
16	0.347601	–	–
17	0.348312	–	–
18	0.348961	–	–
19	0.349573	–	–
20	0.350143	–	–
21	0.350659	–	–

second kind. On the basis of the boundary-value problem given by eqns. A1.12 and A1.13 we shall illustrate the advantages of this method. This formulation will be used as being more convenient than the standard one (see eqns. A1.1–A1.3).

We shall now describe the algorithm. Let $\tilde{\boldsymbol{C}}(t) = \{\tilde{C}_n(t)\}$ denote the solution of the boundary-value problem given by eqns. A1.12 and A1.13 of the order $2M+1$, $M < N$. Elements of the vector function $\boldsymbol{C}(t) = \{C_n(t)\}$ at $M < |n| \leq N$ can be approximately determined through solving the ordinary differential equations:

$$\frac{d^2 \tilde{y}_n}{dt^2} = \Gamma_n^2 \tilde{y}_n + f_n^{(0)}(t) \tag{A1.14}$$

with the boundary conditions given by eqns. A1.13. The functions $f_n^{(0)}(t)$ are determined according to

$$f_n^{(0)}(t) = -\sum_{m=-M}^{M} \Lambda_{nm}(t) \tilde{y}_m \tag{A1.15}$$

The solutions of the 2-point boundary-value problems given by eqns. A1.13–

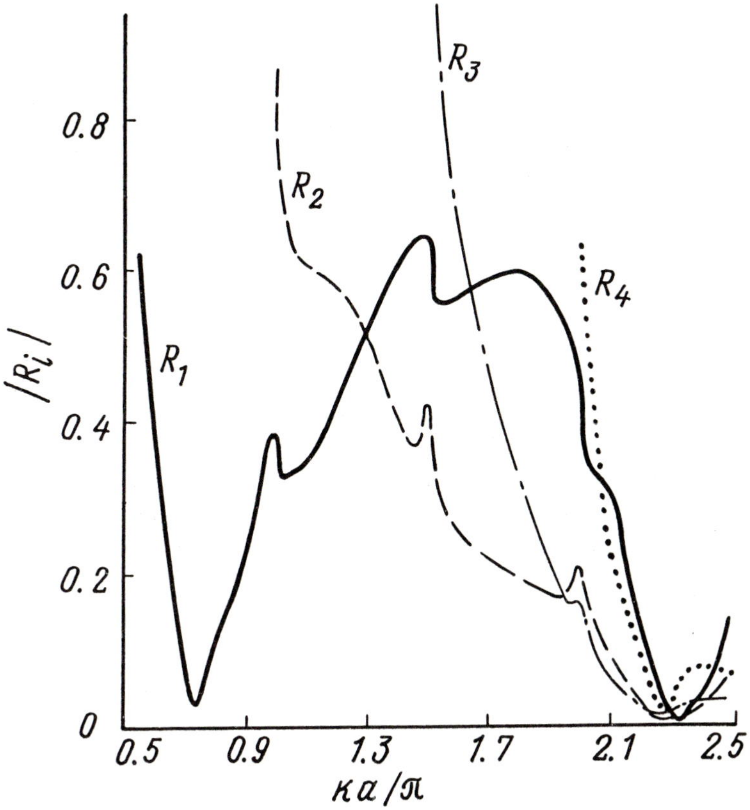

Figure A1.2 *Magnitudes of reflection coefficients against normalised frequency for the waveguide matched load*

A1.15 can be written as

$$\tilde{y}_n(t) = -\frac{1}{2\Gamma_n}\left[\int_{-1}^{t} f_n^{(0)}(u)\exp\{-\Gamma_n(t-u)\}\,\mathrm{d}u + \int_{t}^{0} f_n^{(0)}(u)\exp\{\Gamma_n(t-u)\}\,\mathrm{d}u\right] \quad \text{(A1.16)}$$

Introducing a vector function $\boldsymbol{w}^{(1)} = \boldsymbol{y}(t) - \tilde{\boldsymbol{y}}(t)$, we can easily show that $\boldsymbol{w}^{(1)}$ is the solution of the 2-point boundary-value problem

$$\frac{\mathrm{d}^2 w_n^{(1)}}{\mathrm{d}t^2} = \Gamma_n^2 w_n^{(1)} - \sum_{m=-N}^{N} \Lambda_{nm}(t) w_m^{(1)} + f_n^{(1)}(t) \quad \text{(A1.17)}$$

$$\left.\begin{aligned} \frac{\mathrm{d}w_n^{(1)}(0)}{\mathrm{d}t} + \Gamma_n w_n^{(1)}(0) = 0 \\ \frac{\mathrm{d}w_n^{(1)}(-1)}{\mathrm{d}t} - \Gamma_n w_n^{(1)}(-1) = 0 \end{aligned}\right\} \quad \text{(A1.18)}$$

where

$$f_n^{(1)}(t) = -\sum_{M<|m|\le N} \Lambda_{nm}(t)\tilde{y}_m$$

Let $\boldsymbol{w}^{(1)}(t)$ represent the solution of the 2-point boundary-value problem, given by eqns. A1.17 and A1.18, of order $2M+1$. As for eqns. A1.14–A1.16 the elements of the vector function $\boldsymbol{w}^{(1)}(t)$ at $M<|n|\le N$ can be determined approximately according to the formulae

$$\tilde{w}_n^{(1)}(t) = -\frac{1}{2\Gamma_n}\left[\int_{-1}^{t} \hat{f}_n^{(1)}(u)\exp\{-\Gamma_n(t-u)\}\,du + \int_t^0 \hat{f}_n^{(1)}(u)\exp\{\Gamma_n(t-u)\}\,du\right]$$

$$\hat{f}_n^{(1)}(t) = f_n^{(1)} - \sum_{M<|m|\le N} \Lambda_{nm}(t)\tilde{w}_n^{(1)}$$

For the vector function $\boldsymbol{w}^{(2)}(t) = \boldsymbol{w}^{(1)}(t) - \tilde{\boldsymbol{w}}^{(1)}(t)$ it is also possible to formulate a 2-point boundary-value problem similar to that given by eqns. A1.17 and A1.18 and find its approximate solution $\tilde{\boldsymbol{w}}^{(2)}(t)$. It is thus possible to construct an iteration process at each stage of which the corrections to the approximate solution of the original 2-point boundary-value problem are calculated:

$$\boldsymbol{y}(t) = \tilde{\boldsymbol{y}}(t) + \tilde{\boldsymbol{w}}^{(1)}(t) + \tilde{\boldsymbol{w}}^{(2)}(t) + \cdots.$$

Convergence of the iteration process when M is not very small results from the convergence of Galerkin's incomplete method proved by Ilyinsky and Sveshnikov (1969).

Thus the suggested iteration scheme allows us to reduce the solution of the 2-point boundary-value problem for a system of ordinary differential equations to that of a family of similar problems of a lower order. Then the order of the last ones is such that the simplest modifications of the shooting method can be applied.

Table A1.3 presents the transmission coefficients $T_n = C_n(-1)$ for a grating made by cylindrical dielectric bars. The values $\hat{T}_n$ were obtained by means of direct solution of the 2-point boundary-value problem A1.12 and A1.13 through the DOM, and $\tilde{T}_n$ with the aid of the iteration method of successive corrections. The 2-point boundary-value problems for determining the vector-functions $\boldsymbol{w}^{(i)}(t)$ were solved using the OSM ($N=3$, $M=0$, iteration number $=4$). The

Table A1.3 *Comparison of transmission coefficients calculated by means of various versions of the shooting method*

n	$\hat{T}_n$	$\tilde{T}_n$
−3	0.00092 + *j*0.00025	0.00092 + *j*0.00025
−2	0.00270 + *j*0.00069	0.00264 + *j*0.00067
−1	0.01145 + *j*0.00268	0.01145 + *j*0.00268
0	0.93847 + *j*0.33535	0.93847 + *j*0.33534
1	0.00773 + *j*0.00186	0.00777 + *j*0.00186
2	0.00215 + *j*0.00056	0.00215 + *j*0.00055
3	0.00078 + *j*0.00022	0.00075 + *j*0.00021

numerical results in Table A1.3 correspond to $\vartheta = 30°$, $\lambda/d = 0.2388$, $L/d = 0.2$, $\tilde{\varepsilon} = 4$. As seen from this Table the results obtained through different methods coincide completely.

Computational experiments have shown that for the convergence of the iteration process it is quite sufficient to assume $M = \max(M_+, M_-)$, $M_\pm = [(1 \mp \sin\vartheta)d/\lambda]$, where $[\alpha]$ is the greatest integer in α.

Appendix 2

Expressions for current-density distributions in a microstrip line with a strip of finite thickness

The density of the transverse current on the side surface of the strip is

$$j_y^1(y) = \frac{2}{\pi}\int_0^\infty \left\{\frac{\alpha h(\tilde{\varepsilon}-1)\coth(\beta_1 d_1)}{T_\varepsilon(\alpha)T_\mu(\alpha)\beta_1\beta_2}\cosh(\beta_2 y)\right.$$

$$\times\{\coth(\beta_2 d_2)-\tanh(\beta_2 y)\}\left\{\frac{\hat{N}_1(\alpha)}{T_t(\alpha)\cosh(\beta_2 t)}-\hat{N}_3(\alpha)\right\}$$

$$+\frac{\cosh(\beta_2 y)\hat{N}_2(\alpha)}{(\alpha^2+h^2)T_t(\alpha)\cosh(\beta_2 t)}\left(\beta_1\coth(\beta_1 d_1)\right.$$

$$\left.\times\left\{\frac{\tilde{\varepsilon}\alpha^2}{\beta_1^2T_\varepsilon(\alpha)}+\frac{h^2}{T_\mu(\alpha)}\right\}+\beta_2\tanh(\beta_2 y)\left\{\frac{\alpha^2}{\beta_2^2T_\varepsilon(\alpha)}+\frac{h^2}{T_\mu(\alpha)}\right\}\right)$$

$$+\frac{\beta_2\cosh(\beta_2 y)}{\alpha^2+h^2}\{\coth(\beta_2 d_2)-\tanh(\beta_2 y)\}$$

$$\left.\times\left\{\frac{\alpha^2}{\beta_2^2T_\varepsilon(\alpha)}+\frac{h^2}{T_\mu(\alpha)}\right\}\hat{N}_4(\alpha)\right\}\sin\left(\frac{\alpha W}{2}\right)d\alpha$$

The density of the longitudinal current on the side surface of the strip is

$$j_z^1(y) = \frac{2}{\pi}\int_0^\infty\left[\frac{\cosh(\beta_2 y)\{h\hat{N}_2(\alpha)-\alpha\hat{N}_1(\alpha)\}}{T_\mu(\alpha)T_t(\alpha)\cosh(\beta_2 t)}\right.$$

$$\times\left\{1+\frac{\beta_1\tanh(\beta_2 y)\coth(\beta_1 d_1)}{\beta_2}\right\}+\frac{\cosh(\beta_2 y)}{T_\mu(\alpha)}$$

$$\times\{1-\tanh(\beta_2 y)\coth(\beta_2 d_2)\}$$

$$\left.\times\{\alpha\hat{N}_3(\alpha)-h\hat{N}_4(\alpha)\}\right]\sin\left(\frac{\alpha W}{2}\right)d\alpha$$

The density of the longitudinal current on the lower screen is

$$j_z^s(x) = \frac{2}{\pi}\int_0^\infty \left[1 - \frac{\beta_2 \coth(\beta_2 d_2)}{\alpha^2 + h^2}\left\{\frac{h^2}{\beta_2^2 T_\varepsilon(\alpha)} + \frac{\alpha^2}{T_\mu(\alpha)}\right\}\hat{\mathcal{N}}_3(\alpha)\right.$$

$$- \frac{\beta_1 \coth(\beta_1 d_1)}{(\alpha^2 + h^2) T_t(\alpha) \cosh(\beta_2 t)}\left\{\frac{\tilde{\varepsilon} h^2}{\beta_1^2 T_\varepsilon(\alpha)} + \frac{\alpha^2}{T_\mu(\alpha)}\right\}\hat{\mathcal{N}}_1(\alpha)$$

$$+ \frac{\coth(\beta_1 d_1) \coth(\beta_2 d_2)}{\beta_1 \beta_2} \frac{\alpha h(\tilde{\varepsilon} - 1)}{T_\varepsilon(\alpha) T_\mu(\alpha)}$$

$$\left.\times \left\{\hat{\mathcal{N}}_4(\alpha) - \frac{\hat{\mathcal{N}}_2(\alpha)}{T_t(\alpha)\cosh(\beta_2 t)}\right\}\right] \frac{\cos(\alpha x)}{\cosh(\beta_1 d_1)}\, d\alpha$$

The density of the transverse current on the lower screen is

$$j_x^s(x) = \frac{2}{\pi}\int_0^\infty \left(\frac{\alpha h}{\alpha^2 + h^2}\left[\frac{\beta_1 \coth(\beta_1 d_1)}{T_t(\alpha)\cosh(\beta_2 t)}\left\{\frac{\tilde{\varepsilon}}{\beta_1^2 T_\varepsilon(\alpha)} - \frac{1}{T_\mu(\alpha)}\right\}\hat{\mathcal{N}}_1(\alpha)\right.\right.$$

$$\left. + \beta_2 \coth(\beta_2 d_2)\left\{\frac{1}{\beta_2^2 T_\varepsilon(\alpha)} - \frac{1}{T_\mu(\alpha)}\right\}\hat{\mathcal{N}}_3(\alpha)\right]$$

$$+ \frac{\beta_1 \coth(\beta_1 d_1)}{(\alpha^2 + h^2) T_t(\alpha)\cosh(\beta_2 t)}\left\{\frac{\tilde{\varepsilon}\alpha^2}{\beta_1^2 T_\varepsilon(\alpha)} + \frac{h^2}{T_\mu(\alpha)}\right\}\hat{\mathcal{N}}_2(\alpha)$$

$$\left. + \left[\frac{\beta_2 \coth(\beta_2 d_2)}{\alpha^2 + h^2}\left\{\frac{\alpha^2}{\beta_2^2 T_\varepsilon(\alpha)} + \frac{h^2}{T_\mu(\alpha)}\right\} - 1\right]\hat{\mathcal{N}}_4(\alpha)\right)$$

$$\times \frac{\sin(\alpha x)}{\cosh(\beta_1 d_1)}\, d\alpha$$

The function $P_z(\alpha)$ in eqn. 3.88 is expressed by

$$P_z(\alpha) = \frac{W_0}{k^2}\left\{\frac{k^2 h L_{11}(\alpha) + \bar{h} L_{12}(\alpha) - (\bar{h}^2 + \tilde{\varepsilon})\alpha k L_{13}(\alpha)}{(\tilde{\varepsilon} - \bar{h}^2)^2}\right.$$

$$+ \frac{k^2 h L_{21}(\alpha) + \bar{h} L_{22}(\alpha) - (\bar{h}^2 + 1)\alpha k L_{23}(\alpha)}{(1 - \bar{h}^2)^2}$$

$$\left. + \frac{k^2 h L_{31}(\alpha) + \bar{h} L_{32}(\alpha) - (\bar{h}^2 + 1)\alpha k L_{33}(\alpha)}{(1 - \bar{h}^2)^2} \sin^2\left(\frac{\alpha W}{2}\right)\right\}$$

$$\bar{h} = h/k$$

$$L_{11}(\alpha) = \frac{\tilde{\varepsilon} R_2^2(\alpha)}{2}\left\{(\alpha^2 + \beta_1^2)\frac{\coth(\beta_1 d_1)}{\beta_1} + \frac{(h^2 - k^2\tilde{\varepsilon}) d_1}{\cosh^2(\beta_1 d_1)}\right\}$$

$$L_{12}(\alpha) = \frac{\{R_3(\alpha) - \hat{\mathcal{N}}_4(\alpha)\}^2}{2}\left\{\frac{(\alpha^2 + \beta_1^2)}{\beta_1 \coth(\beta_1 d_1)} + \frac{(h^2 - k^2\tilde{\varepsilon}) d_1}{\cosh^2(\beta_1 d_1)}\right\}$$

$$L_{13}(\alpha) = R_2(\alpha)\{R_3(\alpha) - \hat{\mathcal{N}}_4(\alpha)\}$$

$$L_{21}(\alpha) = \frac{\{R_1(\alpha)\sinh(\beta_2 t) + R_2(\alpha)\cosh(\beta_2 t)\}^2}{2}$$

$$\times \left[(\alpha^2 + \beta_1^2)\frac{\coth\{\beta_2(d_2 - t)\}}{\beta_2} + \frac{(h^2 - k^2)(d_2 - t)}{\sinh^2\{\beta_2(d_2 - t)\}} \right]$$

$$L_{22}(\alpha) = \frac{\{\hat{N}_2(\alpha) - R_3(\alpha)\cosh(\beta_2 t) - R_4(\alpha)\sinh(\beta_2 t)\}^2}{2}$$

$$\times \left[\frac{\alpha^2 + \beta_2^2}{\beta_2 \coth\{\beta_2(d_2 - t)\}} + \frac{(h^2 - k^2)(d_2 - t)}{\cosh^2\{\beta_2(d_1 - t)\}} \right]$$

$$L_{23}(\alpha) = \{R_1(\alpha)\sinh(\beta_2 t) + R_2(\alpha)\cosh(\beta_2 t)\}$$

$$\times \{\hat{N}_2(\alpha) - R_3(\alpha)\cosh(\beta_2 t) - R_4(\alpha)\sinh(\beta_2 t)\}$$

$$L_{31}(\alpha) = \frac{t}{2}(h^2 - k^2)\{R_1^2(\alpha) - R_2^2(\alpha)\} + \frac{\alpha^2 + \beta_2^2}{2\beta_2}$$

$$\times \left[\frac{\{R_1^2(\alpha) + R_2^2(\alpha)\}}{2}\sinh(2\beta_2 \mathrm{t}) + R_1(\alpha)R_2(\alpha)\{\cosh(2\beta_2 t) - 111\} \right]$$

$$L_{32}(\alpha) = -\frac{t}{2}(h^2 - k^2)\{R_3^2(\alpha) - R_4^2(\alpha)\}$$

$$+ \frac{\alpha^2 + \beta_2^2}{2\beta_2}\left[\frac{\{R_3^2(\alpha) + R_4^2(\alpha)\}}{2}\sinh(2\beta_2 t) \right.$$

$$\left. + R_3(\alpha)R_4(\alpha)\{\cosh(2\beta_2 t) - 1\} \right]$$

$$L_{33}(\alpha) = \{R_1(\alpha)\sinh(\beta_2 t) + R_2(\alpha)\cosh(\beta_2 t)\}$$

$$\times \{R_3(\alpha)\cosh(\beta_2 t) + R_4(\alpha)\sinh(\beta_2 t)\} - R_2(\alpha)R_3(\alpha)$$

$$R_1(\alpha) = \frac{1}{\alpha^2 + h^2}\left[\hat{N}_3(\alpha)\coth(\beta_2 d_2)\left\{ \frac{h^2}{k^2 T_\varepsilon(\alpha)} - \frac{\alpha^2}{T_\mu(\alpha)} \right\} \right.$$

$$+ \hat{N}_4(\alpha)\alpha h \coth(\beta_2 d_2)\left\{ \frac{1}{k^2 T_\varepsilon(\alpha)} + \frac{1}{T_\mu(\alpha)} \right\}$$

$$+ \frac{\hat{N}_1(\alpha)\beta_1 \coth(\beta_1 d_1)}{T_t(\alpha)\cosh(\beta_2 t)} + \frac{\hat{N}_2(\alpha)\alpha h \beta_1 \coth(\beta_1 d_1)}{T_t(\alpha)\cosh(\beta_2 t)}$$

$$\left. \times \left\{ \frac{\tilde{\varepsilon} h^2 \beta_2}{k^2 \beta_1^2 T_\varepsilon(\alpha)} - \frac{\alpha^2}{\beta_2 T_\mu(\alpha)} \right\}\left\{ \frac{\tilde{\varepsilon}\beta_2}{k^2 \beta_1^2 T_\varepsilon(\alpha)} + \frac{1}{\beta_2 T_\mu(\alpha)} \right\} \right]$$

$$R_2(\alpha) = \frac{1}{\alpha^2 + h^2}\left[\left\{ \frac{h^2}{k^2 T_\varepsilon(\alpha)} - \frac{\alpha^2}{T_\mu(\alpha)} \right\}\left\{ \frac{\hat{N}_1(\alpha)}{T_t(\alpha)\cosh(\beta_2 t)} - \hat{N}_3(\alpha) \right\} \right.$$

$$\left. + \alpha h \left\{ \frac{1}{k^2 T_\varepsilon(\alpha)} + \frac{1}{T_\mu(\alpha)} \right\}\left\{ \frac{\hat{N}_2(\alpha)}{T_t(\alpha)\cosh(\beta_2 t)} - \hat{N}_4(\alpha) \right\} \right]$$

$$R_3(\alpha) = \frac{1}{\alpha^2 + h^2}\left[\frac{\hat{N}_1(\alpha)\alpha h \beta_1 \coth(\beta_1 d_1)}{T_t(\alpha)\cosh(\beta_2 t)}\left\{ \frac{\tilde{\varepsilon}}{\beta_1^2 T_\varepsilon(\alpha)} - \frac{1}{T_\mu(\alpha)} \right\} \right.$$

$$+\frac{\hat{\mathcal{N}}_2(\alpha)\beta_1\coth(\beta_1 d_1)}{T_t(\alpha)\cosh(\beta_2 t)}\left\{\frac{\alpha^2\tilde{\varepsilon}}{\beta_1^2 T_\varepsilon(\alpha)}+\frac{h^2}{T_\mu(\alpha)}\right\}$$

$$+\hat{\mathcal{N}}_2(\alpha)\alpha h\beta_2\coth(\beta_2 d_2)\left\{\frac{1}{\beta_2^2 T_\varepsilon(\alpha)}-\frac{1}{T_\mu(\alpha)}\right\}$$

$$+\hat{\mathcal{N}}_4(\alpha)\beta_2\coth(\beta_2 d_2)\left\{\frac{\alpha^2}{\beta_2^2 T_\varepsilon(\alpha)}+\frac{h^2}{T_\mu(\alpha)}\right\}\Bigg]$$

$$R_4(\alpha)=\frac{1}{\alpha^2+h^2}\Bigg[\alpha h\beta_2\left\{\frac{1}{\beta_2^2 T_\varepsilon(\alpha)}-\frac{1}{T_\mu(\alpha)}\right\}$$

$$\times\left\{\frac{\hat{\mathcal{N}}_1(\alpha)}{T_t(\alpha)\cosh(\beta_2 t)}-\hat{\mathcal{N}}_3(\alpha)\right\}+\beta_2\left\{\frac{\alpha^2}{\beta_2^2 T_\varepsilon(\alpha)}+\frac{h^2}{T_\mu(\alpha)}\right\}$$

$$\times\left\{\frac{\hat{\mathcal{N}}_2(\alpha)}{T_t(\alpha)\cosh(\beta_2 t)}-\hat{\mathcal{N}}_4(\alpha)\right\}\Bigg]$$

Appendix 3

General formulae for the coefficients $\alpha_{nm}^{(i)}, \beta_{nm}^{(i)}, \gamma_{nm}^{(i)}, \delta_{nm}^{(i)}$

In this Appendix the general formulae for the coefficients in eqn. 7.73 are given. These formulae are applicable for corrugated waveguides with arbitrary shapes of the cross-section and corrugation and have the form

$$\alpha_{nm}^{(1)} = \frac{1}{\omega} \int_{S_\perp} g^{13} \lambda_m^M \Phi_m^M h_{1\rho}^{n*} \, \mathrm{d}S$$

$$\beta_{nm}^{(1)} = \mu_0 \int_{S_\perp} \left\{ \frac{\sqrt{g}}{\rho} h_{1\rho}^{n*} (\tilde{g}^{11} h_{1\rho}^m + \rho g^{12} h_{1\phi}^m) + (\sqrt{g}) h_{1\phi}^{n*} (g^{12} h_{1\rho}^m + \rho g^{22} h_{1\phi}^m) \right\} \mathrm{d}S + \frac{j Z_S}{\omega} \oint_C g^{22} (g g^{11})^{1/2} h_{1\phi}^m h_{1\phi}^{n*} \, \mathrm{d}l$$

$$\gamma_{nm}^{(1)} = -j\mu_0 \int_{S_\perp} \left\{ \frac{\sqrt{g}}{\rho} h_{1\rho}^{n*} (\tilde{g}^{11} h_{2\rho}^m + \rho g^{12} h_{2\phi}^m) + (\sqrt{g}) h_{1\phi}^{n*} (g^{12} h_{2\rho}^m + \rho g^{22} h_{2\phi}^m) \right\} \mathrm{d}S + \frac{Z_S}{\omega} \oint_C g^{22} (g g^{11})^{1/2} h_{2\phi}^m h_{1\phi}^{n*} \, \mathrm{d}l$$

$$\alpha_{nm}^{(2)} = -\frac{j}{\omega} \int_{S_\perp} g^{13} (\lambda_m^M \Phi_m^M h_{2\rho}^{n*} - \lambda_n^E \Phi_n^{E*} e_{1\rho}^m) \, \mathrm{d}S$$

$$\beta_{nm}^{(2)} = -j\mu_0 \int_{S_\perp} \left\{ \frac{\sqrt{g}}{\rho} h_{2\rho}^{n*} (\tilde{g}^{11} h_{1\rho}^m + \rho g^{12} h_{1\phi}^m) + (\sqrt{g}) h_{2\phi}^{n*} (g^{12} h_{1\rho}^m + \rho g^{22} h_{1\phi}^m) \right\} \mathrm{d}S$$

$$+ \frac{Z_S}{\omega} \oint_C g^{22} (gg^{11})^{1/2} h_{1\phi}^m h_{2\phi}^{n*} \, \mathrm{d}l$$

$$\delta_{nm}^{(2)} = \frac{1}{\omega} \int_{S_\perp} g^{13} \lambda_n^E \Phi_n^{E*} e_{2\rho}^m \, \mathrm{d}S$$

$$\gamma_{nm}^{(2)} = -\frac{1}{\varepsilon_0} \int_{S_\perp} \frac{\lambda_m^E \lambda_n^E \Phi_m^E \Phi_n^{E*}}{\omega^2 \sqrt{g}} \rho \, \mathrm{d}S$$

$$+ \mu_0 \int_{S_\perp} \left\{ \frac{\sqrt{g}}{\rho} h_{2\rho}^{n*} (\tilde{g}^{11} h_{2\rho}^m + \rho g^{12} h_{2\phi}^m) \right.$$

$$\left. + (\sqrt{g}) h_{2\phi}^{n*} (g^{12} h_{2\rho}^m + \rho g^{22} h_{2\phi}^m) \right\} \mathrm{d}S$$

$$+ \frac{jZ_S}{\omega} \oint_C g^{22} (gg^{11})^{1/2} h_{2\phi}^m h_{2\phi}^{n*} \, \mathrm{d}l$$

$$\alpha_{nm}^{(3)} = -\frac{1}{\mu_0} \int_{S_\perp} \frac{\lambda_m^M \lambda_n^M \Phi_m^M \Phi_n^{M*}}{\omega^2 \sqrt{g}} \rho \, \mathrm{d}S$$

$$+ \varepsilon_0 \int_{S_\perp} \left\{ \frac{\sqrt{g}}{\rho} e_{1\rho}^{n*} (\tilde{g}^{11} e_{1\rho}^m + \rho g^{12} e_{1\phi}^m) \right.$$

$$\left. + (\sqrt{g}) e_{1\phi}^{n*} (g^{12} e_{1\rho}^m + \rho g^{22} e_{1\phi}^m) \right\} \mathrm{d}S$$

$$+ jZ_S^* \oint_C \frac{\sqrt{g^{11}}}{\omega^3 \mu_0^2 \sqrt{g}} \{(\mathrm{curl}\, \boldsymbol{e}_{1\tau}^m)_z \cdot (\mathrm{curl}\, \boldsymbol{e}_{1\tau}^{n*})_z\} \, \mathrm{d}l$$

$$\delta_{nm}^{(3)} = -j\varepsilon_0 \int_{S_\perp} \left\{ \frac{\sqrt{g}}{\rho} e_{1\rho}^{n*} (\tilde{g}^{11} e_{2\rho}^m + \rho g^{12} e_{2\phi}^m) \right.$$

$$\left. + (\sqrt{g}) e_{1\phi}^{n*} (g^{12} e_{2\rho}^m + \rho g^{22} e_{2\phi}^m) \right\} \mathrm{d}S$$

$$\alpha_{nm}^{(4)} = -j\varepsilon_0 \int_{S_\perp} \left\{ \frac{\sqrt{g}}{\rho} e_{2\rho}^{n*} (\tilde{g}^{11} e_{1\rho}^m + \rho g^{12} e_{1\phi}^m) \right.$$

$$\left. + (\sqrt{g}) e_{2\phi}^{n*} (g^{12} e_{1\rho}^m + \rho g^{22} e_{1\phi}^m) \right\} \mathrm{d}S$$

$$\delta_{nm}^{(4)} = \varepsilon_0 \int_{S_\perp} \left\{ \frac{\sqrt{g}}{\rho} e_{2\rho}^{n*} (\tilde{g}^{11} e_{2\rho}^m + \rho g^{12} e_{2\phi}^m) \right.$$

$$\left. + (\sqrt{g}) e_{2\phi}^{n*} (g^{12} e_{2\rho}^m + \rho g^{22} e_{2\phi}^m) \right\} \mathrm{d}S$$

$$\beta_{nm}^{(3)} = \alpha_{mn}^{(1)*}, \qquad \gamma_{nm}^{(3)} = \alpha_{mn}^{(2)*}, \qquad \gamma_{nm}^{(4)} = \delta_{mn}^{(2)*}$$

where

$$n = \{p, l\}, \qquad m = \{q, r\} \qquad \lambda_n^E = \nu_{pl}^2, \qquad \lambda_m^M = \mu_{pr}^2,$$

$$\tilde{g}^{11} = g^{11} - (g^{13})^2.$$

Index